Franke

Three-Dimensional Molded Interconnect Devices (3D-MID)

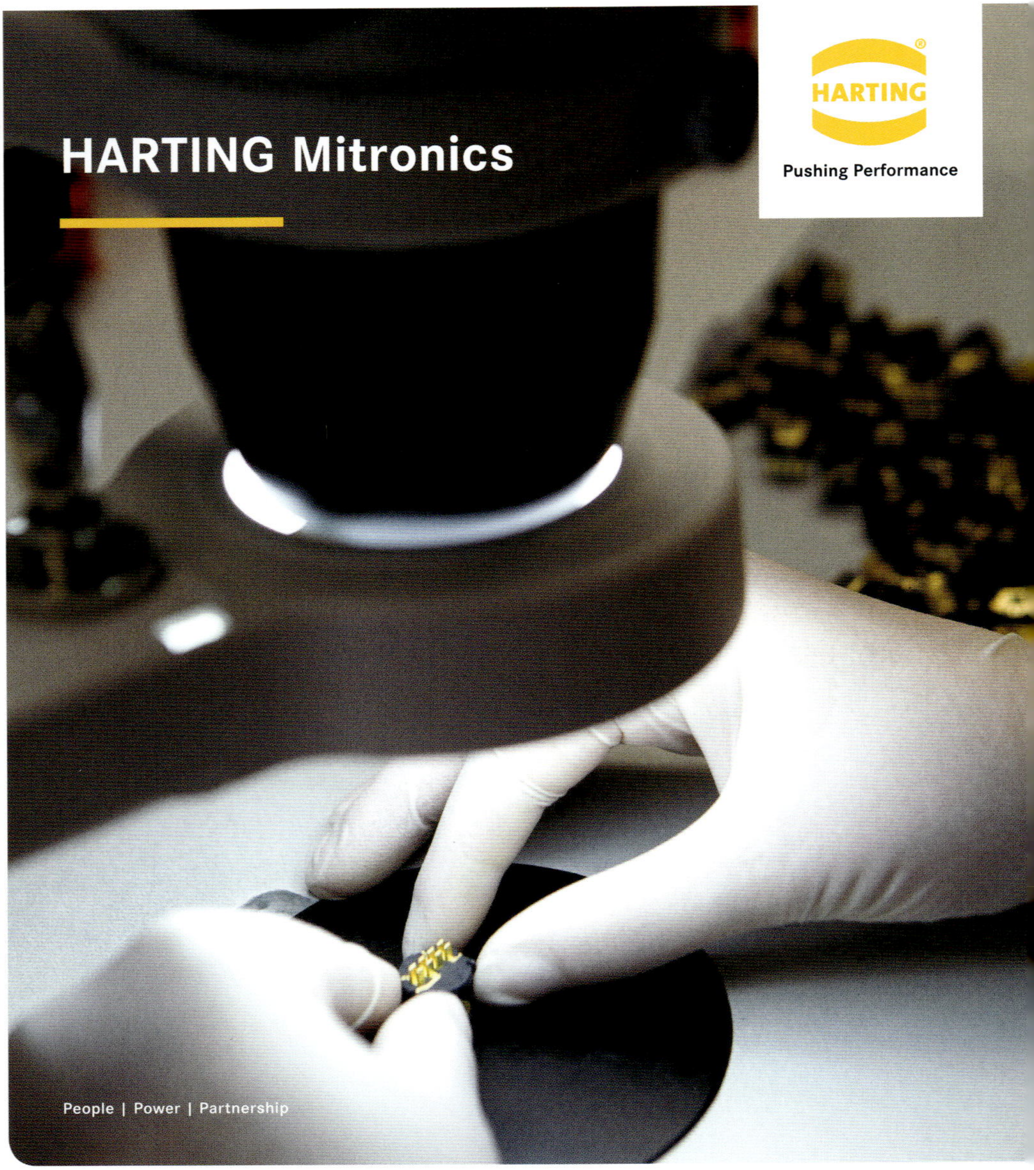
HARTING Mitronics
HARTING
Pushing Performance
People | Power | Partnership

Jörg Franke

Three-Dimensional Molded Interconnect Devices (3D-MID)

Materials, Manufacturing, Assembly, and Applications for Injection Molded Circuit Carriers

HANSER

Hanser Publishers, Munich

Hanser Publications, Cincinnati

The Author:
Prof. Dr.-Ing. Jörg Franke,
Forschungsvereinigung Räumliche Elektronische Baugruppen 3-D MID e.V.,
(Research Association Molded Interconnect Devices 3-D MID)
Fürther Str. 246b, 90429 Nuremberg, Germany

Distributed in North and South America by:
Hanser Publications
6915 Valley Avenue, Cincinnati, Ohio 45244-3029, USA
Fax: (513) 527-8801
Phone: (513) 527-8977
www.hanserpublications.com

Distributed in all other countries by
Carl Hanser Verlag
Postfach 86 04 20, 81631 München, Germany
Fax: +49 (89) 98 48 09
www.hanser-fachbuch.de

Cataloging-in-Publication Data is on file with the Library of Congress

ISBN 978-1-56990-551-7
E-Book-ISBN 978-1-56990-552-4

Bibliografische Information Der Deutschen Bibliothek
Die Deutsche Bibliothek verzeichnet diese Publikation in der Deutschen Nationalbibliografie; detaillierte bibliografische Daten sind im Internet über <http://dnb.d-nb.de> abrufbar.

Production Management: Steffen Jörg
Coverconcept: Marc Müller-Bremer, www.rebranding.de, München
Coverdesign: Stephan Rönigk
Typesetted by Manuela Treindl, Fürth
Printed and bound by Kösel, Krugzell
Printed in Germany

Preface

A molded interconnect device (MID) is an injection-molded thermoplastic part with structured circuit traces. This definition still applies, although the term 3D-MID has also become common. An extension of the meaning to mechatronic integrated device, moreover, allows for the fact that thermoplastics are not the only materials used, and MID are not necessarily produced by injection molding.

There have recently been major advances in MID development, particularly with regard to substrate materials, the methods of producing the interconnect devices with structuring and metallization, and the various connection technologies. Potential areas of application have therefore expanded and impressive advances have been made in viable optical, fluidic, mechanical, electrical, and thermal functionalities and in amalgamating MID with other technologies. The multiplicity of fascinating applications outlined in the MID Survey 2011 illustrated many of these new fields. As in 2003 and 2006, the 2011 survey was commissioned by Germany's Forschungsvereinigung 3-D MID e.V. With its membership now numbering almost 100, the Research Association Molded Interconnect Devices (3-D MID) is an active network bringing together manufacturers, suppliers, users, and research institutes in Germany and elsewhere. The close link between the industrial and scientific communities and intensive cooperation in innumerable projects afforded an excellent basis for the development of this reference volume for MID technology.

This book is not a revision of the manual originally entitled "3D-MID-Technologie: Räumliche Elektronische Baugruppen; Herstellungsverfahren, Gebrauchsanforderungen, Materialkennwerte" (3D-MID Technology: Molded Interconnect Devices, Manufacturing Processes, Requirements for Use, Material Characteristics), published in 2004. The main fields have changed so much since then and so many new areas of interest have arisen as to render a revision of that kind unnecessary. This book aims at presenting the state of the art in 3D-MID technology along the entire process chain. The individual chapters deal with MID-specific terms of reference, merely touching on the topic of guidelines and standards for conventional technologies.

This book, therefore, addresses experts and newcomers to the field of MID, by providing a comprehensive overview of the very latest developments in research activities. For developers and innovation managers it will be an introduction to the subject matter and a source of inspiration.

In-depth knowledge and determined utilization of the integration potential afforded by MID are crucial when it comes to implementing existing ideas in successful MID projects. Consequently, readers will find inspiration in a comprehensive overview of the strengths of MID technology and numerous case studies. Despite many exciting series-production applications, hurdles remain in the form of the as yet unadapted development and the prototyping of 3D-MID. Implementation as a follow-on from production-oriented and function-optimized conceptualization calls for know-how and experience along the entire process chain. The major methods, software tools, substrate materials, and processes for the manufacture of interconnect devices and application of the conductor traces are described in detail in the individual chapters, which also include discussions of the available systems.

Despite the dynamic of recent years, by no means can technological development be considered to have reached its conclusion. Ongoing research is pushing further miniaturization and expansion across the areas of application, for example by reduction in structure size, enhanced qualification of thermoset materials for the LDS process, and productive print technologies for additive conductor metallization, or the manufacture of thermally conductive materials for LED applications.

I would like to take this opportunity to extend my sincere thanks to my assistants Dr.-Ing. Christian Goth and Dipl.-Wirtsch.-Ing. (FH) Thomas Kuhn, who displayed tremendous commitment and strict management in their organization of the book. My thanks also go to all those who submitted contributions, and to the consulting experts who gave the text its final polish. The tremendous support that was forthcoming from industry, particularly from the members of the Research Association, affords this volume the practical relevance it requires. The English-language version of this book is intended as a vehicle to help promote MID technology in the international community. My thanks to Dr. Ingo Kriebitzsch, BMW AG, for organizing the translation.

It is my hope that everyone who reads this book will extract from it new incentives and new ideas for the future development of mechatronic modules in MID technology.

Erlangen, April 2013

Jörg Franke

Contents

1 MID Technology and Mechatronic Integration Potential

C. Goth, T. Kuhn

In line with the variety of applications, the requirements that apply to the functionality, integration density, reliability, and costs of electronic modules have increased enormously, and this in turn has driven the demand for mechatronic solutions. On account of the high degree of design freedom they allow, molded interconnect devices (MID) permit highly integrated systems and, consequently, open up a huge potential for rationalization with regard to the production process. This chapter outlines the technological basics, possible areas of application, and the latest technological developments.

1.1 Technological Basics

1.1.1 Definition and Root Principle

MID is the abbreviation for *molded interconnect device*. On account of the versatility of the injection molding process and with structured metallization, mechanical and electrical functions can be integrated directly into MID parts, as can optical, fluidic, or thermal functions, among others. Widely differing processes can be employed for manufacturing the basic body and for applying or building the conductive structures. The root principle is illustrated in Fig. 1.1.

Extending the significance of the term MID to include *mechatronic integrated devices* takes account of the fact that the three-dimensional carriers do not necessarily have to be injection-molded or indeed made of plastic; other materials such as ceramics can also be used [50]. These mechatronic integrated devices are becoming increasingly important, because as new substrate materials become available they permit the use of new structuring processes, which in turn opens the way to other or further-reaching functionalities and areas of application.

As far as an understanding of MID technology is concerned, however, it is essential to appreciate that besides three-dimensional integration, the material bond between

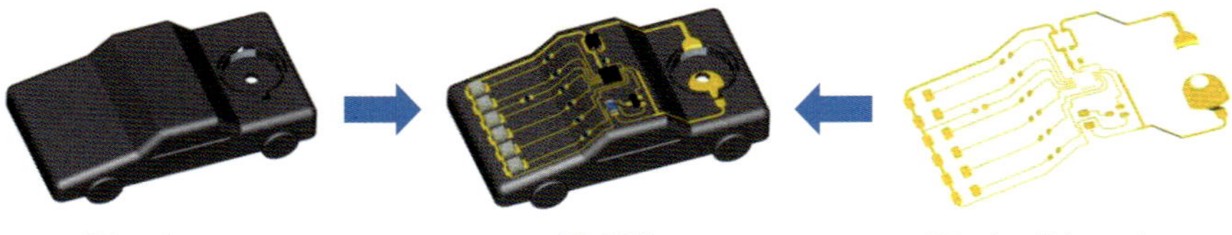

FIGURE 1.1 Underlying principle of MID technology

the component parts is crucial in terms of functional diversity. Mechatronic systems, by contrast, are frequently made by putting together individual components manufactured using conventional build techniques. Function-integrated devices of this nature are constructs made up of modular individual components. The term "three-dimensional circuit board," although commonly used, fails to convey the strengths of MID technology, deriving as they do from the interplay of diverse functions from different disciplines rather than the isolated extension of one particular field. The borders between MID and related areas of technology are drawn in Section 1.1.6. [59, 113, 129]

1.1.2 Geometric Classification

Unequivocal definition of the various dimensional configurations is practical for arriving at the precise geometric classification of molded interconnect devices. One system of geometric classification of the possible modules that is still in use today was introduced in [51]. Figure 1.2 illustrates this system of classification with current examples from the MID world. This system of classification takes into account the layout and the form of the structured process surfaces complete with their complements of electronic components. This is a starting point for deriving the requirements applicable to the production process. The interconnect devices have to be classified as a function of the process step in question. In terms of their placement, it is the distribution of the electronic components on the process surfaces that is crucial, whereas application of the circuit trace is the key factor in terms of functionalization with conductive structures. Modules are classified as belonging to 2½D, n × 2D, and 3D categories.

Conventional circuit boards, for which standard placement machines are used, are flat modules in 2D (class 0) with planar process surfaces. 2½D, on the other hand, has flat or plane-parallel process surfaces plus geometric elements in the *z* direction. Further distinctions can be drawn to yield subdivisions within this class. Class 1A is characterized by flat process surfaces, with 3D elements (e.g., cooling

2D	2½D			n x 2D	3D	
0	1A	1B	1C	2	3A	3B
Planar process surface	Planar process surface 3D elements on the opposite side	Planar process surface 3D elements on the process side	Multiple plane-parallel process surfaces	Multiple process surfaces at angles	Regular surface (e.g., cylindrical surfaces)	Freeform surfaces

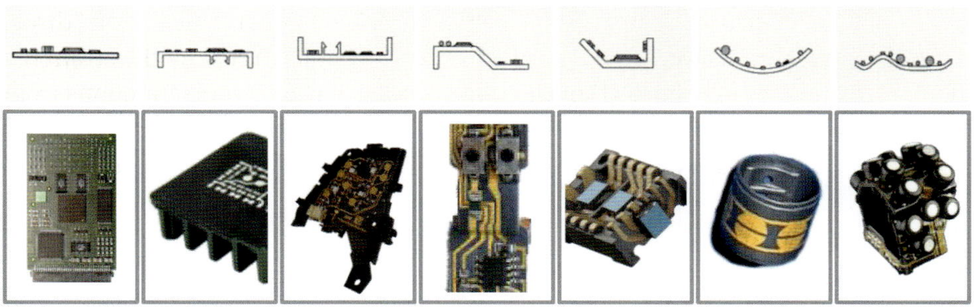

FIGURE 1.2 Geometric classification of 3D-MID [51] (Graphics courtesy of FAPS, TRW Automotive Safety Systems, HARTING Mitronics, Kromberg & Schubert)

fins) on the reverse side. Class 1B has 3D elements (e.g., interconnects) on the process surface; class 1C has two or more plane-parallel process surfaces. Class 2 n × 2D interconnect devices consist of multiple process surfaces intersecting at angles. Applications in 3D are distinguished as having regular surfaces (class 3A) or freeform surfaces (class 3B). [51]

1.1.3 Potential of 3D-MID Technology

The potential of MID technology derives from geometric design freedom in combination with selective structuring and metallization. The 3D layout permits defined angles between components, stacking and precision placement of chips, and the forming of cavities. The versatility of the MID layout therefore enables the integration of contact surfaces for switches or sensors and passive electrical functions (including capacitances, inductances, or resistances) and antennas for transmission or reception of electromagnetic waves. Partial full-coverage metallization of surfaces

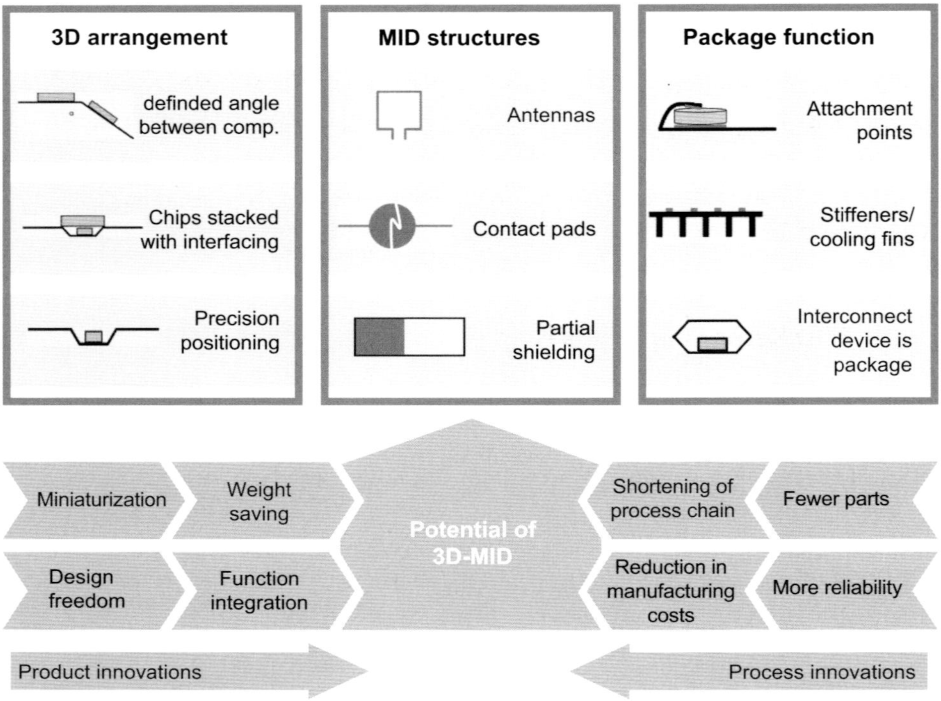

FIGURE 1.3 Potential of 3D-MID

produces shielding to protect against electromagnetic irradiation and emission and heat sinks for dissipating waste heat. Design geometry can be adapted to incorporate attachment points, stiffeners, and cooling fins, for example, directly into the package (Fig. 1.3).

MID technology, in association with production-perfect and function-oriented product development, can thus be utilized to tap into tremendous potential for rationalization and to optimize both product structuring and the production process. In this context product innovations follow on primarily from miniaturization and weight savings and from the function integration mentioned above. This, in turn, is keyed by the virtually unlimited freedom of design in 3D.

Shortening the process chain, reducing the number of parts, cutting production costs, and boosting yield all contribute to optimization of the production process. The reduction of the number of interfaces by system integration is a prime advantage in this respect, streamlining production while at the same time boosting reliability. Yield can be increased by integrating poka-yoke constraints, particularly where the MID interfaces with its periphery, although much the same effect is achieved when the number of parts is reduced.

The environmental compatibility aspect is another advantage. MID are generally made of recyclable thermoplastics and pose no difficulties in terms of EOL disposal. Materials consumption and materials diversity, moreover, are reduced. [51]

On the one hand, these advantages help boost the competitiveness of companies manufacturing electronic or mechatronic systems. On the other hand, they also contribute to increasing customer benefit and maximizing the options open to users and system manufacturers for optimizing the overall system.

1.1.4 MID Reference Process

MID manufacturing processes are many and varied. In any given case the most appropriate has to be chosen on the basis of the stated criteria. The most important processes include two-shot molding, laser structuring (additive and subtractive), hot embossing, and film insert molding. At this time, too, printing technologies and plasma structuring are of growing significance. Other technologies such as masking and primer technology, by contrast, are of minor significance in the MID environment, as are the physical processes of metallization.

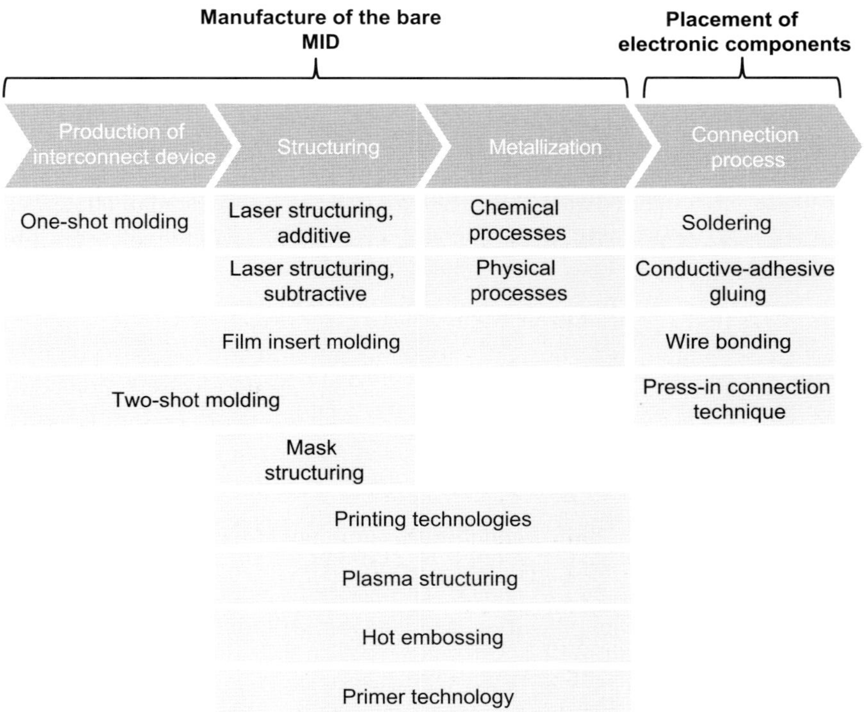

FIGURE 1.4 Reference process, MID [51, 58, 63]

All these processes are oriented toward the higher-order MID reference process (Fig. 1.4). A three-step process for production of the MID blank, it has a follow-up step in which any of the various connection techniques for completing the MID are applied [58]. The primary connection techniques for MID are soldering, conductive-adhesive gluing, wire bonding, and press fitting. The various manufacturing processes for MID parts are described in Chapter 3; Chapter 5 focuses on interconnection technologies. References 51, 58, and 63 are among the studies that set out other, similar reference processes for MID technology. As a standard reference volume for MID technology, this book pursues the aim of comprehensively covering all possible process chains, with the production processes of relevance for MID technology being examined in greater detail.

1.1.5 Factors Influencing Choice of Technology

The decision in favor of a particular technology or material invariably reflects the trade-off between the dictates of the market and the capabilities of any given option. The difficulty at the crux of optimum utilization of the technological, economic, and ecological potential of MID technology resides in an integrated product and design approach, which is also invariably a conceptual approach. Frequently more complex than those associated with conventional technologies, the decision-making processes this entails pose relatively high requirements for individuals and for corporate structures alike. Arriving at the ideal design for an MID necessitates taking a multiplicity of influencing factors duly into account. [135]

The challenges for the production of optimum MID assemblies are the constraints of the technical requirements (current-carrying capability, chemical and thermal resistance) that apply to the product, balanced against the opportunities afforded by the MID production processes and the materials used (Fig. 1.5). There is a library of design guidelines and data sheets for MID materials, some of which are developments specific to this purpose. Economic perspectives also factor into design considerations, because in many instances costs and, in turn, price are crucial in terms of the success of a given innovation.

The choice of materials and a structuring process that fits the applicable requirements are crucial for the implementation of any MID-based electronic component. Invariably, the substrate material must be suitable for the conditions of use and for the connection technology of choice. An important factor in this respect is the adhesive strength of the metallization, because this will undoubtedly have a major effect on long-term reliability.

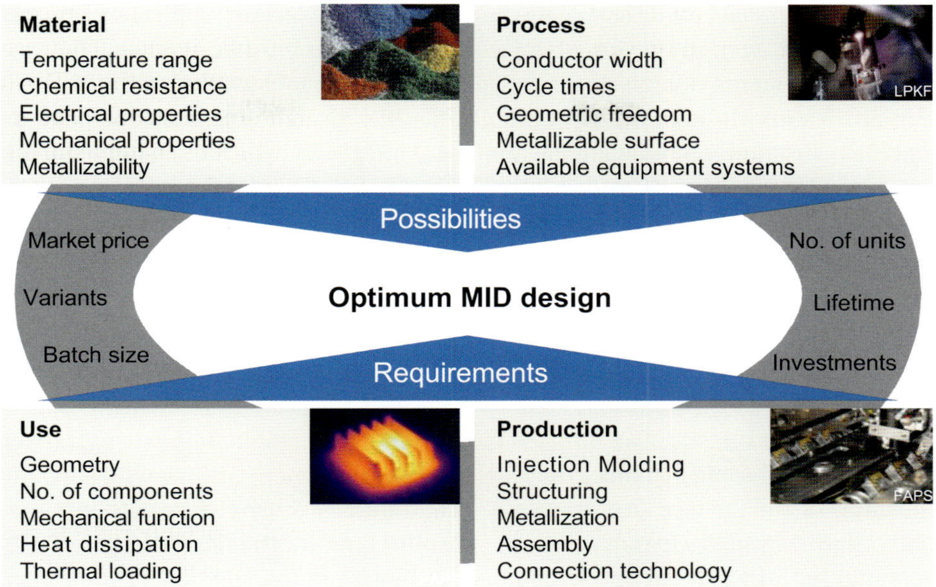

FIGURE 1.5 Decision-taking for MID in the context of possibilities and requirements [129]

1.1.6 Differentiation from Related Technology Fields

MID technology stands clearly apart from leadframe, printed-circuit board, and film technologies on account of the specific potential discussed in Section 1.1.3. None of the other candidates can fully replicate this potential (Fig. 1.6).

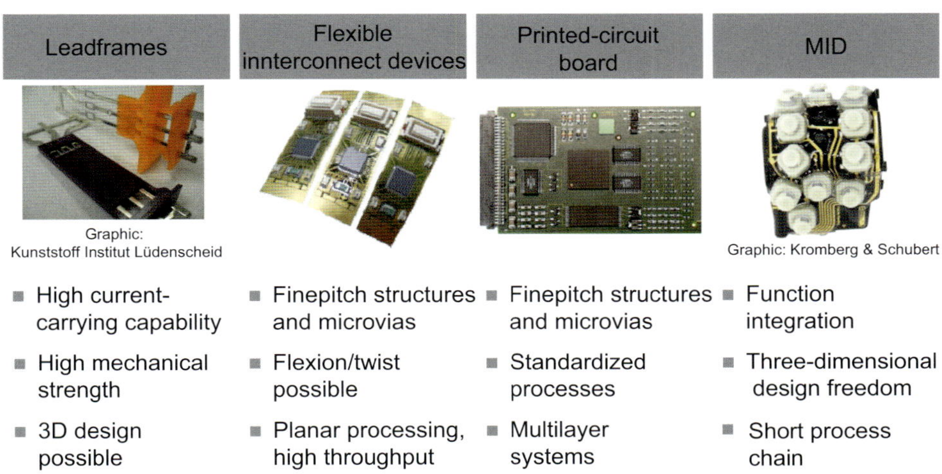

FIGURE 1.6 MID vis-à-vis conventional technologies

These conventional build technologies, moreover, are subject to technical constraints that MID can circumvent, for example with regard to miniaturization. The case studies in Chapter 9 depict current applications demonstrating overall-system miniaturization through 3D layouting. Difficulties in handling limp components and the restrictions on reducing the number of interfaces, which is a potential weak point for reliability, are further constraints to which the established technologies are prone. Conversely, MID technology is only conditionally suitable for areas of application in which these technologies are very much at home. Choosing the most suitable technology is always a matter of weighing up requirements, the intended area of use, and the complexity of the application.

In leadframe production, a die and punch set is used to emboss an electrically conductive structure from flat, strip-feed sheet material. The leadframe itself is then jacketed in plastic. The finished components are combination wiring carrier and package. In some instances they are also the interconnect device with galvanic finishing of the surface (e.g., tinning). Leadframes are used primarily in situations requiring high current-carrying capability and mechanical strength. Their potential for miniaturization is limited, even though individual lands can be very narrow (100 µm) [59, 129]. Flexible and rigid printed-circuit boards are suitable for mounting highly integrated components with minute connection structures and high packing density and for stacking into multilayer systems. Detailed standards and test criteria have been formulated. Substrate materials, connection mediums, and components are intensively qualified and closely matched to each other. Highly efficient systems engineering for high throughput rates in planar production is available. Film technology is often the first choice whenever the finished interconnect devices are likely to be subjected to bending or twisting action or shock and vibration loading [63, 153].

1.2 Relevant Industries and Areas of Application

An outline of published MID applications reveals a high level of diversification in terms of products available, the industry sectors serviced, and the areas of application (Fig. 1.7). Successful MID applications have been realized in all markets for conventional electronic modules. Chapter 9 presents in-depth case studies illustrative of many different projects. Current issues and developments in the MID-relevant industries and areas of application are discussed in more detail below. A great deal of this information was sourced from [50], with the addition of the latest findings from publications and press releases and the expert knowledge base.

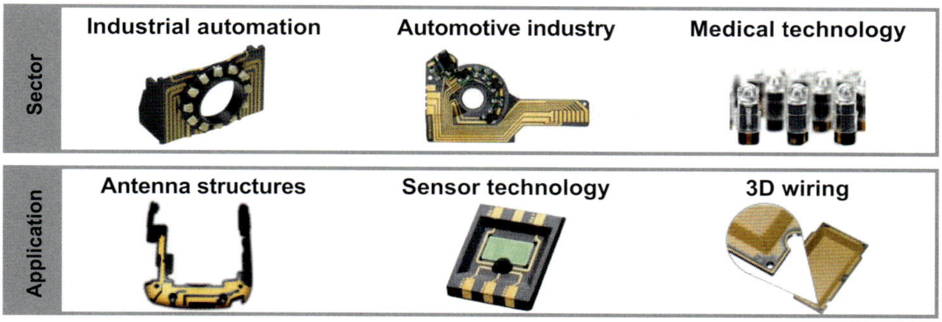

FIGURE 1.7 Typical MID applications from various industries and different applications (graphics courtesy of HARTING Mitronics, LPKF, 2E mechatronic)

Two industries in particular show considerable potential for high-end MID applications and can be graded as technology drivers. On account of the large number of high-profile projects in Europe in particular, these two sectors are the automotive industry and medical technology. The IT and telecommunications market is characterized by the mass production of MID antennas in Asia. Industrial automation and aerospace technology are other interesting sectors.

Among the multiplicity of potential applications, sensor technology and antenna technology are of particularly high significance at this time. In the future there is a strong possibility of optical functions leading to many new applications, for example in combination with LED technology.

1.2.1 MID-Relevant Industries

The following is an overview of developments and current trends in the relevant industries.

Automotive

In automotive engineering, the number of electronic systems is increasing constantly on account of the demand for enhanced road safety, access to the latest trends in communications and IT, and environmental compatibility. Miniaturization is essential and functional diversity continues to expand, and both these factors have a positive effect on demand for MID integration. At the same time the high pressure of costs in the car-making industry dictates that over and above technical advantages, MID parts have to be an economically viable alternative to conventional technologies. The solutions as implemented generally afford the makers an edge on costs, but one that does not emerge until the complete system is taken into account instead of the MID on its own. Reliability is crucial in the automotive industry, and many experts are of the opinion that this requirement can be satisfied. MID materials (substrates, connection mediums) have been developed for applications extending to elevated-temperature environments, such as the conditions inside an engine compartment. Further development in the field of fully automated production will be necessary for other high-volume MID applications in the automotive industry.

Medical Technology

On account of demographic development and the booming interest in healthy lifestyles, medical technology is experiencing a rising demand for miniaturized, low-cost, high-efficiency medical devices. Important MID applications in this sector include, for example, systems for audiology (e.g., hearing aids). Microphone directional precision within a tiny package is essential in this application, to enable the user to pinpoint audio sources. Single-use, disposable products are another possible application. Sterile sample carriers have to be used for bioanalysis. Two-shot molding is an ideal process for implementing the functions of a sample carrier, because the method enables a microfluidic channel to be integrated into the carrier body along with analysis electronics.

IT and Telecommunications

Tremendous use has been made of MID technology in recent years for antennas, for example in smartphones. Production volume of MID antennas to date extends to millions of units. The benefits lie in the combination of 3D design freedom with electromagnetic transceiver properties for different wireless standards, including for example WLAN, Bluetooth, UMTS, and LTE. Antennas up to 300 mm in length are being used increasingly in mobile computers, and relocating the antenna modules directly into the casing makes the overall package both slimmer and lighter [107].

Industrial Automation

High flexibility, availability, and cost efficiency are the crunch factors for success in automation technology. Smart sensor concepts and lighting modules based on 3D arrays of LEDs or sensor chips are major possibilities for MID. MID can also be used for long-range RFID antennas, because three-dimensional antenna structures enable noncontacting signal transmission. This in turn opens the way for optimized logistics control and process monitoring.

Others

The aerospace industry and the military sector are other important fields for MID. These applications focus on the three-dimensional arrangement of sensors, but to date very little information has been put into the public domain. The household-goods sector is of no more than secondary importance at this time, but this is not to say that potential for MID applications might not unfold in future, for example within the major-appliances part of the domestic-equipment market.

1.2.2 Areas of Application

The areas of application of relevance for MID are outlined below.

Sensor Technology

Current sensors, light sensors, pressure sensors—these are only some of the many established sensor applications that demonstrate the strengths of MID technology. Sensor build and positioning lay the groundwork for development of function integration and miniaturization in these applications. In many metrological processes, the optimum arrangement or precision alignment of the mechanical and electrical components is crucial, and this is one important consideration. Another is that the measurement process should be sited as close as possible to the medium. These spatial dependencies are at the crux of precision in measurement. MID can be made much smaller than conventional sensors.

Antennas

MID antennas have become mainstream in recent years, and nowhere more so than in the mobile-communications sector. Units of this kind are manufactured in very high numbers in Asia. Molex announced the production of its twenty millionth LPKF-LDS® antenna [76] back in 2009, and a large proportion of MID antennas are two-shot moldings. The sector is further expanding to include tablet PCs, with demand leaning more toward larger antenna structures. RFID antennas for very widely varying applications constitute yet another sector.

3D Wiring

The design freedom afforded by MID technology removes the planar constraints of a conventional circuit board, so conductors can be tracked virtually anywhere in the available 3D space. The three-dimensionalism attained in this way gives us highly integrated, multifunctional constructs. In complicated layouts with many electronic components, conventional techniques with planar placement can be combined with the 3D conductor tracking singular to MID technology.

Interconnect Device/Package

It is not possible to define clear borders around the application potential of MID as interconnect devices in or as packages for electronic modules. This is because, seen purely as interconnect devices, MID parts can compete with conventional printed-circuit boards only subject to certain constraints. Certainly, both individual electronic components and multifunctional chips can be surface-mounted and wired, but in general the functions involved are of the signal-processing or data-processing kind. The voltage supply is limited on account of the current-carrying capability of MID parts. Consequently, the greatest potential is in using MID parts as packages with directly integrated conductor tracks. The simplest example is the MID protective cap: an MID package with meandering conductor structures used as an electronic surveillance device to detect attempted burglary. Component integration can also be interesting for certain applications (e.g., seat-positioning switches).

Plug Connectors

The use of MID parts as plug connectors is set to become much more important in future. Every MID application has to be connected to its periphery, so a connector has to be integrated into the MID component. The pins for this purpose can be either metal or metallized plastic. This use of MID as plug connectors will necessitate standardization and the production of a code of practice for the design of an MID plug connector.

Optical Function

Optical functions can lead to new applications in various sectors. For example, MID technology can be used in lighting, a field where development is now keenly led by advances in LED engineering. The requirements for the positioning and alignment of LEDs are similar to those for sensors. Using MID as reflectors with optical coatings is another possible avenue, although surface roughness and heat dissipation both look challenging in this respect. Yet another interesting aspect is the integration of fiber-optic waveguides into MID.

1.3 The MID Market, a Global Comparison

Pioneering work on MID technology was undertaken by companies in the 1970s, mostly in the United States but also increasingly in Europe. In this early phase attention was focused primarily on developing substrate materials and the various metallization processes. By the 1990s companies and research institutes were combining their development capacities and forming interdisciplinary project teams. The section below describes the development of the technology and points out the current main areas of attention and the preferred processes in the individual regions of the world.[1]

[1] This information was collected jointly with Dr. Wolfgang John.

1.3.1 Historical Development

A patent filed by E. W. Lehtonen back in 1969 mentioned the possibility of manufacturing electrical interconnects by molding thermoplastic materials [51]. Further thoughts about integrating electronic and mechanical functions in a molded part date back to patents from the late 1970s and early 1980s. It was during these years that the first design and feasibility studies were undertaken as joint projects by AT&T Bell Labs, Union Carbide, and General Electric Plastics in the United States. US company Circuit-Wise linked up with Leicester Circuits in England to examine the possibilities of producing conventional printed-circuit boards as injection-molded parts. As long ago as 1984 an early German MID pioneer, Volker Zippmann, was nurturing his company Buss-Werkstofftechnik's first steps toward the new technology with the new material PEI (Ultem), a development of General Electric Plastics. In the years that followed, large international players took up the challenge of developing high-specification plastics and the accompanying specific metallization processes. Companies from many different walks came up with tentative solutions for implementing mechatronic product concepts and, in turn, provided input for evolution of the technology and its emerging processes (Fig. 1.8). But when Eastman Kodak and Mitsui Petrochemical & Pathtek developed new two-shot molding techniques, from about 1986 onward the fledgling technology got the push it needed with the hugely increased freedom of the third dimension. As it turned out, primarily on account of economic aspects, the dominant notion of the day—that molded interconnect devices would eventually supersede conventional printed-circuit boards—proved impractical.

The leading processes of those early days derived from the printed-circuit board industry and included a modified form of masking, laser etching, two-shot molding, and hot embossing. A raft of evolutions and modifications of these established processes emerged during this time. For example, two-shot molding was streamlined by using core-catalyzed plastics for the plateable component, so it was no longer necessary to eject the first shot from the mold and seed it before laying on the second shot.

FIGURE 1.8 Early MID applications: Solder base for an airbag system, Buss-Werkstofftechnik GmbH (left); SPS AG 95 control unit for PLC, Siemens AG (right)

By the early 1990s the term Molded Interconnect Device (MID) had displaced the various other nomenclatures used until then (e.g., molded printing wiring board, molded electronic assembly). The term and the abbreviation were ratified by MIDIA (Molded Interconnect Devices International Association) at its inaugural meeting in March 1993. The notion of forming this association was first mooted in the fall of 1992 at a meeting of the two largest MID manufacturers Mitsui-Pathtek and Circuit-Wise in the United States, and the declared aim was to unite the national and international activities of the big MID players. In the United States these were the three companies: Mitsui-Pathtek, Circuit-Wise, and ufe. Their ranks were later joined by other companies, including Fuba Hans Kolbe, Siemens, and Buss-Werkstofftechnik from Germany and a handful of Japanese specialists such as Sankyo Kasei.

In Germany too, interdisciplinary workgroups had been brought into being in the early 1990s to tackle cross-industry problems. In 1992 these developments culminated in the founding of the Forschungsvereinigung Räumliche Elektronische Baugruppen 3-D MID e. V. (Research Association Molded Interconnect Devices), and today this body's activities are still coordinated by the Institute for Factory Automation and Production Systems (FAPS) of the Friedrich Alexander University Erlangen-Nuremberg. The stakeholding companies saw in this form of joint research with a close university tie-in the ideal platform for the launch and evolution of MID technology in Germany. The approach is unique in international terms, and it has brought the technology the high level of importance it enjoys in the country today (see Section 1.6 for more information). By 1995 a national association for the development of MID technology had been formed in Japan. The member companies of this association also became affiliated members of MIDIA in the United States.

1.3.2 MID Focuses by Region

Until 1996, the US market was very transparent. It bore the stamp of the three big players Circuit-Wise, Mitsui-Pathtek, and ufe, companies that had brought out a raft of products, including many for the US automotive industry. From the European or German perspective, it looked at times as though projects of this nature were being pursued with a great deal more vigor in the United States. In 1996, however, Circuit-Wise took over Mitsui-Pathtek's MID business and moved energetically toward concentration in the sector. Market response to the move was reticent and skeptical, and income dropped off sharply. Recovery did not follow until increasing acceptance for the MID business of Circuit-Wise emerged across the industry. Circuit-Wise founded a spin-off company that it called MID LLC to handle MID-related business.

In 2000, Rochester-based MID LLC was acquired by Tyco Electronics. The new owner continued MID operations in the town of Rochester until 2008, concentrating heavily

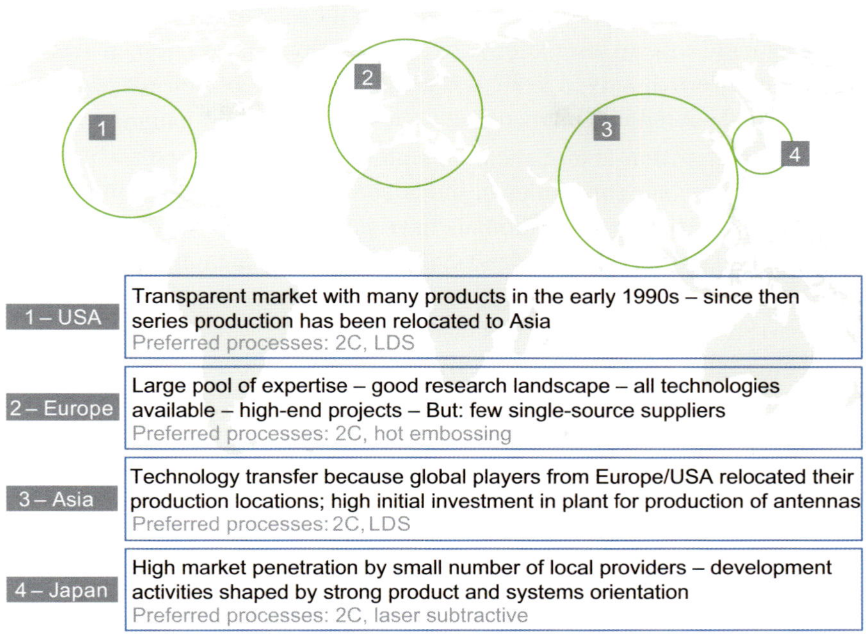

FIGURE 1.9 Preferred processes and focuses per region [50]

on mobile-communications antennas on a two-shot molding basis and adding laser direct structuring (LDS) as a second process in 2006, before relocating production to China. The biggest EMS service providers, especially for the mobile communications market, have all relocated production to Asia. The preferred processes are two-shot molding and, increasingly, LDS (Fig. 1.9).

Virtually the entire set of production processes is available in Europe. With a very well established and broad-based scientific background, the LDS process has greatly enlivened the European MID market in recent years. The number of high-end component-set MID projects that reached series-production status, including many for the automotive sector, is proof of the high level of MID development in Europe. It also demonstrates the leading position in the international rankings for the production of mechatronic components, with a raft of product-matched MID technologies. There is a downside to the further propagation of the technology in that, until now, only a few single-source providers have emerged to handle the entire process chain, and this applies to two-shot molding and also to LDS, including the buildup and connection technologies.

Regarding process development in particular, Japan's MID history is characterized by a very strong focus on the product-oriented use of the technology by a very small number of MID manufacturers. Sankyo Kasei and Matsushita are the two lead players. Japan looks primarily to function integration and the miniaturization of products developed by the country's own industry. Preferred production processes are two-shot molding and laser subtractive structuring. Japan's official association of MID stakeholders dissolved itself at the end of the 1990s and was refounded in 2002 with the declared aim of encouraging joint activities.

Following the expansion in globalization since the late 1990s and the pressure of costs confronting major companies in industries such as mobile communications, the big suppliers of cellphone antennas have put all of their production facilities in Asia. This in turn has meant high-volume production of antennas in Asia from early on. Two-shot molding was the process of choice at the beginning, but LDS has also become increasingly common on account of its flexibility for accommodating design changes. Cellphone antennas can scarcely be compared to complex, highly integrated MID modules such as those developed and used for automotive applications, for example. In Asia, nonetheless, high-volume production has engendered a huge process knowledge base that can be put to good use in the development of higher-specification products.

1.4 Main Fields of MID Research

Broadly speaking, the main fields of MID research can be broken down into five principal sectors (Fig. 1.10):

1. substrate materials
2. interconnect-device manufacture
3. 3D placement and interconnection technology
4. quality and reliability
5. planning and development

Other activities focus on end-to-end optimization of the process chain (e.g., reel-to-reel production, variable-output production solutions with modular plant configurations) and MID as part of the overall system (e.g., integration by standardized interfaces). Chapters 2 through 7 discuss the state of the art in these areas.

Substrate Materials

Thermally conductive plastics will significantly expand the potential use of MID. High-current applications call for new materials to deal with heat generation. Characterization of the key between metallization and plastic substrate is another important factor. Pigmentizing MID plastics to produce colors other than gray or black makes MID a much more attractive solution for packages. Biocompatible materials can extend the range of possible applications in the medical-technology sector.

Substrate materials	Thermally conductive plastics
	High-temperature thermoplastics
	Bio-compatible, transparent or colorized materials
	Thermosets as substrate material
Interconnect-device manufacture	Printing technologies/plasma structuring
	Multilayer constructs
	Increased current-carrying capability
	Rapid prototyping
3D assembly and interconnection technology	Multiple workpiece carriers for standard SMT process chains
	Chip-on-MID placement technologies
	Embedded components
	MID housing
Quality and reliability	Test and inspection criteria for MID
	Standardization and specification
	Forecasts on long-term reliability
	Simulation models
Planning and development	Design tools
	Development methodologies
	Heat-removal concepts
	Recycling

FIGURE 1.10 Main areas of research in MID technology [50]

Interconnect-Device Manufacture

Printing technologies (Aerosol-Jet®, inkjet) and plasma structuring (Flamecon®, plasmadust®) have major potential for future use in series production. Sustaining long-term stable functional reliability for MID under extreme operating conditions necessitates protecting the electronic components against medium exposure and high thermal and mechanical loads. One avenue currently being explored in this respect is that of enclosing MID modules by assembly injection molding. Another point of importance in the manufacture of interconnect devices is increasing current-carrying capability. Limited by chemical metallization in the case of some technologies, this necessitates practical extension by galvanic processes.

3D Placement and Interconnection Technology

Automating 3D placement is another main focus of research activities at this time. Consideration of the process chain in its entirety is the central issue in this respect, along with appropriately versatile systems and high throughput rates. In this respect multiple workpiece carriers can be a key to integration into standard production systems. Connection technology will have to be brought more closely into line with the developments in stock electronics production. Studies exploring chip placement on MID are in the foreground in this respect, but this also applies to progress in miniaturization with regard to component size (e.g., resistors and capacitors with 0201, 01005 packages) in surface mounting.

Quality and Reliability

MID-specific test criteria are of significance for further development of the technology. Common practice now is to select test criteria for MID on the basis of existing codes that are not MID-specific (e.g., IPC). MID manufacturers want to see specifications for test criteria and for criteria selection that will enable optimum quality tests to be put in place for MID-specific applications.

Planning and Development

There is a need for design tools and development methodologies that will help identify and open up new areas of application. MID technology offers tremendous potential for LED carriers, for example, and also for optoelectronic applications. In particular, this will necessitate the development of efficient heat-removal concepts. Simulation is another important point for a better understanding and estimation of thermomechanical effects.

1.5 Key Factors for Successful Projects

The MID Survey 2011 set out the key factors for the implementation of successful MID projects (Fig. 1.11). These factors are the complexity of the process chain, correct cost calculation, MID-compatible design, and willingness to take risks.

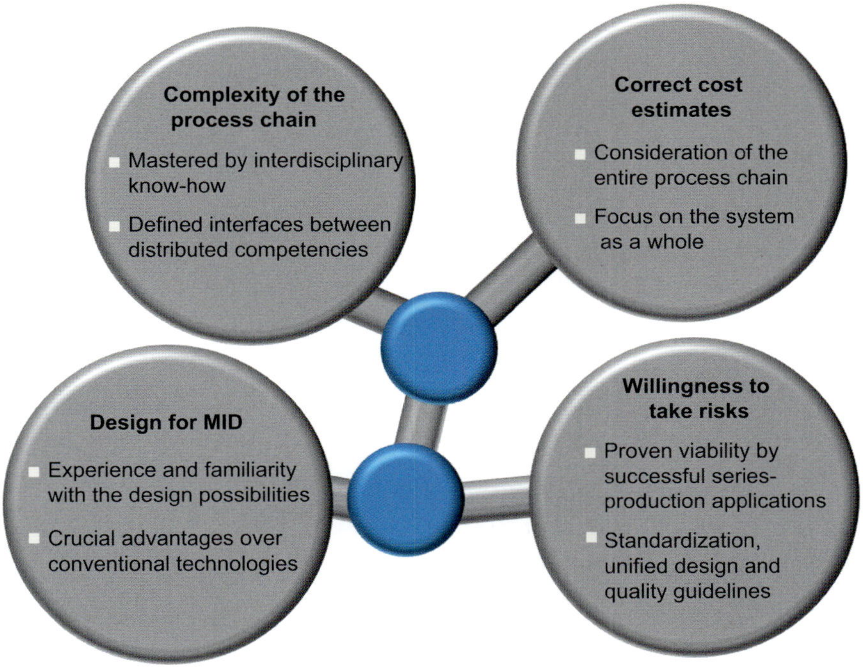

FIGURE 1.11 Key factors for successful MID projects

Complexity of the Process Chain

Companies looking to implement successful projects need interdisciplinary know-how that has to be brought together in the individual projects. Defined interfaces between the stakeholding companies in MID projects are essential whenever areas of responsibility are divided.

Correct Cost Estimates

A well-grounded estimate of costs is a prerequisite in the lead-in to a project. Costs cannot be calculated correctly unless the entire process chain or the overall system is taken duly into account. This can prevent the MID technology being erroneously identified as the reason for costs going beyond those proposed for other, more conventional solutions.

MID-Compatible Design

Recognizing the potential of the technology, many companies are able to come up with ideas for new products. Successful implementation requires appropriate experience in assessing the design-based solutions that can be implemented with MID and deciding which production process should be chosen for a given case.

Willingness to Take Risks

Reasons for a lack of sufficient willingness to take risks have included the shortage of proof of long-term reliability and the lack of MID-specific design and quality guidelines. One factor now expanding the willingness of individual companies to take the associated risks is the growing body of reference projects and successful stock applications.

In line with these four points, there is a need for know-how along the entire process chain from formulas for thermoplastics through injection molding, structuring, and metallization to final placement of the electronic components. Correspondingly, in-process quality assurance with defined interfaces between stakeholders is of tremendous importance. It is in this context that this book addresses each individual process step with the purpose of avoiding possible mistakes in MID projects.

1.6 Network-Oriented Cooperation within the Research Association Molded Interconnect Devices (3-D MID)

To an increasing extent, innovative products and production processes are the result of cross-company cooperation. Customers, suppliers, and research institutes are primary movers in technology-oriented cooperation, and to some extent the same can also be said of competitors.

One initiative of note in this respect is the cooperation between companies and research institutes along the process chain, which, at a very early juncture, led to the establishment of Germany's Forschungsvereinigung Räumliche Elektronische Baugruppen 3-D MID e. V., the Research Association Molded Interconnect Devices. Membership in this network affords the affiliated companies crucial access to external technological know-how. At this time, in 2013, the association networks almost 100 companies and research institutes from the fields of plastics and injection-molding technology, structuring/metallization, and 3D assembly with placement, soldering, and testing. System providers and users of MID technology extend the network even further.

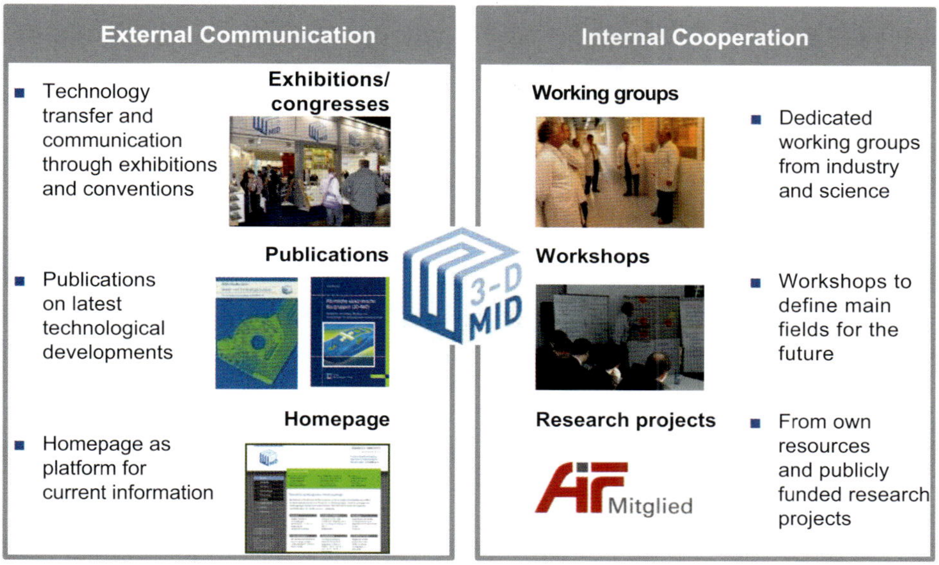

FIGURE 1.12 The declared aims of the Research Association 3-D MID are the promotion and further development of MID technology

Numerous joint research projects are in progress to promote and develop MID technology (Fig. 1.12). Joint information activities for technology transfer and supplementary marketing measures further contribute to driving the market penetration of this attractive technology.

All research activity is undertaken in close alignment with industry. The primary focus is on areas of research from end to end along the process chain: planning and development of 3D-MID, development and modification of substrate materials, interconnect-device manufacture, 3D placement and connection technology, and the entire field of quality and reliability.

2 Materials for 3D-MID

M. Fuchs

Materials are products that integrate in a number of production steps into an end product, or that can form the end product by themselves. Modern materials science broadly classifies materials as belonging to any of five major groups:

1. metals (e.g., aluminum or magnesium)
2. nonmetals (including the noble gases such as helium, the halogens such as chlorine, and a small number of other nonmetals that are not included in the other groups of substances)
3. organic raw materials (e.g., carbons and wood)
4. inorganic, nonmetallic raw materials (glass and ceramic are two examples)
5. semiconductors (e.g., tellurium and germanium)

Yet another group is formed by the composites, which consist of at least two aggregated materials. A composite has material properties differing from those of its component constituents. The properties of composites depend to some extent on substantive properties and on the geometry of the components. The connection can be of the cohesive-bonding or positive-locking kind, or a combination of both. In terms of geometry, composites are distinguished as

- particle composites,
- fiber composites,
- laminar composites, or laminates as they are also known,
- infiltration composites, or
- structural composites.

The components of a composite can themselves be composites. Particle composites have particles (and fiber composites have fibers) embedded in another component of the composite, which is known as the matrix. The fibers in fiber composites can be nonoriented or oriented in a preferential direction or directions. Fiber composites can have a multilayer build-up, but they are not considered laminates if the layers in the sequence are of the same kind. Laminated composites, or laminates, consist of layers of different kinds. A sandwich composite is a special case and consists of two identical outer layers sandwiching a thicker core. The individual components of an infiltration composite each form a separate but cohesive open-pore material. This classification by polymers (plastics), metallic, ceramic, and organic materials produces the basic possible combinations. Materials science classifies these materials primarily on the basis of their chemical structure, although production and workability and practical properties of the materials are also taken into consideration (Fig. 2.1).

The objective is to combine the different advantages of the individual materials in the end product for a given application, while excluding the disadvantages.

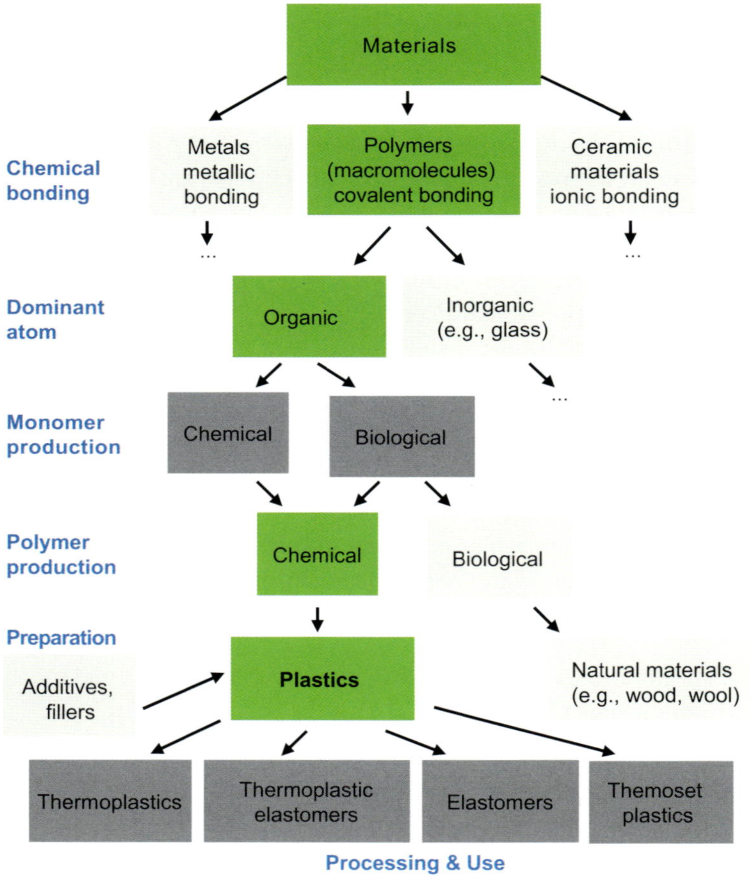

FIGURE 2.1 Distinguishing features for the classification of materials [8]

2.1 Introduction to MID Materials Classes

A class of materials used in the manufacture of three-dimensional molded interconnect devices (MID) is plastics. They are the substrate materials of preference for the basic body of MID. Vis-à-vis ceramics, the advantage of plastics is that on account of high fracture toughness they are better able to dissipate local peak stresses under mechanical loading by elastic and plastic deformation. Ceramic is better at retaining its shape (high specific rigidity) and has less thermal expansion, but the technological complexity of working ceramics increases with the size and shape of the part. A uniform, homogeneous microstructure has to be sustained throughout a large volume, even though heat can be introduced only from the outside in the firing process. Quite simply, plastics offer more production flexibility, from macro- through to micro-MID.

Plastics are defined as organic or semiorganic materials with high molecular weight. They are made up of very large, long-chain molecules (macromolecules) themselves synthesized from recurring basic building blocks by chain, or successive, reactions (polymerization, polycondensation, or polyaddition). The nonmetallic elements carbon (C), hydrogen (H), and oxygen (O) are the main constituents in the atomic makeup of organic plastics. But nitrogen (N), fluorine (F), and sulfur (S) are also frequently present in the chemical structure of plastics [8]. The polymer profile is sharply characterized by the makeup, in other words the constitution and the configuration, of the macromolecules. There is also the morphological structure deriving from the conformation and aggregation of the macromolecules, which also influences polymer properties (softening, melting, and crystallization). The structural diversity of plastics is ultimately the net outcome of chemistry and morphology. In terms of their structural features, organic polymers are characterized by the following properties in particular [8, 35]:

- By comparison with other materials, plastics are *specifically light substances* (densities up to about 2.0 g/cm^3).
- Plastics have *limited thermal stability* (bond breakage is irreversible).
- Plastics are *electrical insulators* and have *poor heat conductivity*.

Plastics are commonly further subdivided in terms of the degree to which their molecular chains are crosslinked, and by their composition and internal structure, as shown in Fig. 2.2.

Molding compounds based on epoxy or phenolic resins mixed with fillers are used for the conventional printed-circuit board technologies widely employed in electronics production. These molding compounds belong to the group of thermoset plastics. They consist of tightly crosslinked macromolecules, as reflected in their extremely hard and brittle material behavior at room temperature.

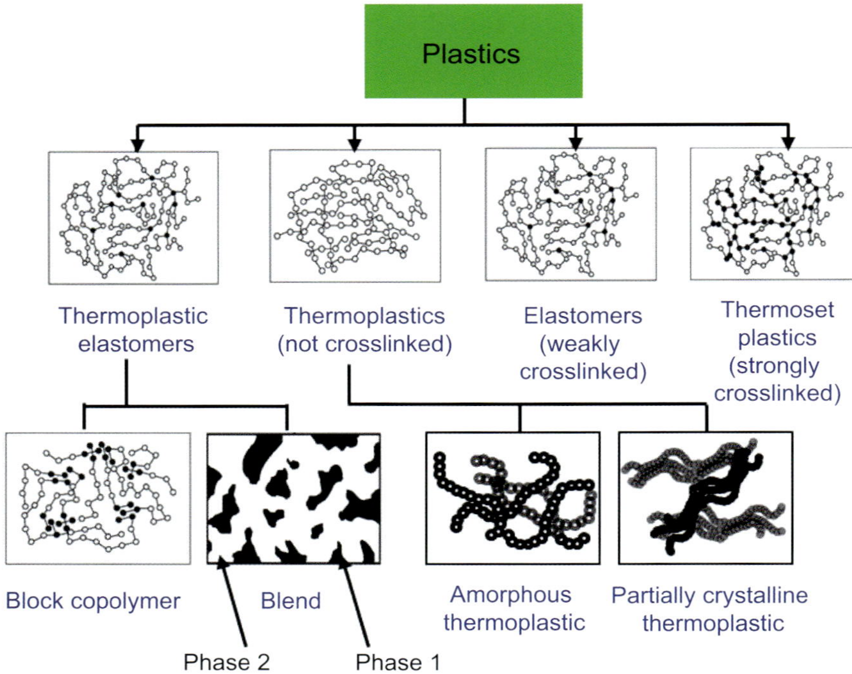

FIGURE 2.2 Classification of plastics

Depending on their basic cured condition, they can become harder when heated, but they do not soften. Thermoset plastics do not melt, so they cannot be welded [48].

Conventional assembly technologies that combine separately developed and manufactured mechanical and electronic components to form an interconnect device are invariably trade-offs between the factors of quality, cost, and time. Planar wiring boards made of laminated epoxy resins (EP resins), for example, both hamper progressive miniaturization and unnecessarily stretch the process chain, for example because they necessitate the inclusion of diverse assembly steps. The connecting elements are potential weak points limiting the electronic interconnect devices in terms of quality and reliability [51]. The MID industry endeavors to orient interconnect-device production toward the aspects of design freedom, flexibility, and rationalization. Given the concept underlying their design and since their manufacture is based on injection-moldable thermoplastic materials, MID offer tremendous potential in all these respects. In this context and also in light of the various routes and strategies for processing granulate into a finished injection-molded interconnect device, it is worth referring once again to Chapter 1 of this book.

2.2 Materials Properties and Determination of Characteristic Values for MID

Users can select from a wide offering of plastics with different properties profiles (Fig. 2.3). Over the course of time, the compelling search for more and more high-capability materials has extended the range of standard plastics and engineering plastics to include many high-specification polymers.

Obviously, therefore, the expectations that apply to any base material impinge on decision-making processes at several different levels. Alongside purely technical aspects such as intrinsic loadability and compatibility and the basic workability properties, ecological factors (e.g., suitability for return to the natural materials cycle) and of course economic factors (e.g., procurement costs and processing costs) influence materials development and selection.

The properties that characterize plastics are presented below, along with the test methods and processes for gauging these properties. The information is offered here with the objective of developing and expanding awareness for this group of materials.

Broadly speaking, plastics characteristic values are measured

- for initial assessment of new products,
- for comparison with known products,
- for preselection of plastics for a given load case, and therefore for a given use case, and finally
- for measuring moldings.

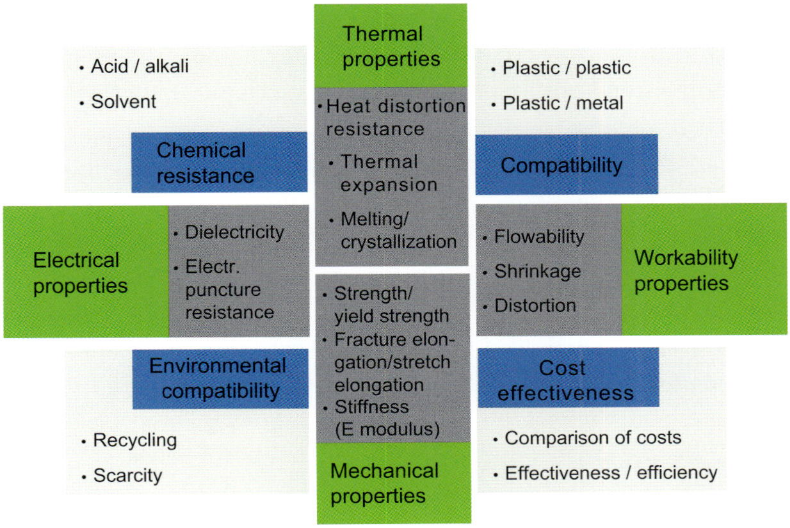

FIGURE 2.3 Requirements criteria for polymer materials for MID

To this extent, therefore, these characteristic values are also definitive for MID, the properties of which are ultimately based on the thermoplastic or thermoset behavior of different base materials.

The requirements outlined in Fig. 2.3 can be very diverse. In order to satisfy these requirements, characteristic values have to line up with certain basics:

- They have to be *meaningful.*
- They have to be *comparable.*
- They should be *rationally measurable.*

In regards to plastics, it is particularly important to distinguish between materials characteristic values as such and processing-related characteristic values. The latter in particular are coresponsible for the properties and manufacturing costs of moldings and therefore of MID as well—and to a much more significant extent than is the case with metals or ceramics. In terms of materials selection and MID preproduction, knowledge of the following is very important:

- rheology (i.e., flowability)
- hardening
- degradation
- shrinkage and tolerances, especially in the case of partially crystalline plastics
- anisotropy (e.g., reinforced plastics)
- crystallization and melting (partially crystalline plastics)

In line with the common ISO standards, these characteristic values are measured by test methods standardized for thermoplastics (TP) and thermoset plastics (TS). Table 2.1 lists the standards for determination of rheological characteristic values in accordance with the DIN ISO 10350 data catalog [8, 66]. The literature can be referred to for the standards for determination of mechanical, thermal, and electrical characteristic values.

There are also unified standards in place to regulate the production of the test specimens for determining these characteristic values.

TABLE 2.1 DIN ISO 10350 data catalog - rheological characteristic values [8].

<table>
<tr><th>Property</th><th>ISO (IEC) standard</th><th>Test specimen (dimensions in mm)</th><th>Unit</th><th>Comment</th></tr>
<tr><td colspan="5">Rheological properties</td></tr>
<tr><td>Melt volume rate</td><td>1133</td><td>Moldable compound</td><td>$cm^3/10$ min</td><td></td></tr>
<tr><td>Working shrinkage (parallel)</td><td rowspan="2">294-4 (TP)
2577 (DP)</td><td rowspan="2">60 × 60 × 2</td><td rowspan="2">%</td><td rowspan="2">ISO 294-3 (TP)
ISO 10724-2 (DP)</td></tr>
<tr><td>Working shrinkage (vertical)</td></tr>
</table>

2.2.1 Mechanical Characteristic Values of Plastics

The mechanical behavior of plastics, in other words the relationship between *strain* σ and the resulting *elongation* ε and the force F acting on a molding and the resulting *deformation* ΔL is best characterized by tensile testing. Unlike the flexural test, tensile testing can also be carried out with the softer plastics [8, 66].

In terms of the characteristic values measured by mechanical testing, a broad distinction is made between

- short-term behavior (e.g., tensile test),
- shock-loading behavior (e.g., impact test),
- static long-term behavior (e.g., tensile creep test), and
- dynamic long-term behavior (e.g., fatigue test).

The tests and test techniques for ascertaining the short-term behavior of MID materials and impact behavior are dealt with in more detail below, because they are particularly relevant for sizing and designing MID.

Short-Term Behavior: Tensile Test

In accordance with ISO 527-1 and -2, dog-bone-shaped test specimens (ISO 3167 multipurpose test specimens) are stretched at a constant rate specified in the test standard.

Force F after length change ΔL of the basic test section L_0 is recorded for each specimen. Appropriate division by the specimen's original cross-section A_0 returns *strain* σ in N/mm^2, while division of the length change by the original length of test section L_0 yields *elongation* ε expressed in percent. The *modulus of elasticity* E as the ratio of strain to elongation can be determined from the gradient of the force/deformation curve. The modulus is heavily dependent on elongation, so the gradient of the strain/elongation curve between 0.05% and 0.25% is the section often used to determine the modulus of elasticity as a characteristic value [8].

The relevant characteristic values are shown again in diagrammatic form in Fig. 2.4. Fig. 2.5 shows typical strain/elongation diagrams for selected plastics by way of example. The reliability of the MID is determined by the mechanics of the polymer basic body, so materials characterized by strength (elongation strain) and ductility are eminently suitable (see curves (b) and (c), Fig. 2.4).

Strength is definitively influenced by the following criteria:

- the thickness of the test specimen
- the point of removal of the test specimen from the sample (e.g., close to or remote from the sprue, boundary layer, or middle layer)
- most importantly of all, the direction of removal of the test specimen from the sample (influences of orientation, e.g., in fiber-reinforced plastics)

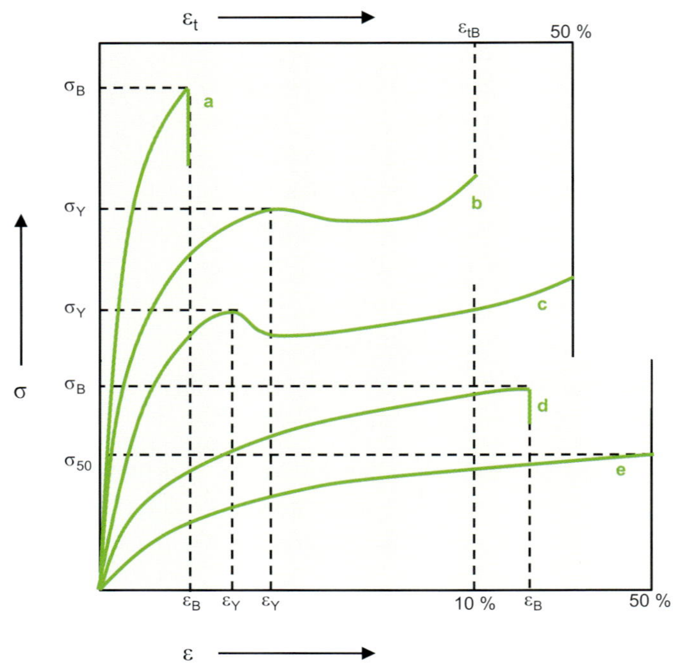

FIGURE 2.4 Schematic σ/ε diagrams with characteristic values from tensile testing
a) brittle plastic; b) tough plastic; c) stretchable plastic;
d) and e) softened plastic with E = tensile modulus of elasticity;
σ_B = fracture strain; σ_y = elongation strain; σ_{50} = strain at 50% elongation;
ε_B = fracture elongation; ε_y = stretch elongation; ε_{tB} = nominal fracture elongation

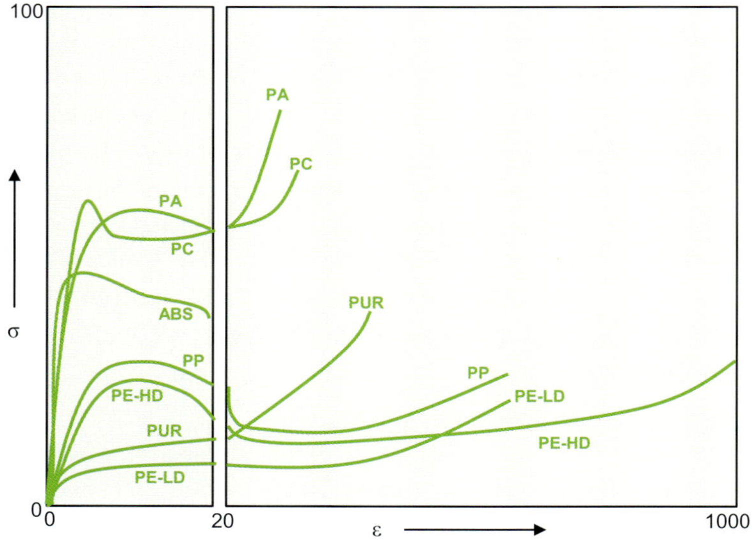

FIGURE 2.5 Sample strain/elongation transients for selected plastics from tensile testing

Short-Term Behavior: Flexural Test

ISO 178 flexural tests are conducted on beam-shaped test specimens. The preferred dimensions for these test specimens are (l) 80 mm × (w) 80 mm × (h) 4 mm, and the specimens are supported at each end and a bending load is applied by a die at the midway point between the supports. Measured forces and deflection are used to calculate characteristic values of flexural strain and outer fiber elongation, for example up to a defined maximum outer fiber elongation or up to fracture of the specimen [8, 66].

Short-Term Behavior: Hardness (Indentation Test)

Indentation testing is carried out to determine the resistance of the surface of the plastic to penetration by defined indenters. To a certain extent there is a relation between hardness and modulus of elasticity, insofar as the proportion of elastic deformation is predominant. The methods outlined below are suitable for measuring hardness [8]:

- *Ball-pressure hardness H (ISO 2039-1,2)*

  Characterizes the quotient of test force F, applied to the surface of the plastic by a ball 5 mm in diameter, and the area of the indentation produced by the loaded ball in the surface. The area is calculated from indentation depth measured under the load 30 s after application.

  The load stages applied as a function of the hardness of the plastic are as follows: 49, 132, 358, or 961 N.

- *Rockwell α hardness R_α (ISO 2039-2)*

  A ball 12.7 mm in diameter acts on the surface with a force F of 588.4 N.

  Indentation depth is measured after 15 s (d_h) and 10 s (d_s) to obtain the figure for Rockwell α hardness R_α:

$$R_\alpha = 150 - \left(d_\mathrm{h} - d_\mathrm{s}\right) \tag{2.1}$$

- *Rockwell HR (ISO 2039-2)*

  Calculated from indentation depth e measured after removal of the major load (after being applied for 1 s):

$$\mathrm{HR} = 130 - e \tag{2.2}$$

  The four hardnesses R are distinguished by test force F and ball diameter d as follows (F = 588 N, d = 2.7 mm), L (F = 588 N, d = 6.35 mm), M (F = 980 N, d = 6.35 mm), and E (F = 980 N, d = 3.175 mm).

- *Vickers hardness HV (EN ISO 6507)*

  A diamond in the shape of a square-based pyramid is applied to the surface with test loads of 100, 300, and 600 N for 40 s. On removal of the load the length of the indentation diagonals d in mm is measured and used to calculate Vickers hardness HV:

$$\mathrm{HV} = \frac{0{,}189}{d^2} \cdot F \tag{2.3}$$

- *Shore A and D hardnesses (ISO 868/DIN 53505)*

  Soft plastics can be tested by handy instruments to ascertain these properties. The test measures the resistance of the material to the penetration of a truncated cone presser (Shore A) or a rounded-tip cone presser (Shore D) as deformation after 3 s or, in the case of plastics with pronouncedly plastic response, 15 s after the hardness tester is pressed into the surface.

Behavior under Shock Loading

Behavior under shock loading, which indicates the toughness of plastics, is determined by standardized methods in impact or puncture tests. Impact testers or drop testers are used to determine the work absorption capability of bar-shaped or plate-shaped test specimens, in most cases until failure or in the case of puncture tests also until initial damage occurs. The test specimens can be notched. Common test procedures include [8]

- Charpy impact bending test,
- Izod impact bending test,
- tensile impact test, and
- puncture test.

Nowadays there are three preferred procedures set out in DIN EN ISO 10350 (single-point data catalog), with the choice depending on the toughness of the plastic:

- Charpy impact toughness for brittle plastics, carried out in accordance with ISO 179
- Charpy notched impact toughness for tough plastics, carried out in accordance with ISO 179
- tensile impact toughness of plastics that are not fractured by the above methods, carried out in accordance with ISO 8256

2.2.2 Thermal Characteristic Values

The properties of plastics depend on temperature to a very significant extent. Consequently, it is essential to determine the characteristic values that are definitive for use for the temperature ranges corresponding to the MID application, including thermal loading occurring during the applicable MID assembly and connection processes. Permissible top and bottom temperature limits for the use of plastics define the usable temperature range. Major factors for this range are [8]

- reversible softening or embrittlement when exposed to temperature over a short period of time, and
- behavior when exposed to temperature over an extended period of time with additional loading excluded.

2.2.2.1 Short-Term Exposure to Temperature

Ideally, short-term behavior under thermal loading is described in terms of mechanical properties as a function of temperature. Useful pointers when determined as a function of temperature are the modulus and damping values, modulus of shear G, and mechanical loss factor $\tan\delta$.

These thermally relevant characteristic values can be obtained by dynamic mechanical analysis (DMA). This analysis returns information about the way mechanical properties change under slight, usually sinusoidal dynamic loading as a function of temperature, time, and/or frequency. The periodic force (input signal) causes a corresponding response signal, namely deformation or strain in the viscoelastic test specimen, which can be evaluated for dynamic-elastic characteristic values in terms of amplitude and phase offset [36].

The resulting real and imaginary parts of the resulting complex module E^* can be analyzed. The storage modulus E' (real component) reflects the rigidity of the viscoelastic material and can therefore be used to assess the elastic properties. Given low load and reversible deformation, it corresponds roughly to the E modulus under single, rapidly imposed stress. The loss modulus E'' (imaginary component), by contrast, reflects the viscous properties of the test specimen and is proportional to the work dissipated in the material while the stress is applied. It characterizes the portion of an oscillation that is transformed into heat, for example, and therefore in energy terms that cannot be recovered and in mechanical terms that is no longer present. Establishing a relation between loss modulus and storage modulus yields a unit of measure for the mechanical damping or inner friction of a viscoelastic system. This is defined as the loss factor $\tan\delta$. Consequently a material with a mainly elastic deformation portion is characterized by a low $\tan\delta$. The loss factor can be seen as a measure of work applied mechanically beforehand and no longer

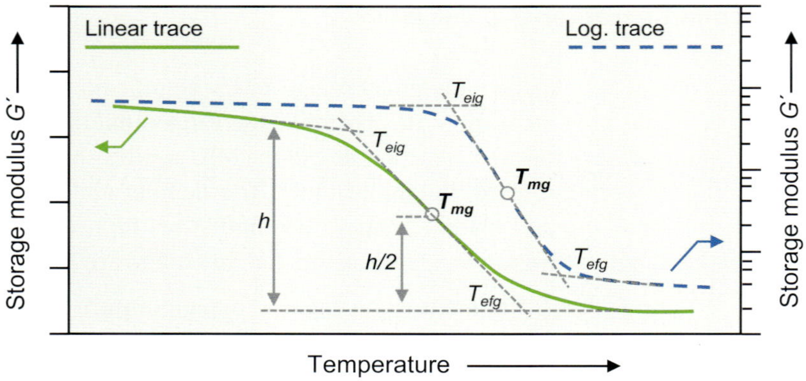

FIGURE 2.6 Temperature-dependent mechanical plastic behavior

recoverable when strain is relieved. This nonrecoverable proportion of work is converted directly into heat. G and $\tan\delta$ as a function of temperature can therefore be considered as something akin to fingerprints of the various plastics [8, 36].

DMA curves are also the point of departure for determining the glass transition temperature T_g. T_g is determined by evaluating the step-like drop evinced by the storage modulus with linear curve imposition across temperature by the half-step height method (Fig. 2.6).

The dimensional stability of a plastic when exposed to heat is another crucial property. The Vicat method of determining a plastic's softening point and the heat deflection temperature (HDT) method are two ways of rapidly determining which polymers have suitable characteristic values. In both methods the test specimens are heated under defined load with a certain heating rate while deformation is measured. Vicat softening temperature and HDT are each defined as the temperatures at which deformation reaches a given value [8].

DIN EN ISO 75-1,-2 (successor to DIN 53461) sets out a method of determining heat deflection temperature (HDT), which is in widespread use, so it is outlined below. The standard test specimens are rectangular in section and should preferably be laid flat and subjected to three-point flexion at constant load. A force F that can vary depending on the thickness of the specimen is applied by weights and/or springs to achieve an outer fiber elongation of 1.80 (method A), 0.45 (method B), or 8.00 N/mm^2 (method C). The specimens loaded in this way are then heated at a constant rate of 120 K/h (or 50 K/h). When deflection of the specimen reaches an outer fiber elongation of 0.2%, the temperature at which this occurs is the *heat deflection temperature* (also known as *heat distortion temperature*).

2.2.2.2 Long-Term Exposure to Temperature

Behavior under conditions of long-term exposure to temperature cannot be ascertained in short-term mechanical or thermo-mechanical tests, because time-dependent aging and relaxation processes are involved.

As set out in DIN EN ISO 2578, aging can be expressed in terms of temperature indices (TI). Selected properties are examined and compared before and after test specimens are kept under elevated temperatures in heating chambers. If there are other mechanical strains or deformations that still apply, the behavior of the plastic can only be gauged on the basis of creep tests and stress relaxation tests [8].

2.2.2.3 Relevant Thermal Characteristic Values for MID

Of the many thermal characteristic values measurable for polymers, two that are highly relevant for molded interconnect devices are glass transition temperature T_g (for amorphous and partially crystalline thermoplastics) and melting temperature T_M (for partially crystalline thermoplastics). They can be determined by dynamic mechanical analysis (DMA) or differential scanning calorimetry (DSC). Other characteristic properties of primary importance are (short-term and long-term) service temperature and heat deflection temperature, as are temperature response and heat conductivity and the thermal expansion of the thermoplastic substrate materials.

The substance and significance of the thermal expansion of plastics is discussed in more detail below. The reason is that MID are multimaterial systems, consisting of polymer substrate (thermoplastic, and in the future thermoset plastics as well), metal (e.g., Cu-Ni-Au), and electronic devices. Fig. 2.7 illustrates this, taking an MID conductor structured by laser direct structuring by way of example.

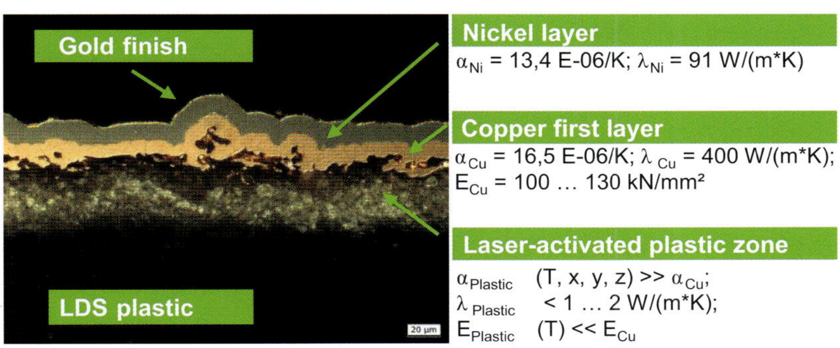

FIGURE 2.7 Microscope image of a laser direct structured, wet-chemically metallized Cu/Ni/Au conductor on a PET/PBT blend

Thermal Expansion

The expansion of plastics in response to temperature is much more pronounced than that of metals. The thermal expansion of the plastic is superimposed with reversible and irreversible effects that lead to a change in the linear transient with temperature.

They include

- softening at the glass transition or at the melting point (reversible),
- shrinkage due to relaxation of orientations (irreversible), and
- temperature-related effects of late crystallization, crosslinking, or thermo-oxidative degradation.

Many MID-relevant partially crystalline plastics, including for example PA, PBT, and PET, have glass transition temperatures in the range from –20 to 100 °C; in other words they are situated in the range that corresponds to that of the potential area of usage. This glass transition temperature is overshot and undershot by short-term and sometimes prolonged temperature fluctuations during production and in service (heating and cooling). These plastics, therefore, experience dramatic changes in their coefficient of thermal expansion (CTE) under practical conditions. In the long run the differences in thermal expansion in a composite will have a negative effect on the dimensional stability of the MID and thus on the adhesion of the metallization.

Development and Dissipation of Thermal Stresses

If there is anything to prevent deformation, the thermal expansion and contraction occurring in all materials when the temperature changes produce thermal strains. These strains can be all the more acute when plastics are positively locked to other materials such as metals, which show radically different expansion under the effects of temperature change. The thermal expansion of plastics over temperature is nonlinear, producing strains in the molding. Plastics generally have a much higher coefficient of expansion than metals or ceramic materials, and this can lead to difficulties with complicated assemblies and composites. The strains that can occur when a flat plastic-metal composite is heated act particularly as shear stresses in the interfacial areas. The metal hampers expansion, producing compressive strains in the plastic. These strains, in turn, act as tensile stressors on the metal layer. These latter strains in an MID can crack the conductors, bringing about total failure of the part. Shear stresses at the interface, moreover, can impair adhesion, and under thermal (cyclic) loading this can result in delamination of conductive functional layers. However, there is also the possibility of thermal strains in the plastic relaxing with temperature and time.

2.2.3 Electrical Characteristic Values

Plastics in general are good to excellent insulators, although this also means that they have a tendency to become electrostatically charged. Electrical behavior differs widely from one plastic to another; the differences are just as pronounced as those between the mechanical and thermal properties of individual plastics. This behavior is equally dependent on geometry and on loading and process parameters that apply in injection molding [8]. Broadly speaking, electrical characteristic values can be grouped as follows:

- insulation behavior (volume resistivity/surface resistivity)
- resistance behavior (puncture/creep/arcing resistance)
- dielectric behavior (dielectric index/dielectric loss factor)
- electrostatic behavior

Insulation Behavior

Insulation behavior is characterized by the resistivity that the base material can offer to the flow of an electric current. A distinction is drawn between volume resistivity (taking only the current flowing inside the plastic into account) and surface resistivity (measurement across electrodes touched to the surface) [8]. Conductivities can be introduced selectively by compounding electrically conductive fillers with the base material.

Volume resistivity and conductivity

The volume resistivity R of a body is defined as the ratio of applied voltage U to internally flowing current I:

$$R = \frac{U}{I} \quad \text{in } \Omega \tag{2.4}$$

Conversely, electrical conductivity S is a measure of a material's ability to conduct current. Electrical conductivity, then, is identical to the reciprocal value of resistivity R.

Surface resistivity

Surface resistivity R_0 is the ratio of a direct voltage U applied across two electrodes on the surface of a test specimen to current intensity I between the electrodes:

$$R_0 = \frac{U}{I} \quad \text{in } \Omega \tag{2.5}$$

Resistance Behavior

Electrical puncture resistance

High electrical voltages can cause electrical burn-through or puncture. Electrical puncture resistance E_d is the quotient of the voltage U_d at which puncture occurs and the distance between the conductive parts to which the voltage is applied (i.e. thickness d of the test specimen):

$$E_d = \frac{U_d}{d} \text{ in } \frac{\text{kV}}{\text{mm}} \tag{2.6}$$

Electrical creep resistance

The figures for short-term electrical puncture resistance are merely pointers with regard to the behavior of a material subjected to long-term application of an electric voltage. Over periods up to about 1 s electric puncture resistance is virtually constant; the occurrence of an electrical burn-through is caused by electrons. Thermal puncture can occur if the voltage is applied for longer, and the phenomenon can also affect plastics that warm up either on account of sufficient conductivity or in the alternating field due to dielectric losses. The reason is that electrical puncture resistance diminishes as temperature rises.

Dielectric Behavior

In some areas of application, including for example high-frequency engineering and antenna design, the dielectric constant ε_r and the dielectric loss factor $\tan\delta$ must be known to gauge component performance.

The dielectric constant is an expression of the ratio of capacitance of a capacitor with plastic as the dielectric to the capacitance of the same capacitor with air as the dielectric. The loss factor, by contrast, is a measure of the energy that an insulating substance in an alternating field transforms into heat and is therefore lost as electrical energy. The dielectric characteristic values are material constants. They are dependent on temperature and frequency range and also on ambient conditions such as humidity, for example [8].

2.3 Materials for MID Technology

The thermoplastics that are of most interest for the MID market or most commonly encountered as substrate materials are introduced in this section. Certain properties of these materials are explained, although it should be noted that with the development of new materials and the evolution of existing materials continuing apace, this list and the choice of materials from the MID portfolio are by no means comprehensive. The section closes with some notes on thermoset, injection-moldable compounds as an outlook on new high-performance plastics. Although they are not yet state of the art in MID technology, on account of their attractive properties profiles (e.g., high dimensional stability or low thermal elongation), these moldable compounds would appear to be suitable future candidates for substrate materials.

2.3.1 Thermoplastics for MID

In principle MID manufacture is not restricted to any given base material. In fact, very different polymers can be used. The choice of material and production process is primarily linked to the mechanical, thermal, and electrical requirements of the end purpose of the finished MID. The materials group of thermoplastics is normally subdivided on the basis of heat distortion resistance or long-term service temperature on the one hand, and the price of the raw materials on the other [93].

The plastics pyramid (Fig. 2.8) is a widely accepted mode of graphically presenting the differences between three major subgroups of thermoplastic materials:

- standard thermoplastics
- engineering thermoplastics
- high-performance plastics

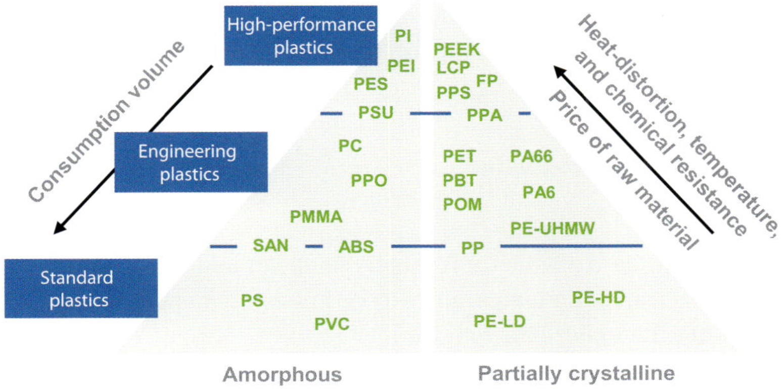

FIGURE 2.8 Plastics pyramid including worldwide consumption figures

In comparison to the overall market for thermoplastics, in figures high-performance plastics are a very small but extremely interesting sector in terms of application engineering. Despite extraordinary properties profiles that qualify them for much wider areas of application than say engineering plastics, high prices and costly processing are the primary factors restricting sales.

Consequently, many MID are still made of engineering-grade thermoplastics, because they are available at low cost and present virtually no difficulties whatsoever as regards their workability in existing production facilities. In most instances existing MID materials can be adapted reasonably well for a given purpose within economically acceptable bounds

- by selectively combining materials,
- by blending thermoplastics,
- by including particular additives (e.g., nucleating agents or flame retardants) or conductive fillers (e.g., metallic or ceramic particles), or
- by post-production modification independent of the production process as such (e.g., by electron bombardment).

Typical MID substrate materials are briefly characterized below within the broad system of classification shown in Fig. 2.8. This characterization deals primarily with important material properties (thermal limits of use, for example), but it also gives some indication of how the material behaves for metallization, along with its chemical resistance and solder resistance. The overview affords a point of initial access to users of MID technology. Ultimately, however, each user must be aware that material characteristic values can be much more complex than they appear from the outlines below. Moreover, a material like polyamide 6 (PA 6) from one raw-materials supplier does not necessarily have exactly the same properties as another manufacturer's PA 6, due primarily to differences in the manufacturing process and formulations.

When selecting materials it is always advisable to consult the data sheets issued by the manufacturers of the raw materials and obtain detailed information from databases such as CAMPUS and compare these data closely with the requirements set out in the specifications for the intended product.

2.3.1.1 Standard Thermoplastics

Polypropylene (PP)

PP is one of the standard plastics. It is largely nonpolar and has a crystallinity of 60 to 70%. The glass transition temperature of PP homopolymers is about 0 °C. Above this temperature the material is tough.

If not placed under mechanical loading, parts made of PP have a maximum long-term service temperature of about 110 °C. A temperature of 140 °C is permissible for short-term exposure without external loading. Melting temperature is in the range from 155 to 160 °C. For this reason the material is not suitable for standard soldering processes. PP has high chemical resistance on account of its nonpolar character.

PP has no inherent flame-retardant properties. High shrinkage means that dimensional stability is poor. This shortcoming can be countered only by using talc-reinforced, mineral-filled, or glass-ball-filled PP for reduced and more isotropic shrinkage. Talc reinforcement reduces heat distortion resistance [8, 40, 48].

Syndiotactic Polystyrene (sPS)

sPS is a partially crystalline plastic that can withstand temperatures up to 265 °C for short periods. Long-term service temperature can reach 150 °C. Maximum crystallinity is approximately 52% and depends on processing conditions, on account of the relatively slow rate of crystallization. HDT is determined by degree of crystallinity. sPS has a melting point of 275 °C and a glass transition point of 105 °C. It generally has 10 to 40% glass fiber reinforcement. sPS is extremely chemical-resistant but easily damaged by oxidizing mediums such as chlorine, ozone, or niter, particularly at high temperatures.

sPS can be metallized by wet-chemical or catalytic processes. Low isotropic shrinkage makes sPS eminently suitable for precision moldings. sPS absorbs virtually no water at all. Suitable additives can render sPS flame-retardant (up to UL94 V-0 at 0.8 mm).

The types highly reinforced with glass fiber can be soldered with all the common soldering processes and even with lead-free solders [8, 40, 48].

2.3.1.2 Engineering Thermoplastics

Modified Polyphenyl Ether (modified PPE, also PPO)

Modified PPE, also known as polyphenylene oxide (PPO), is a blend of PPE and polystyrene. It belongs to the group of amorphous plastics. It has good mechanical properties and its electrical properties are virtually unaffected by temperature and frequency. Long-term service temperature is in the range from 100 to 110 °C. Infrared soldering processes cannot be used, although some selective soldering methods can. Modified PPE can be chemically metallized.

It is not resistant to hydrocarbons and is susceptible to stress cracking on account of its styrene content. There are types containing different proportions of glass fiber and with varying degrees of flame retardation and also blends with PA, PBT, and PPS (polyphenylene sulfide) [8, 40, 48].

Polycarbonate (PC)

PC is an amorphous, transparent plastic that combines high strength, heat distortion resistance, and impact toughness (even at low temperatures). The material's intrinsic flame resistance can be rendered even higher by environmentally compatible means in accordance with the applicable regulations. The material also has good electrical insulating properties that are virtually independent of temperature and moisture. The limit for heat distortion resistance is between 135 and 145 °C; the glass transition temperature is between 140 and 150 °C.

Along with the standard types there are glass-fiber-reinforced and flame-retardant versions, foam and extrusion products, and injection-molding grades with particularly good flowability properties. Shrinkage is relatively low and uniform. In combination with low thermal expansion, the net result is a high level of dimensional accuracy and stability and a low distortion tendency [8, 40, 48].

As with most amorphous thermoplastics there is a certain susceptibility to stress cracking, which can generally be improved by certain crosslinking agents and solvents.

Polycarbonate HT (PC-HT)

The product is an evolved version of polycarbonate. It retains the usual properties, but the limit for heat distortion resistance is up from 160 to 240 °C.

Acrylonitrile Butadiene Styrene (ABS)

ABS belongs to the group of amorphous plastics. It is impact-resistant and it also exhibits good hardness and scratch-resistance properties. However, it can be used only at temperatures up to 90 °C, and it is therefore not suitable for standard soldering processes. The glass transition temperature is in the range from 85 to 100 °C. Standard processes suffice for chemical metallization.

ABS evinces relatively good resistance to stress cracking. The proportions of all three constituent components can be varied to adapt the material to very widely differing requirements. Virtually all the different grades are available with an antistatic setting. The material is not inherently flame-resistant. Glass fiber reinforced and flame-retardant types are available. Special galvanically workable types can be metallized using standard processes [8, 40, 48].

Polycarbonate + Acrylonitrile Butadiene Styrene Blend (PC+ABS)

This amorphous blend of PC and ABS combines the properties of both products. These include high rigidity, dimensional accuracy and dimensional stability, and low, uniform shrinkage, which in turn means only slight susceptibility to distortion.

The individual types have limits of heat distortion resistance between those of PC and ABS, increasing with the proportion of polycarbonate in the blend. Flame-retardant and glass-fiber-reinforced types are available along with the standard types. In particular the products with a high ABS content can be metallized by galvano-chemical means much like ABS [8, 40, 48].

Polyamides (PA)

The exceptionally wide range of polyamides constitutes one of the most important groups of materials within the partially crystalline engineering plastics. Some polyamide moldable compounds show properties on a par with those of the high-temperature plastics (in particular PPA, PA 6T/X, PA 46, and PA 6/6T). High mechanical loadability is characteristic, regardless of structure. They also have a high damping capability, high wear resistance, and good slip and emergency-running properties. It is important to bear in mind that in particular the mechanical and electrical properties of the molding depend on moisture content, and the same applies to the dimensions. Water absorption can be as high as 10%. The severe bubbling this can cause has negative effects for soldering. PA 6 and PA 66 are suitable for metallization.

Polyamides generally are resistant to solvents, fuels, and lubricants. However, they are not inherently flame-retardant. The extensive range of types, which also includes flame-retardant settings, permits a great deal of adaptation to the applicable sets of requirements [8, 40, 48].

Polybutylene Terephthalate (PBT)

PBT is one of the partially crystalline engineering plastics. It is used primarily on account of its good dimensional stability; hardness, rigidity, and strength; good temperature resistance; and good electrical properties.

PBT cannot be chemically metallized. Hot embossing, in contrast, is possible, and so too is physical vapor deposition (PVD) coating. The material is highly resistant to many solvents, but it is not inherently flame-retardant. There are types containing different proportions of glass fiber and with varying degrees of flame retardation, and impact-toughness-modified settings and blends are also available [8, 40, 48].

2.3.1.3 High-Performance Thermoplastics

Polyphenylene Sulfide (PPS)

PPS belongs to the group of partially crystalline and nonpolar high-temperature plastics. Its structure lacks branches, so it is highly crystalline. The degree of crystallinity and therefore the properties of the molding depend heavily on processing. PPS is generally available only with an elevated glass fiber or glass fiber/mineral content.

The material is very hard and stiff and is suitable for service temperatures up to 240 °C (short-term peaks up to 300 °C). The melting temperature is 285 °C, and the glass transition temperature is between 85 and 100 °C. The common soldering processes can be applied to PPS. Resistance to chemicals is excellent, rendering chemical metallization very difficult. Water absorption is very low; the figure is 0.02%. PPS is inherently (requiring no additives) flame-resistant (UL 94 V-0) [8, 40, 48].

Polyethersulfone (PES)

PES belongs to the group of amorphous, transparent, polar high-temperature plastics. PES is available in both reinforced and nonreinforced versions. It exhibits high strength, rigidity, and hardness across a wide temperature range from −100 to +200 °C (and can briefly withstand a peak of 260 °C). The glass transition temperature is 225 °C. Water absorption up to 2.1% increases ultimate elongation. An elevated moisture content can cause bubbling during soldering. Polyethersulfones can be chemically metallized as follows:

1. Surface treatment by printing and metallization under vacuum
2. Galvanization after appropriate pretreatment

The material is not chemically resistant to ketones, esters, hydrocarbons, aromatics, and highly polar solvents. In common with all amorphous polymers, PES is in danger of stress cracking if brought into contact with certain organic solvents (e.g., ketones, esters, chlorinated hydrocarbons). PES is inherently flame-retardant without the addition of flame-retarding agents, as evinced by its UL 94 V-0 classification at a section thickness of only 0.5 mm [8, 40, 48].

Polyetherimide (PEI)

PEI also belongs to the group of amorphous high-temperature plastics. PEI has high strength (even without reinforcement), rigidity, and hardness and a maximum long-term service temperature of 170 °C. Its glass transition temperature is 210 °C. PEI is easily solderable by reflow soldering processes.

It can be chemically metallized and is not resistant to ketones, chloroform, ethyl acetate, and methyl ethyl ketone. Stress cracking resistance to some mediums is only conditional at best. It can absorb up to 0.6% water. PEI is inherently flame-retardant (UL 94 V-0 at 0.4–0.8 mm) [8, 40, 48].

Liquid Crystal Polymers (LCP)

LCPs consist of rigid, rod-shaped macromolecules that self-parallel in the melt and form liquid-crystalline structures. LCP molecules “self-reinforce” in the flow direction when the melt solidifies, so the products are strongly anisotropic in their mechanical properties.

LCPs can be used to manufacture extremely delicate moldings with thin wall sections and long runner channels. Weld lines, however, constitute a major weakness. Other characteristic properties besides high strength and rigidity in the flow direction include the long-term service temperature of 185 to 250 °C for a melting temperature between 280 and 355 °C [8, 40, 48].

LCPs are solderable with common methods and galvano-workable types are available. Only concentrated acids and alkalies at very high temperatures can attack these materials. There is no danger of stress cracking. Along with low shrinkage, LCPs also have very low water absorption and are inherently flame-retardant [8, 40, 48].

2.3.2 Modified Thermoplastics for MID

The higher temperature requirements due to the changeover to lead-free solder processes required by law constituted one of the main reasons why high-performance thermoplastics are virtually the only unmodified products that can be used adequately as substrate materials.

Engineering thermoplastics such as PA 6 or PA 66, often chosen as base materials on account of their good metallization and service behavior, have melting points of approximately 220 °C and 260 °C, respectively, so they are right at the thermal limits of use. Vis-à-vis economical engineering thermoplastics, therefore, this decision correlates high cost intensity for procurement and processing (e.g., increasing energy consumption on account of high process temperatures) with more difficult metallization.

So the search is on for approaches that will render the cheaper and relatively more workable engineering thermoplastics still usable for mainstream assembly and connection technology processes such as lead-free reflow soldering, for example.

Today, these thermoplastics are enriched with up to 40% by weight of reinforcing additives to increase their mechanical serviceability limits. Glass fibers, glass balls, and a wide variety of mineral substances are used as reinforcing additives. As far as extending the thermal service limits is concerned, MID research in recent years has brought about two main approaches for modifying thermoplastics capable of satisfying the higher requirements—particularly those deriving from the production process:

- **use of electron-beam crosslinking** to extend thermal service limits of standard and engineering thermoplastics
- **use of fillers or filler systems** in engineering thermoplastics for improved thermo-management and for reducing hotspots and shadowing effects

Both approaches necessitate a certain modification of the thermoplastic molding compounds, described in more detail below in terms of basics, interactions, and the resulting end properties.

2.3.2.1 Radiation Crosslinked Thermoplastics

The radiation-induced crosslinking of polymers is a technology that can be traced back to its origins in basic research undertaken during the 1950s [71]. Large-scale usage of radiation-induced crosslinking to produce high-temperature cable insulation and to crosslink shrink-on tubing and polyolefin polymers dates back to the 1970s [96, 97]. In the past, its use was mainly restricted to the group of polyolefins.

Recent endeavors to transfer experience gained in radiation crosslinking to injection moldings and extrusion films made of engineering plastics are a new departure into relatively unknown territory. Successes have already been achieved in research involving crosslinked PA 6 and PBT moldings. The areas of application this opens up are extremely interesting for the electrics and electronics industry in particular. Despite their excellent mechanical and electrical properties, until now engineering thermoplastics have not been adopted for high thermal loadprocesses (such as reflow soldering) and applications (e.g., engine blocks) primarily on account of their low heat distortion resistance and poor temperature resistance. This thermal weakness can be overcome by radiation crosslinking. The outcome is an economically viable alternative to the established high-performance plastics, which are invariably more costly in terms of both procurement (costs of raw materials) and processing (tooling and process costs) [89, 97, 146].

Chemistry and Structure of Thermoplastic Materials

As stated above, plastics are polymerization, polycondensation, or polyaddition products of low-molecularweight monomers. The formation reactions are controlled by suitable reaction conditions such as temperature and pressure and by activating and inhibiting additives (e.g., catalysts, activators, or inhibitors) [8, 35].

The chemical or radiation-induced crosslinking of a plastic leads directly to changes in morphology. The macromolecules are differently sensitized for crosslinking reactions of this nature in the various phases. In the amorphous phase in particular the molecules are more mobile, so this is where most crosslinks are formed. The molecular chains become less mobile as the degree of crosslinking increases, so there is a subsequent increase in the characteristic glass transition temperature. Behavior in the crystalline zones is the opposite. The crystals are not as mobile, so they can participate only to a limited extent in crosslinking. In a crystalline structure, moreover, the binding forces in formation are invariably very strong. Crosslinks in the crystalline structure are therefore directly interpretable as disruptions that enable even slight energy or heat input to dissolve the physical compound of macromolecules. In these cases, consequently, the formation of the crosslinked mesh is associated with a reduction in crystallite melting temperature of the partially crystalline polymer [18].

Crosslinking Processes

Plastics can be crosslinked chemically, for example in the presence of peroxides, but also by radiation chemistry. Broadly speaking, radiation-induced crosslinking has a number of advantages over peroxide-induced crosslinking in terms of both process and quality, including the following [96]:

- Instead of being tied to the temperature limits of peroxide decay, crosslinking can take place at room temperature (→ virtually no detrimental effects of heat).
- The parameters for crosslinking can be modified in many ways (→ high level of flexibility).
- The rate of crosslinking can be accelerated (up to 1000 m/min, depending on type of material, cross-section, and wall thickness).

This explains the predominance of radiation-chemical reactions as a means of initiating crosslinking in thermoplastic materials. In principle the effect can be induced by electromagnetic waves such as X-rays or gamma rays or by corpuscular radiation (beta rays) such as accelerated electron beams, for example. The differences between

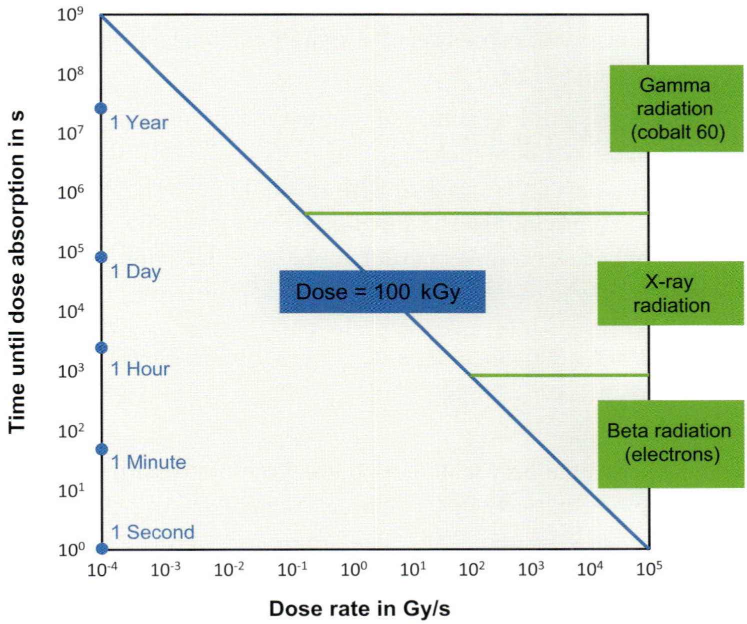

FIGURE 2.9 Duration of irradiation required depending on type of radiation for absorption of a defined energy dose D = 100 kGy [71]

the individual types of radiation relate primarily to dose rate and penetration range and consequently to their specific properties with regard to absorption in material, as graphed in Fig. 2.9 [71].

Gamma rays generally have a low dose rate (10^1 to 10^3 Gy/min) but high penetration range. A high dose rate necessitates sufficiently long irradiation over a period of several hours; see Fig. 2.9. The use of this process is of interest for parts with complex structure.

By contrast, *beta rays/electron beams* have a high dose rate (10^3 to 10^6 Gy/min) and low penetration range. High energy doses can be introduced into the material within a fraction of a second to trigger radiation-chemistry reactions. Consequently, thin-walled plastic moldings and films could be considered main applications for electron-beam crosslinking.

Advantages of Radiation-Induced Crosslinking

In principle, many of a thermoplastic material's properties can arise from radiation-chemistry crosslinking (Table 2.2). Even minute quantities of absorbed energy can produce considerable changes in the properties of the plastic. Improved stability of shape at elevated temperature is a radiation-modified property that merits particular mention. The engineering thermoplastics that are meltable between 180 and 260 °C and are used for MID lose their thermoplasticity when crosslinked and become unmeltable. They can then sustain thermal loading spikes in excess of 400 °C, which means that lead-free processes of assembly and connection technology can be used without difficulty [18, 71].

Radiation-induced crosslinking is also of interest if different properties have to be established within a (complex) molding. The areas of the MID that should remain soft, elastic, or weldable can be masked with metal. Energy-rich radiation does not penetrate the metal covering, so the molding is only partially crosslinked in the uncovered areas, where it becomes harder, more heat distortion resistant, and less susceptible to cold flow.

TABLE 2.2 Improvement in MID Properties by Radiation-Induced Crosslinking

Mechanical properties	
▪ Increase in strength ▪ Rise in moduli	▪ Improvement in creep behavior ▪ Improvement in abrasion resistance
Thermal properties	
▪ Increase in temperature resistance ▪ Reduction in thermal expansion	▪ Improvement in solderability ▪ Improvement in flame retardation
Chemical/physical properties	
▪ Improvement in chemical resistance ▪ Increase in adhesion	▪ Reduction in solubility ▪ Improvement in hydrolysis resistance

2.3.2.2 Highly Filled Thermoplastics

Compounding special fillers with the polymer has proven to be a good way of integrating functions such as thermal and electrical conductivity into thermoplastics. These plastic compounds benefit from material synergies derived from the good workability of the plastics and the additional properties that the plastic gains from the fillers [31]. Tailoring thermally conductive compounds to requirements would entail complying with the following criteria for selection and composition of the fillers and additives:

- degree or maximization of filling
- native conductivity of the filler
- filler shape and size
- filler mixtures
- additivity

Technically adequate thermal conductivity figures are in the region of 1 to 20 $Wm^{-1} K^{-1}$ and require the admixture of weighted quantities of fillers and additives. Filler systems can be ceramic materials (thermally conductive but electrically insulating) and metallic materials (thermally and electrically conductive) and diverse modified forms of carbon [4, 31]. The high proportion of fillers alters behavior radically compared to that of nonmodified thermoplastics; see Fig. 2.10.

Flow behavior and mechanical properties in particular are changed significantly by the addition of fillers. Fracture strain generally diminishes, whereas rigidity increases. Strength depends largely on the bonding of the filler to the embedding

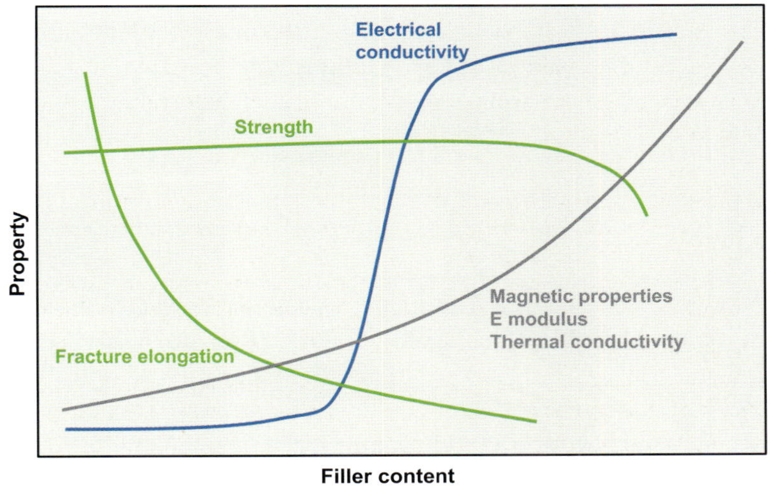

FIGURE 2.10 Dependency of material properties on filler content

matrix. Special additives and other fillers such as glass fibers can be used to improve the bond between matrix and filler, increasing strength and fracture strain [4, 31].

In principle, compounds for MID can be manufactured with a very wide variety of fillers from the groups of metallic (only for MID processes that do not require wet-chemical conductor metallization) or ceramic materials. At this time some fillers are coming into widespread use on account of their verifiably excellent thermal properties, offering considerable benefits for MID production (e.g., reduction in hotspots for lead-free soldering processes) and MID use (e.g., integrated thermal management in highly loaded MID such as LED modules). The most important of these fillers are

- graphites,
- conductive carbon black,
- copper,
- aluminum oxides, and
- most recently and very importantly, boron nitrides.

MID with heat conductivity ratings up to about 20 $Wm^{-1} K^{-1}$ are possible these days, depending on filler content and also on particle shape and size. In comparison, unmodified thermoplastics (see Section 2.3.1) have thermal conductivity values in the region of 0.15 $Wm^{-1} K^{-1}$, so selective modification with fillers up to about 60% by volume clearly benefits the processing and use phases. Thermal conductivity shows an overproportional increase as filler content rises (Fig. 2.11). Over and above filler content, the fillers themselves can be classified in a number of different ways.

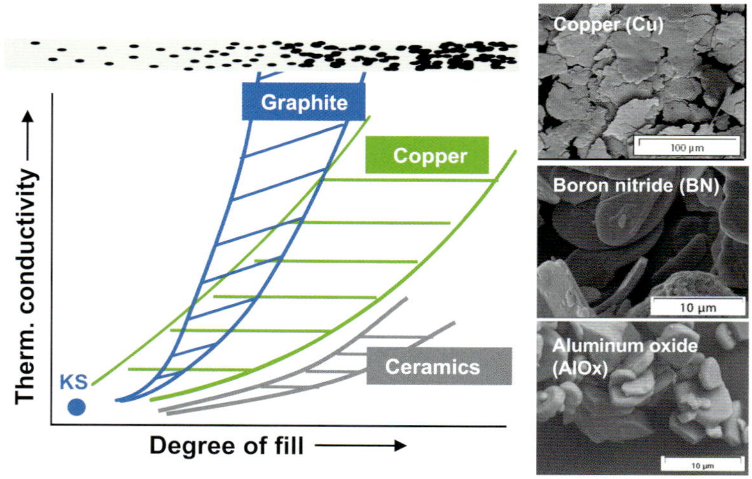

FIGURE 2.11 Overview of fillers used in compounds [55]

Filler geometry affords one system of classification based purely on the mechanics of reinforcement:

- one-dimensional fillers (fibers)
- two-dimensional fillers (platelets)
- three-dimensional fillers (powders)

Glass fibers in particular, because of their geometry, cause good force introduction and lead to a considerable improvement in mechanical characteristic values. Three-dimensional fillers such as glass balls or chalk usually increase only the E modulus; far from benefiting tensile strength and toughness, they often have the effect of lowering these values. Classification of fillers by geometry is also crucial in terms of the thermal conductivity achievable in the molded component. Anisometric, one- and two-dimensional fillers lead to higher thermal conductivities than three-dimensional, spherical fillers. Flow-induced filler orientation tends to layer the fillers through the cross-section of the component, which in turn leads to a preferred direction of thermal conductivity in the molding. Thermal conductivity, therefore, is a process-dependent and component-dependent material property, particularly when the fillers used are anisometric.

As regards the modification of thermoplastics, however, it is important to bear in mind that the proportion of fillers cannot be increased arbitrarily without crucially affecting the wetting essential for processing. The proportion of filler such as glass fiber reinforcement is generally between 15 and 50 percent by weight. Below 15% by weight there is virtually no reinforcing effect, and above 50% by weight it becomes difficult to wet the fillers with the uniformity necessary for smooth surfaces. Moreover, highly filled, thermally conductive plastics require adaptation of the process parameters for dependable injection molding. This is because of the higher melt viscosity and the rapid cooling of the melt while the mold cavity is filling. Consequently, MID design for highly filled thermally conductive compounds is always a trade-off between the modified properties and the workability of the materials [55, 135].

As degree of fill increases, there is generally an overproportional increase in the thermal conductivity of MID. The higher the anisometry of the filler, the higher the contact probability and thus the formation of conductive paths. This is why filler geometry is so significant in terms of the property profile. Anisometric particles such as copper or boron nitride platelets lead to high thermal conductivity of the compound, largely because of their geometry. This is confirmed by computational models. However, particle geometry can also lead to a strongly anisotropic property in injection moldings, as Fig. 2.12 illustrates by graphing as examples the results of nanoflash measurements of PA 6 injection molded plates 4 mm thick.

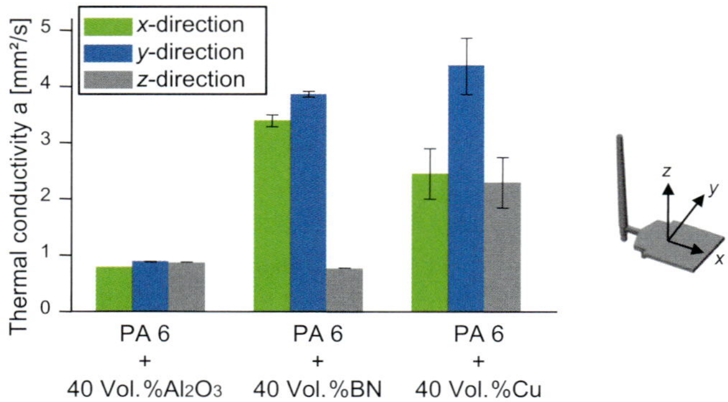

FIGURE 2.12 Directional thermal conductivity a of various thermally conductive compounds at room temperature; thickness d = 4 mm [77]

Figure 2.12 also shows that thermal conductivity is always a directional characteristic value, an aspect that has to be taken duly into account in the framework of integrated MID product development with highly filled plastic compounds [77, 135].

This directionality of the property can vary widely, and it is due to characteristic filler layering with a preferred direction imposed on the molding by the injection-molding process. Conductivity paths of different thicknesses formed in the solidified MID are the cause (Fig. 2.13). When a suitable layer model is superimposed, it becomes clear that it is in the y direction in particular that all three layers can contribute to optimum transport. In the x direction only the top and bottom layers afford suitable arrangements. In the z direction the middle layer alone is oriented for good thermal transfer [77].

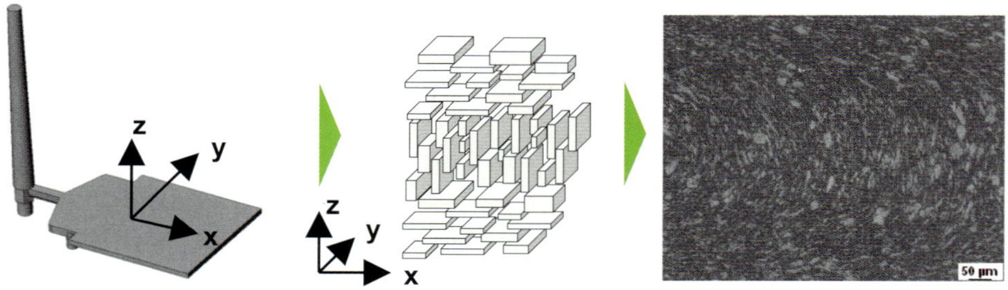

FIGURE 2.13 Filler orientation model to describe anisotropic thermal conductivity in injection-molded test specimens [77]

2.3.2.3 Thermoplastics for Selected Technologies of MID Metallization

Methods including two-shot molding, laser-based structuring (laser direct structuring, laser subtractive structuring), hot embossing, and film insert molding are discussed next as common processes in MID production. A basic distinction is drawn between subtractive structuring and additive metallizing processes. The ultimate choice of a technology for any given application depends on many criteria. The choice can also depend to a very large extent on the technological possibilities and restrictions of the individual processes, including high design flexibility, the ability to produce microfine structures, and suitability as the point of departure for subsequent assembly and connection techniques to be applied to the MID moldings.

The following sections set out the purely materials-related and tool-related prerequisites for the two processes now most commonly used in MID production: two-shot molding and laser direct structuring (LDS). Chapter 3 of this book deals in depth with the workflows, possibilities, and boundary conditions associated with these manufacturing processes. It also contains information about the production methods such as hot embossing and film insert molding, neither of which is considered in more detail here from the materials point of view.

Two-Shot Molding

In the past, two-shot molding or two-shot injection molding generally referred to a process in which a first shot of material is subsequently enclosed within a second shot. In MID technology, however, the process is one in which the second-shot plastic compound is applied on top of the first shot. In the course of time two main versions of this basic process have established themselves for MID technology [48]:

- PCK process (PCK = printed circuit board Kollmorgen)
  Palladium is mixed with the metallizable plastic to act as a nucleus for metallization. Other processes, following the outline of these versions 1 and 2 as described below, generally use inherently metallizable plastics.
- SKW process (SKW = Sankyo Kasei wiring board)
  Metallization is based on a catalyzing step inserted between the two shots in injection molding.

In version 1 of the PCK process (Fig. 2.14, left) a metallizable, electrically nonconductive plastic is used for the first shot, and molding produces a raised structure corresponding to the subsequent conductors of the MID. The second shot then covers the areas between the conductor tracks with a nonmetallizable plastic.

The alternative is to mold the conductor structure as a recessed structure in the first shot of nonmetallizable plastic and fill these recesses with the metallizable compound in the second shot. This is version 2 of the PCK process for producing an MID (Fig. 2.14).

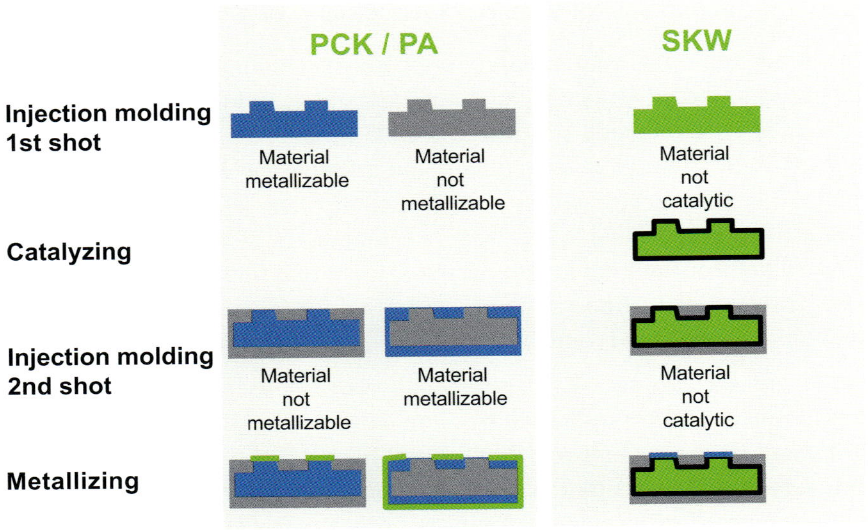

FIGURE 2.14 Two-shot molding processes for the production of MID [48]

After the second shot the MID has achieved its final geometrical shape, and in the subsequent steps suitable metals can be deposited on the surface of the metallizable plastic. The desired thickness of galvano-chemical copper is deposited on the activated surface. Widely differing finishes can then be applied to suit the given application, with Cu, Ni, and Au all being standard at this time [48].

Two-shot molding is one of the MID manufacturing processes offering the most geometric freedom, as indicated in Chapter 3. The conductor tracks are structured in the injection-molding process, so conductor widths and insulation spacing down to 0.25 mm are possible (Table 2.3), depending on the flowability properties of the plastic compound and the runner lengths in the tool.

TABLE 2.3 Geometric Limits of Two-Shot Molding (in mm) [48]

<table>
<tr><th>Design limits</th><th>LCP/SPS</th><th>PEI</th><th>PPA</th><th>PES</th><th>PA</th></tr>
<tr><td>Conductor track width</td><td rowspan="2">0.25</td><td rowspan="2">0.6</td><td rowspan="2">0.4</td><td rowspan="2">0.6</td><td rowspan="2">0.5</td></tr>
<tr><td>Conductor track spacing</td></tr>
<tr><td>Min. hole diameter</td><td colspan="5">0.6</td></tr>
<tr><td>Aspect ratio, length/diameter</td><td colspan="5">< 6 : 1</td></tr>
<tr><td>Min. pin diameter</td><td colspan="5">0.8</td></tr>
<tr><td>Pin length for pin diameter</td><td colspan="5">2</td></tr>
<tr><td>Section thickness (1st shot)</td><td rowspan="2">0.25</td><td colspan="4">0.8</td></tr>
<tr><td>Section thickness (2nd shot)</td><td colspan="4">1</td></tr>
</table>

In terms of tool design, it is important to note that different cavities are required for the first shot and the second shot. As regards the materials, the two thermoplastics must either both evince good melt compatibility (for good adhesion when the second shot is molded on to the first) or else the design must include mechanical anchorage points (e.g., back-cuts, or surface roughness for a good key). Consequently, the choice of material combinations for two-shot molding is limited. The MID-relevant materials systems include combinations based on LCP, sPS, PA, PPA, or PEI [48].

The metallization of two-shot moldings is one of the issues discussed in detail in Chapter 3.

Laser Direct Structuring

Two-shot molding, hot embossing, and laser subtractive structuring are all established processes in MID series production, and in recent years they were joined by a new, innovative technology: laser direct structuring (LDS) developed by LPKF Laser & Electronics AG, a manufacturer of industrial laser technology headquartered in Garbsen, Germany. The LPKF-LDS® process is an additive method of producing 3D conductor structures on a plastic interconnect device. Based on mechanisms different from those of other methods, it is a process described in detail in Chapter 3.

Interactions between Laser and LDS Additive

Laser direct structuring is based on the notion of laser activation of specially additivated plastics. The additives that enable the LDS process and that are finely distributed in the MID basic body are transformed into catalytically effective species during laser structuring and must satisfy different requirements [6]:

- good chemical compatibility and homogeneous distribution in the polymer matrix
- no impairing of electrical and mechanical properties of the substrate
- thermal stability in the processing-temperature range for the plastic
- no catalytic activities in the unactivated state

The preferred selective additive is a metal oxide or mixed metal oxide. Its temperature resistance is extremely high, so no nucleation takes place while the plastic melt is being processed. It is only the beam from a Nd:YAG laser (γ = 1064 nm) that activates the additive, now enclosed in its organic matrix. The laser light also excites the polymer molecules, causing them to oscillate. When a minimum energy threshold is reached, the bonds between the molecular chains degrade and the polymer matrix is removed (photo-chemical ablation) or evaporated (relaxation) (Fig. 2.15). The exposed particles of additive activated by the laser subsequently act as the nuclei for reductive copper plating in the electroless metallization bath [6, 73].

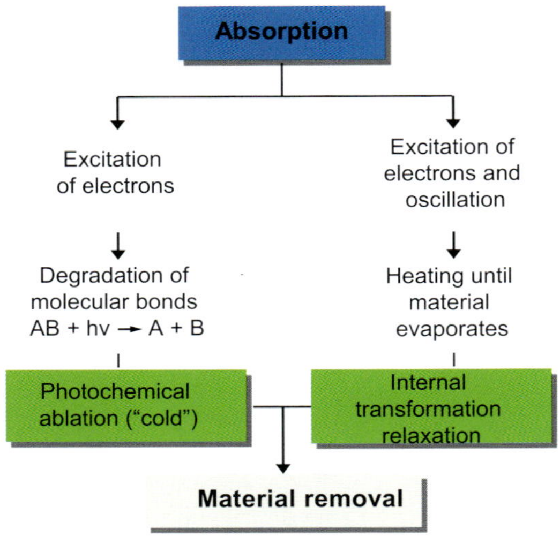

FIGURE 2.15 Laser/plastic interaction in the LDS process [73]

Alongside sufficient activation, metal-to-plastic adhesion is another important parameter for qualification of electronic interconnect devices. An initial adhesive strength of 0.6 to 1.1 N/mm is a specification for printed-circuit boards (DIN IEC 326).

Since surface roughness is crucial for adhesive strength of the metallization subsequently applied, the polymer matrix is modified by suitable, radiation-resistant fillers (mostly inorganic). The fillers in the plastic are not ablated by the laser. This produces microscopically small caverns and back-cuts in which the copper can anchor itself firmly, so the metallization acquires a secure key to the base material without complicated post-treatment (Fig. 2.16) [73, 75, 172].

Laser structuring triggers nucleation and it also produces the necessary surface roughness. Within certain limits, this roughness can be optimized by the choice of filler type, particle size, and laser parameters [172].

Process Chain of Laser Direct Structuring (LDS)

The laser direct structuring process developed by LPKF can produce high-resolution circuit layouts on complex three-dimensional interconnect-device structures in a multistep process sequence.

The remarks below concern themselves only with the main points of materials selection and aspects of substrate manufacture. All other aspects are discussed in depth in Chapter 3.

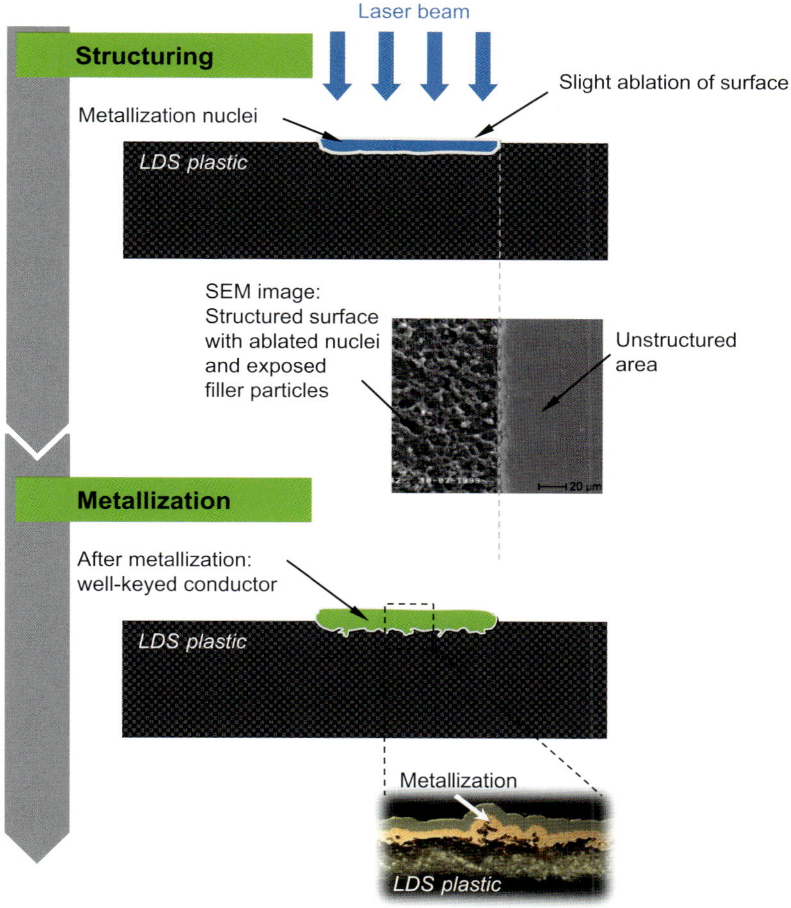

FIGURE 2.16 Laser-induced metal/plastic adhesion [172]

Materials Selection

The correct choice of materials to suit the assignment is one of the basic steps in series production of 3D-MID. It is the job of the developer to decide which of the materials available make the best fit to the properties profile set out in the specifications. In principle any thermoplastic that is suitable for the production of molded interconnect devices can be modified for laser direct structuring. This is why almost all the big-name manufacturers now offer LDS variants of their polymers for use in the manufacture of molded interconnect devices. The range available for the wide gamut of uses extends from standard thermoplastics through engineering thermoplastics to high-performance thermoplastics suitable for use in the reflow soldering process. The materials currently qualified and available on an industrial scale are shown by way of example in Fig. 2.17 [73, 75].

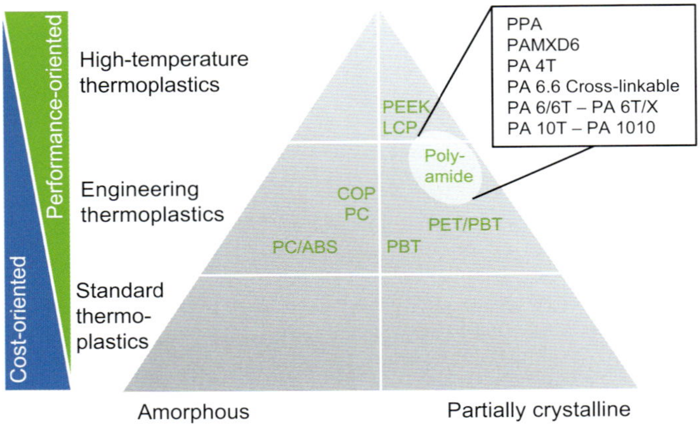

FIGURE 2.17 Current portfolio of materials compatible with the LPKF-LDS® process (graphics courtesy of LPKF)

Substrate Manufacture

The moldings for laser structuring are one-shot molded from commercially available thermoplastics containing special additives. The dried and preheated plastic granulate is forced under high pressure into the mold and demolded once it has cooled and solidified. The process does not require special workability properties. The molding process is consequently faster and the tool required is simple in design, without the dual cavities necessary for two-shot molding. Each manufacturing process requires suitably adapted materials, so in all instances due consideration has to be given to the central material properties such as flowability, working temperature, mechanical and electrical properties, heat distortion resistance, and costs.

2.3.3 Thermoset Plastics for MID

To date the only materials used in MID technology have been thermoplastics for processes such as two-shot molding and laser direct structuring, for example. Soldering with lead-free soldering narrows down the choice considerably, excluding virtually everything except the commercially available materials PET/PBT, PA 6/6T, PPA, and LCP. These materials differ in their properties and, consequently, in terms of their suitability as well. Table 2.4 provides an overview of important thermal properties of commercially available thermoplastics. Given the current importance of the LDS method among the MID manufacturing processes, by way of example the overview lists the main thermal properties of plastic molding compounds suitable for laser direct structuring.

TABLE 2.4 Important Properties of Thermoplastics Suitable for Laser Direct Structuring [7, 22, 147]

Polymer type	Elongation	Heat distortion temperature		
	ISO 11359-1/-2	ISO 75-2		ASTM D648
		HDT-A (1.8 N/mm^2)	HDT-B (0.45 N/mm^2)	(45.5 N/cm^2)
	[10^{-6}/K]	[°C]	[°C]	[°C]
LCP	12–27	227	–	–
PPA	–	–	–	279
PA 6/6T	30–50	–	265	–
PET/PBT	36–56	–	255	–

The differences in thermal expansion between copper and LDS plastic, particularly the partially crystalline types such as PET/PBT, PA 6/6T, and PPA, some of which exhibit abrupt changes at the glass transition, have long-term negative effects on the adhesive strength of the metallization and therefore on the serviceability of the MID. LCP is a material well suited for MID applications for which only a slight mismatch between the coefficients of thermal expansion (CTE) of the plastic and the conductors is admissible.

However, the relatively high price of LCP remains a motivator for the development and production of new high-performance materials so that MID suitable for lead-free soldering can still be produced economically. The thermoset, moldable compounds that are not yet the accepted state of the art in MID technology have properties profiles that appear to offer more than enough points of departure for use as substrate materials for molded interconnect devices.

Thermoset Materials

Thermoset moldable compounds can be mixed with a very wide variety of fillers to modify their properties to meet the requirements for a given application. Adding suitable fillers can produce coefficients of thermal expansion and elongation behavior virtually identical to those of copper. Once thermoset materials have cured, moreover, their three-dimensional molecular network structure gives them a very high level of dimensional stability. Consequently, in terms of temperature resistance to soldering many of the materials in this class are potentially suitable as MID substrates. Table 2.5 summarizes some of the important thermal properties of commercially available thermoset moldable compounds. Thermoset moldable compounds, moreover, have economic potential because in some cases the cost of the material is low. Phenolic resin moldable compounds in particular are available at a price of less than about \$7 (EUR 5) per kg and could therefore be considered an economical alternative to LDS high-temperature thermoplastics.

TABLE 2.5 Important Thermal Properties of Selected Thermoset Moldable Compounds [136, 167]

Polymer type	Elongation	Heat distortion temperature	
	ISO 11359-1/-2	ISO 75-2	
		HDT-A (1.8 N/mm^2)	HDT-C (8.0 N/mm^2)
	[10^{-6}/K]	[°C]	[°C]
Phenolic resin	16–24	> 250	160–180
Epoxy resin	15–25	> 250	170–200
UP resin	10–20	> 250	> 200
DAP resin	10–20	> 250	> 200

Thermoset Injection Molding

On account of the high degree of design freedom and cost efficiency it offers, thermoset injection molding has established itself as a process for manufacturing thermoset components on a mass-production scale (Fig. 2.18).

Compression molding tends to restrict design freedom, but high-precision parts of complicated geometry can be produced fully automatically with short cycle times in thermoset injection molding. Other processes such as casting and methods of rapid prototyping are less suitable on account of the extended process times, which render them less economical for mass production. Pulverulent moldable compounds can be used for thermoset injection molding (Fig. 2.18).

There is no settling of the functional fillers. The materials can be handled by standard machines. In terms of process parameters, however, thermoset injection molding

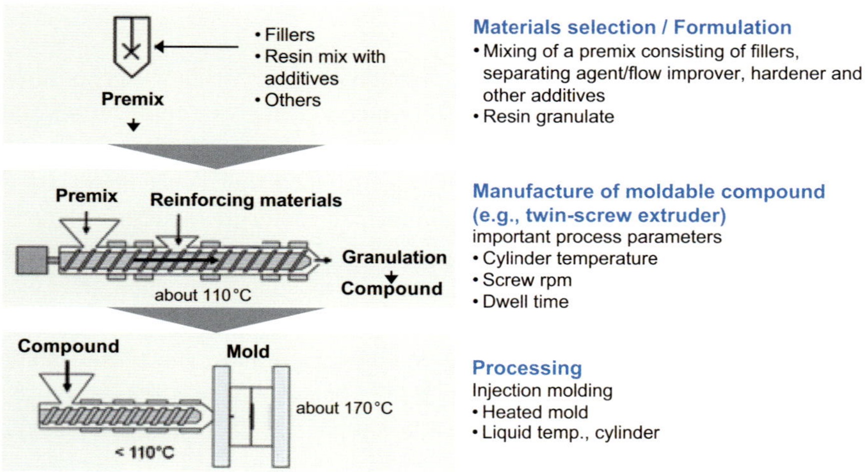

FIGURE 2.18 Process chain for thermoset injection molding

differs radically from thermoplastic injection molding, which is currently being used successfully for the production of 3D-MID. In the thermoset process the material is first plastified at temperatures up to about 100 °C in the plastification unit and then injected into the heated mold, where the viscosity of the material initially drops further on account of the temperature rise. At the same time the material starts to set in the mold, until after a certain heating time the fusion point of the compound is exceeded and the part can be demolded.

3 Structuring and Metallization

R. Schramm

There are various plastics shaping processes that can be used in MID production. Shaping by one of the three mainstream plastics forming processes, one-shot injection molding, two-shot injection molding, or insert molding, is followed by structuring and metallization, the next steps in the MID process chain. This chapter begins by looking at the various structuring technologies before moving on to metallization and its process variants, characteristics, and possibilities.

3.1 Structuring Processes

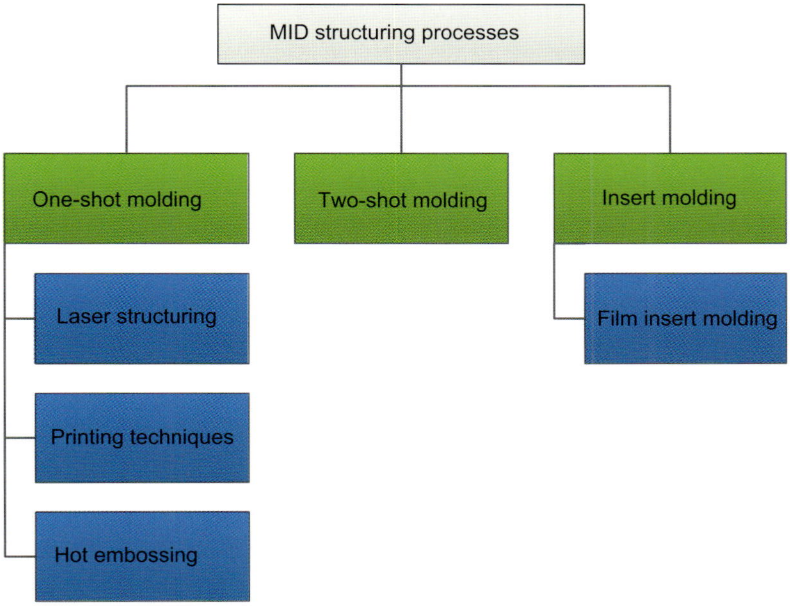

FIGURE 3.1 Classification of MID structuring processes by shaping

It is common to categorize MID structuring processes as belonging to three shaping categories: one-shot injection molding, two-shot injection molding, and insert molding (Fig. 3.1).

At this time, there is an established process suitable for series production available in each of these categories. The individual processes differ widely, primarily in terms of process chain, outlay, time factor, and cost.

In recent years, alternative production methods have been developed alongside the more long-standing conventional processes indicated in Fig. 3.1. These newer processes are dealt with in more detail in Section 3.1.4.

3.1.1 One-Shot Injection Molding

One-shot injection molding is characterized by high-volume production at moderate prices per unit and short cycle times. The thermoplastic material, generally a granulate, is conveyed by a screw feeder from a hopper to the injection-molding tool. The shear forces and the application of additional heat melt the granulate in the screw feeder. The liquefied plastic cools inside the tool and assumes its final shape. The tool is then opened, the molded part is ejected, and the next cycle starts. Throughput can be increased by using multipart molds designed to produce multiple injection-molded parts per cycle.

The plant actually required for one-shot molding of MID depends on the plastic used and on the size and complexity of the part. For example, the nature of the plastic affects the shear forces required for the injection-molding tool. Tools might have to be designed and built to specification, and suitable injection-molding machines are required. The complexity and size of the part affect determining costing for tool design and manufacture. It is easy to see, therefore, that the costs for one-shot injection molding depend on numerous factors, all of which have to be taken into consideration for any given application.

3.1.1.1 Laser Structuring

The main laser-structuring techniques for MID are derivatives from the world of printed-circuit board engineering. For this reason the various techniques are commonly classified as additive, semiadditive, or subtractive. Laser direct structuring is an additive technique, and its most important manifestations are LPKF-LDS® and ADDIMID. MIPTEC, a process developed by Panasonic, is one of the semiadditive techniques. Subtractive laser structuring techniques are commonly used on ceramic substrate materials. Figure 3.2 is an overview of the individual structuring techniques and the steps involved.

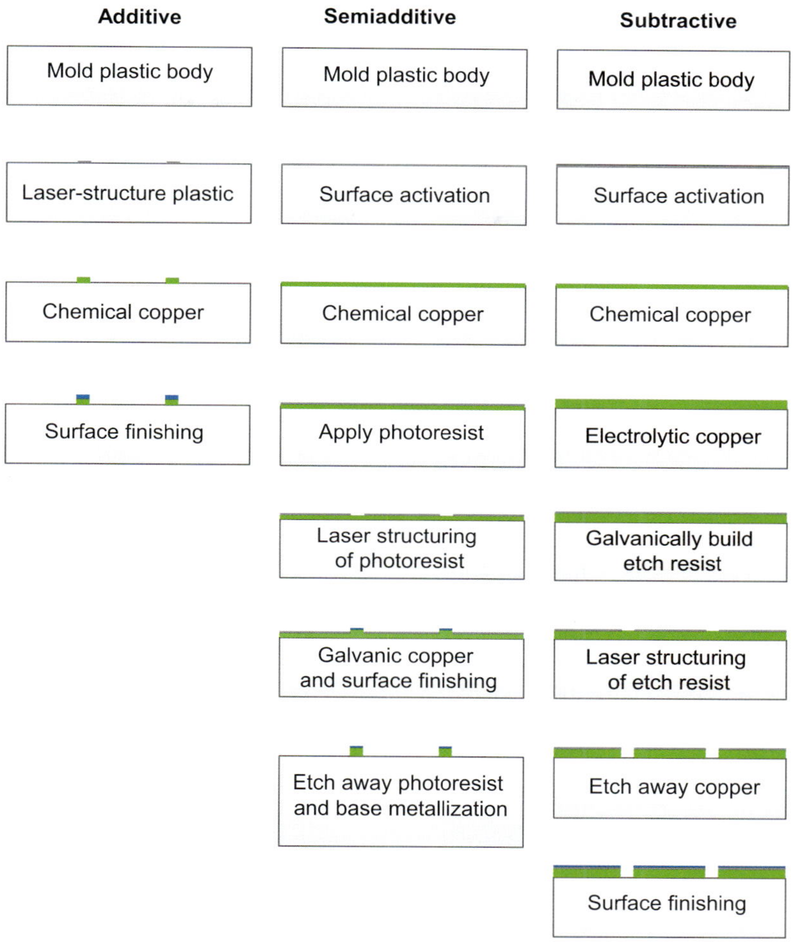

FIGURE 3.2 Process steps in additive, semiadditive, and subtractive laser structuring as in [48]

3.1.1.1.1 LPKF-LDS® Process

Laser direct structuring (LDS) currently accounts for more than 50% of the market in MID structuring. LDS has become a beacon process in MID production over the last decade. The reasons for this development are many and varied. On the one hand, LDS offers a high level of versatility, possibilities for prototyping and series production, low tool costs, microfine structures, and high 3D design freedom. On the other hand, it is now also possible to work a very wide range of thermoplastics with LDS. Particularly as regards color and type of plastic, materials diversity has increased enormously over recent years. This development is set to continue in the future. This section contains a brief overview of the materials currently available, followed by a more in-depth introduction to the LDS process as such.

LDS Substrates

Numerous different types of plastic are now available for the various areas of application and the corresponding sets of requirements. The choice ranges from run-of-the-mill plastics such as ABS through high-specification polyamides to high-temperature plastics such as LCP and PEEK. The difference in the prices of plastics is linear to the properties of the materials. PC/ABS is the most common material, mostly because it is used so much in the telecommunications sector. The low price of the plastic is of course one of the most significant contributory factors in this respect.

At this time there is also a very wide choice of plastic materials for the LPKF-LDS® process. The available materials are listed in Chapter 2. In the vast majority of cases, the plastics usable for laser direct structuring require a specific additive. It is added to the blend during compounding. The extreme heat resistance of this additive makes compounding possible, and it also prevents nucleation in the injection-molding process. The specific properties of the plastics are only slightly affected. The additive contains inactive chemical substances that are activated by laser irradiation and exposed by ablation of the molding skin [15].

It is also possible to prepare plastics for laser direct structuring without the specific LDS additive. LPKF has a product akin to an LDS painting system for this purpose. ProtoPaint LDS is a two-component primer/curing agent system for coating plastic injection moldings with an LDS-compatible skin. Application is by standard spray gun or with a recently developed and more user-friendly aerosol. A coat approximately 30 to 40 μm thick has to be applied in two steps and then cured. The plastic parts prepared in this way can then be structured; the results are comparable to those achieved with LDS plastics. This process, however, is intended primarily for prototyping because the constant-use properties are not comparable with those of plastics with LDS additive in the compound [15].

LPKF-LDS® Process

The LPKF-LDS® method is a four-step process: injection molding, laser structuring of the plastic, metallization, and surface finishing (Fig. 3.2). Plastics compounded with the LDS additive are fully compatible with the injection-molding process.

Laser direct structuring is based on the principle of ablation and nucleation by laser irradiation. The patterning Nd:YAG laser ablates about 1 to 2 μm of material from the surface and simultaneously activates the additive necessary for metallization (Fig. 3.3). This creates a microscopically rough surface. This surface consists of catalytically active nuclei embedded in the microcavities produced by the laser. These tiny pits form a good key between the plastic and the metallized layer without any need for additional, postablation treatment. Processing rates as high as 4000 mm/s are possible at this time, diminishing with increasing three-dimensionality and part complexity because scanning speed for focus tracking in z is limited. The

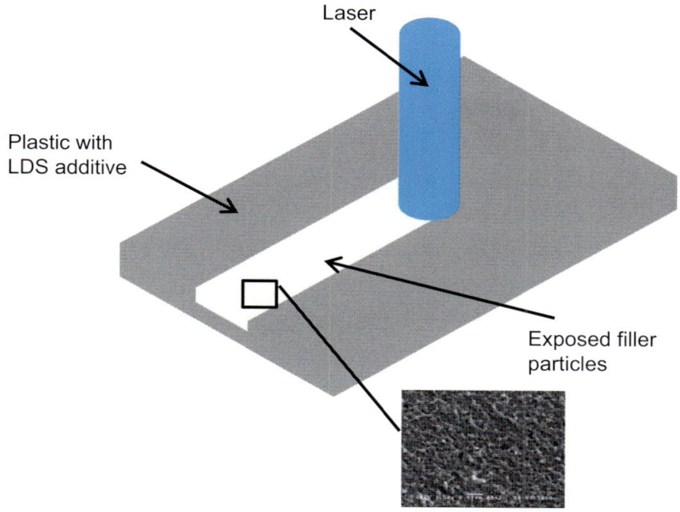

FIGURE 3.3 Laser ablation with simultaneous additive activation

wider the angle to the target surface, the more rapidly the focus has to be adjusted in the *z* direction. Maximum scanning speed is down to 1500 mm/s by the time this angle reaches 45°. LPKF currently recommends a minimum conductor track width of 150 µm at 200 µm minimum spacing. Technically speaking, even thinner conductors of 100 µm for example are possible, as is closer spacing. However, this necessitates optimizing the combination of plastic, structuring, metallization, and component placement. At the moment the technological limit is a conductor width of 50 to 55 µm [15].

Laser Systems

The oldest laser structuring machine from the LPKF company for LDS is the MicroLine3D 160i as a single-head laser system complete with rotary indexing table for series production. [103]

LPKF currently has a range of four different laser systems for LDS. The economical, entry-level model is the Fusion3D 1100. The basic-configuration machine has a large working surface manually adjustable along the *z* axis. In combination with user-specific workpiece adapters, the machine is ideal for prototyping and also suitable for small-volume and medium-volume production runs. The Fusion3D 1500 is an evolved version of the Fusion3D 1100 designed specifically for long parts. The working zone measures 400 mm × 78 mm × 80 mm ($L \times W \times H$). The system can also be used with a multiple laser array. Up to three laser heads help minimize process times and boost throughput. The Fusion3D 6000 was designed specifically for 24/7 operation in large-volume series production. With an array of up to four laser heads, the machine enables simultaneous structuring of several sides of the component. Automatic component manipulation ensures precision alignment and high throughput. The LDS process has numerous advantages to offer. Extracting maximum benefit from them, however, entails complying with a series of design rules. The most important of these rules are discussed below. More information is obtainable by consulting the LDS-MID design rules from LPKF [100, 101, 104, 108].

Scan Volume

The size of an MID part for LDS depends on the maximum scan volume of the laser system. In the case of the MicroLine3D 160i, for example, this is a truncated pyramid with a bottom-face diameter of 160 mm, a height of 24 mm, and a shell-to-base angle of 77.4° [104]. The counterpart truncated pyramid of the entry-level Fusion3D 1100, by contrast, measures 120 mm across its base and 50 mm in height [108].

Vertical Interconnects

The potential of vertical interconnects is enormous. In combination with three-dimensional structuring and metallization, they can help optimize layout in terms of both packaging and cost efficiency. For example, a vertical interconnect can obviate the need to track the conductor over an edge, and this saves on conductor surface area. Conversely, minimizing surface area for metallization contributes to a reduction in cost. LCP is the only LDS plastic that can be holed by a laser to produce vertical interconnects. If some other plastic is used the holes for vertical interconnects have to be engineered into the design of the tool for injection molding. The rules illustrated in Fig. 3.4 apply to structural design of vertical interconnects.

The thicker the section the wider the inside diameter of the vertical interconnect has to be for the laser beam to reach all the inside surfaces in the hole.

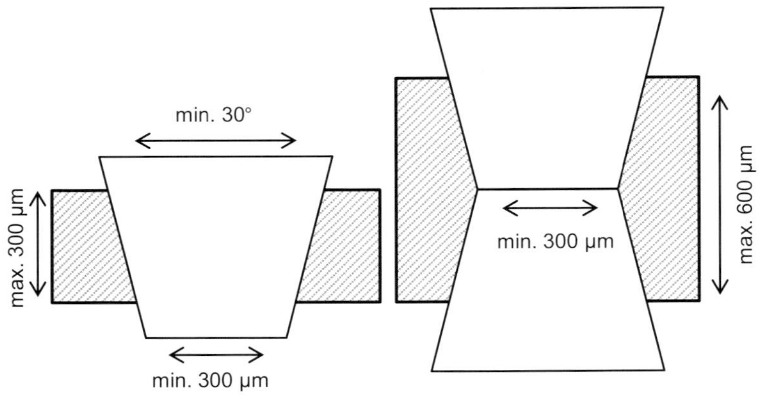

FIGURE 3.4 Design rules for vertical interconnects as in [104]

Edge Radius and Conductor-to-Wall Spacing

Recommended minimum edge radius is 150 µm (Fig. 3.5), even though a radius of 100 µm is technologically possible. As with conductor width, all the components in the process have to be matched to each other. The same applies to conductor-to-wall spacing. Ablation deposits during structuring or precipitation from the metallization process are possible, so the recommended clearance is 150 µm from a 45° wall or 250 µm from a 70° wall (Fig. 3.5). Once again, closer spacing is possible but necessitates optimum intermatching of all the process components.

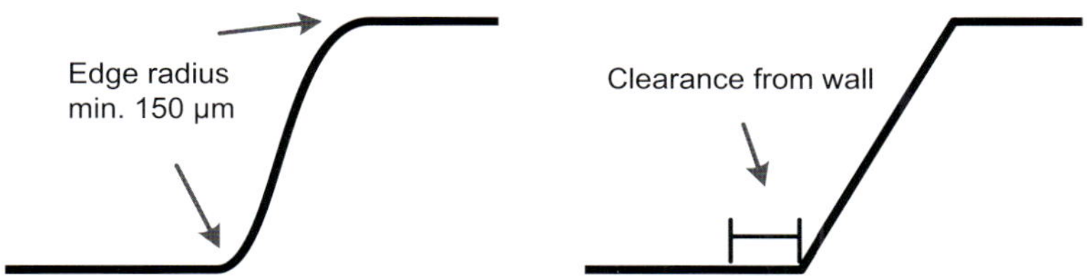

FIGURE 3.5 Design rules for edge radius and conductor-to-wall spacing as in [104]

Angle of Incidence

The structuring process is not dependable unless the angle of incidence of the laser beam with the surface of the plastic is pegged at its maximum. This is on account of energy transfer to the surface for activation. Energy input has to be sufficient to ablate the plastic at the surface and activate the additive. As soon as the beam is off-normal its energy is spread across a larger area, so the design rules recommend a maximum angle of 70° between the orthogonal on the process surface and the laser beam. It is also important to bear in mind that the angle between the horizontal and the laser beam increases with distance from the center. Therefore 13° is the maximum angle of incidence in this case (Fig. 3.6). Depending on the equipment used, the time saved simply by adhering to this design rule can be considerable.

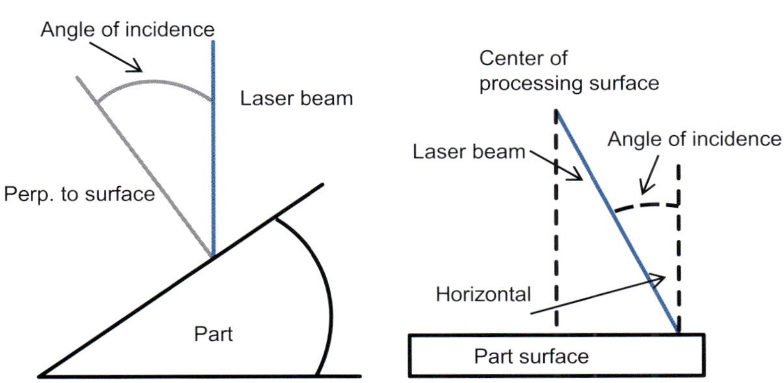

FIGURE 3.6 Design rules for angle of incidence according to [104]

Fitting component rotators to the MicroLine3D 160i is one way of adjusting the angle of incidence. Synchronized operation of the Fusion3D 6000's multiple laser heads arranged around the component for structuring minimizes time for rotating the components, significantly boosting productivity in the structuring process. [101]

3.1.1.1.2 ADDIMID Technology

ADDIMID is another additive laser structuring technique. It can be employed to structure stereolithographic components or injection-molded plastic bodies. The process was qualified within the framework of a research project at the Bayerisches Laserzentrum, the laser applications research facility of the Friedrich-Alexander University, Erlangen-Nuremberg. The similarities between the LPKF-LDS® and the ADDIMID processes are firstly in the resemblance of the process chains and secondly in modification of a plastic by the addition of special fillers. At this time the technology is under development, but initial results of research are indicative of potential for future applications. [3]

ADDIMID technology involves upgrading the plastic with an additive consisting of a metal powder with an electrically insulating cladding material. The materials used in trials to date are microfine Cu, Ni, and Al powders with particle diameters in the range from 0.5 to 1.2 μm. As in the LPKF-LDS® technique, the plastic bodies are injection-molded after the raw material has been compounded. ADDIMID is again similar to LDS in that it uses a Nd:YAG laser with a wavelength of 1064 nm for structuring. A CO_2 laser with 10.6 μm wavelength has also been used for this purpose. Both methods exploit the duality of lasering as a crucial process advantage. The surface for structuring is partially ablated while at the same time the fillers mixed through the plastic matrix are activated and subsequently act as the catalyst for chemical metallization [3, 32].

In addition to high flexibility, suitability for prototyping and series production, low tooling costs, and high 3D design freedom, the ADDIMID technique offers the advantage of low material costs for the additive. Aluminum in particular is priced at around $22 (EUR 16) per kg and can be compounded with a multiplicity of low-cost, high-performance plastics, contributing to the huge economic potential of this process. [3]

Nothing definite can be said as yet about the structure widths of conductor traces and spacing. Research to date has focused primarily on reproducible laser direct structuring in combination with adequate mechanical bonding strength of the chemical metallization. In this context, too, very promising results have been obtained. They are dealt with in more detail in Section 3.1.4.3.

3.1.1.1.3 Alternative Laser Structuring Processes

MIPTEC

Panasonic's semiadditive MIPTEC (microscopic integrated processing technology) method of structuring differs widely from LPKF-LDS® and ADDIMID in terms of its process chain. As Fig. 3.7 shows, injection molding of the plastic body is followed by allover plating with copper. The unwanted metal is then removed by UV or IR lasering. The copper layers can then be galvanically reinforced. The unwanted copper is etched off and the body is ready for surface finishing. As a rule, multiple parts are produced in a single sheet and are then diced.

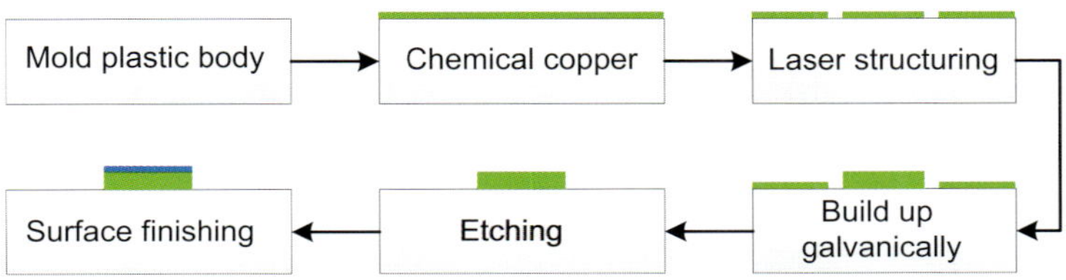

FIGURE 3.7 MIPTEC process chain [123]

Removal of the metallization means that this process is capable of producing very fine conductor structures of 50 µm, with interconductor spacing of the same order. Because full-coverage plating is followed by laser structuring, metallization produces a high surface quality with very low roughness, so bare semiconductor chips can be bonded with no postworking steps. The only materials currently compatible with the MIPTEC process are PPA as a thermoplastic substrate and the two ceramics Al_2O_3 and AlN. But there are radical new horizons opening up, particularly for ceramics. For example, a multilayer ceramic with vertical interconnects and internal conductor structures has already been implemented. With very good thermal conductivity and minimal thermally induced expansion, ceramics have a crucial advantage for MID LED applications in particular. Conductor structures generated by sintering can be sized down to 30 µm for both conductor width and spacing [123, 124].

The UV and IR lasers for the MIPTEC process are subject to design rules similar to those for Nd:YAG or CO_2 lasers in laser direct structuring. PPA has a minimum diameter of 250 µm for vertical interconnects, and the corresponding figure for the ceramic materials is 400 µm. Furthermore, in both the plastic and the ceramic substrates, the aspect ratio for vertical interconnects is 2. The maximum scan area also depends on the substrate material. The proprietary recommendations for PPA are 30 × 65 mm, 50 × 70 mm, and 63 × 55.7 mm. The advice for the two ceramic compounds is two squares with an edge length of 40 mm and 50 mm or a rectangle measuring 30 × 65 mm. Total height of the components can vary. The maximum difference in *z* between two structuring planes is 10 mm. The laser's angle of incidence for patterning in the MIPTEC process is 90° (Fig. 3.8) [123, 124].

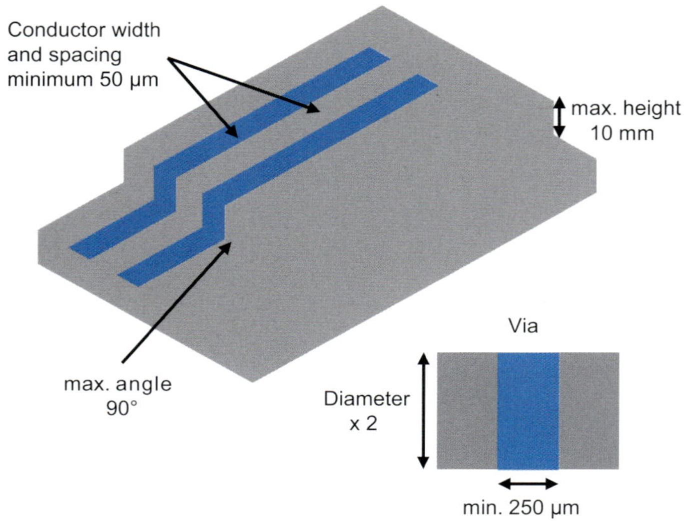

FIGURE 3.8 Sample design guidelines for the MIPTEC process according to [124]

Laser Subtractive Structuring

There are more process steps involved in laser subtractive structuring (LSS) than in the additive or semiadditive processes (Fig. 3.2). Injection molding is followed by short surface activation to permit electroless copper or nickel plating. This chemical premetallization is followed by a galvanic process to build up the plating to target thickness. The next step is structuring as such, by application of an activatable etch resist.

Etch resists can be of the photoresist or galvanoresist type. Photoresists react chemically to UV energy input, and galvanoresists are removed by laser energy. Photoresists can be further categorized as positive or negative. A positive photoresist is soluble after exposure and washes off, whereas exposure decreases a negative resist's solubility (Fig. 3.9).

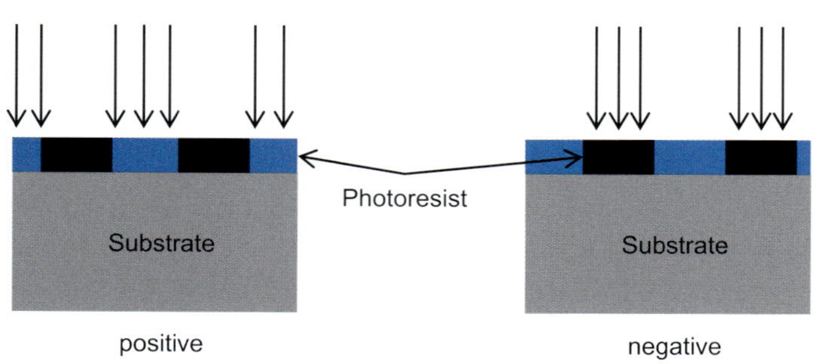

FIGURE 3.9 Process differences between positive and negative resists

Resist technology can be used to produce structures with a 30 μm conductor width at a patterning speed up to 2000 mm/s. The postexposure etching processes can cause these very fine structures to exhibit breakage of the resist at the edges, which in turn can lead to high fluctuations in the width of the insulating channels between the tracks. Contaminants in the resist can also be a problem. The presence of a single grain of dust is enough to prevent activation of the resist it shadows. Shorting as the result of incomplete insulation is one possible fault [48].

Galvanoresists are a good alternative for avoiding these problems, but the downside is that patterning speed is generally limited to 600 mm/s. Higher patterning speeds lead to incomplete ablation of the resist, with the resulting risk of shorting. Postworking of areas not ablated is possible in principle, but inevitably adds further steps to the process. Another alternative is repeated exposure of the same areas at a higher throughput rate. This, however, is always a trade-off between higher speed and the extra time needed for the repeat passes. One final option is chemical or electrochemical finishing of the parts, but this too adds an extra, complicated step on to the process [48].

Drawbacks of the subtractive technique are the complicated and extended process chain, plus the severe fluctuations in the thickness of the plating. Thick buildups in particular can give rise to problems with the width of the insulating channels. Underetching, too, can have a negative effect on the degree of miniaturization, on account of localized weak bonding strength.

Conventional Masking

The technique of masking is one of the oldest for MID structuring. The technology was borrowed from planar wiring board production. Figure 3.10 shows the extensive process chain it involves. Up to the point at which the mask is applied, the procedure is much the same as semiadditive laser structuring. The difference is

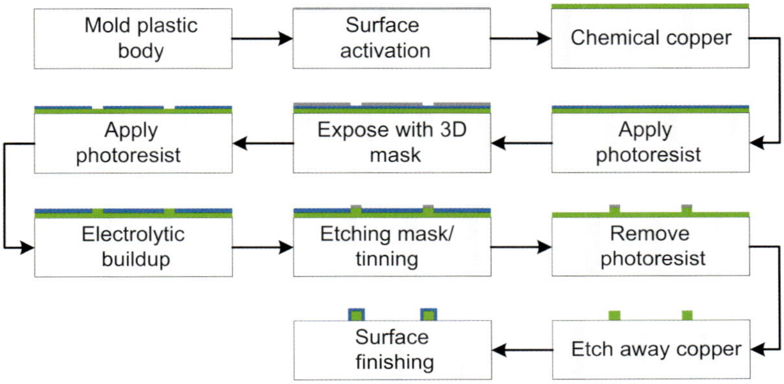

FIGURE 3.10 Steps in the masking process

in application of the photomask, which is generally in two parts, with subsequent exposure of the photoresist to UV light. Development of the photoresist is followed by buildup of the copper plating. The photoresist is then removed along with the unwanted copper. Surface finishing is the last step in the process chain.

The durability of photomasks is limited on account of the high intensity of the UV radiation. On the other hand, laser-patterned photomasks make layout changes a relatively simple and cost-effective matter, assuming that there is no need for alterations to the injection-molding toolset. The benefits of this flexibility always have to be weighed against the higher costs of consumption [48].

Conductor tracks in the 150 μm range present no difficulty. Even smaller structures are possible with suitable stencils, resists, and base materials, and the same applies to vertical interconnects. The only drawback is that the two-part photomasks impose constraints regarding three-dimensionality, for example because conductor geometry cannot be overly complex [48].

3.1.1.2 Printing Techniques

In terms of general process flow (Fig. 3.11), the Aerosol-Jet® and inkjet printing techniques differ radically from laser structuring. The process chain as such is very straightforward and compact, and print technologies also offer new possibilities for contacting, wiring, and integrating electronic components. The inks are applied directly to the substrate surface after injection molding and then cured in a sintering process. The process-specific differences between Aerosol-Jet® and inkjet printing are set out in detail below.

FIGURE 3.11 Process chain for Aerosol-Jet and inkjet printing

3.1.1.2.1 Aerosol-Jet® Printing

Aerosol-Jet® is a noncontacting, maskless, vector-based printing technique. The aerosol can be generated from low-viscosity (0.7 cP) or normal-viscosity (2500 cP) coating substances. Atomization is by either ultrasonics or pneumatic suction effect. The print mediums are many and varied, ranging from conductive inks through semiconductive to dielectric or insulating materials. The particles of the solids are nanosized (approx. 50 nm), so in combination with optimized focusing the technique can produce superfine conductor structures. The coatings can be cured by conventional thermal sintering processes in a curing oven or by selective sintering by light- or laser-soldering systems.

Printing Process

Figure 3.12 shows the pneumatic Aerosol-Jet® printing process in diagrammatic form. The ink, a suspension with a solids content of approximately 60 to 70%, is pneumatically atomized and mixed with an inert carrier gas. The suction effect is the basis for atomization. Plastic tubes deliver the aerosol to the print head. In the print head a metered feed of clean inert gas aerodynamically collimits the aerosol, which is then applied to the substrate surface in a tight jet by a nozzle.

The focused aerosol particles are 1 to 5 μm in size and permit microfine conductor structures down to widths of 10 μm and 20 μm spacing. Ink, substrate, and print parameters all have to be closely matched in order to achieve structures as fine as this. On account of the parallel alignment and high exit speed of the aerosol, the contour sharpness of the conductor track geometry remains virtually constant even when the nozzle-to-substrate gap is as wide as 3 mm. Consequently, small irregularities in the substrate surface are easy to print without modification of the print head's *z*-axis alignment.

Printing can be effected by moving the nozzle relative to the substrate, or by moving the substrate and keeping the nozzle in a fixed position. Creation of the aerosol is a continuous process. This in turn means that the aerosol flowing constantly to the nozzle has to be interrupted in order to print distinct structures. This is done with a shutter mounted directly at the nozzle outlet. The simplest arrangement is a curved metal rod. Machine programming commands control the shutter, interrupting or clearing the aerosol jet, as appropriate.

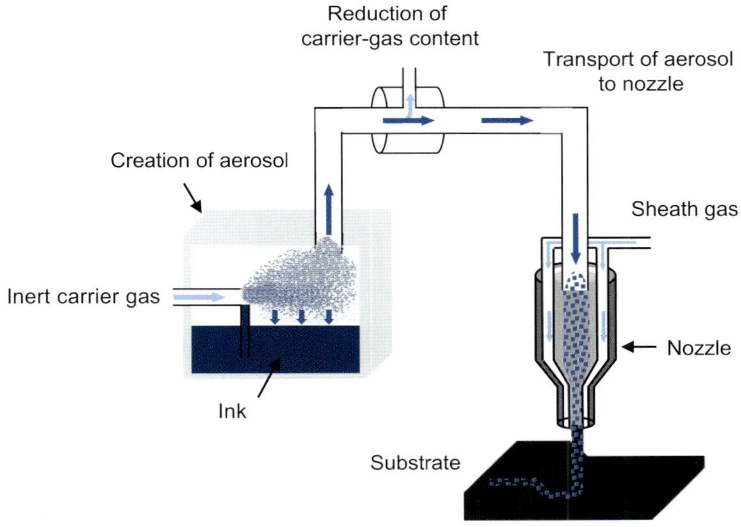

FIGURE 3.12 Diagrammatic view of the Aerosol-Jet® process (pneumatic atomization) as in [91]

Inks and Substrates

Ink composition varies very widely in terms of concentration, shape, size, substance of the particles, and the viscosity of the ink suspension, so the print medium can be optimized for any given substrate. A broad distinction is drawn between solutions and dispersions. A solution contains solids dissolved in a solvent. A dispersion is a heterogeneous mixture containing undissolved, fine particles of a solid in suspension.

Inks can be water-, solvent-, or polymer-based. Until now, polymer-based inks have been produced only for certain solids, and the technology has yet to advance past the research stage. An aqueous ink needs considerably more drying time than a solvent-based ink. Compared with that of solvent-based products, the drying process of aqueous inks is dependent on relative humidity. Adhesion to smooth substrate surfaces, moreover, is trickier and can cause runs in the conductor tracks. Aqueous inks have the advantage of better environmental compatibility. Polymer-based ink contains nanoparticles and what is known as a polymer template, an adhesion agent dissolved in water. The polymer template is also responsible for the dimensional stability of the nanoparticles, keeping processing properties unchanged throughout the entire process [163].

Figure 3.13 shows the composition of a typical ink in diagrammatic form. Solid particles, solvent, stabilizers, and modifiers all have to be carefully matched to each other. Good printability is based on dynamic viscosity and particle sizes within the working range of the atomizer used. The ink must also evince a certain stability, durability, scratch-resistance, and adhesive strength. The sintering temperature should be as low as possible, in order to avoid excessive thermal strain on the substrate. Economical production, clearly, is a primary criterion for aerosol inks intended for industrial use [9].

As with inks, there are many different substrates to choose from. Plastics, metals, semiconductors, and glass can all be printed, and in theory paper and textiles as well. Three-dimensional structures are just as printable as planar structures.

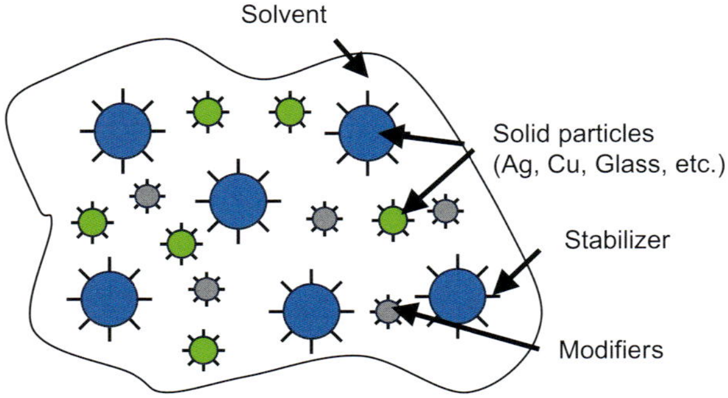

FIGURE 3.13 Schematic of a typical ink composition for aerosol printing [65]

In practical terms, however, it makes more sense to use the aerosol process for three-dimensional bodies than for flats. The vector-based procedure clearly limits working speed. Conversely, the variability of the nozzle-to-surface gap can be advantageous for printing complex 3D parts.

The main concern is the chemical compatibility of the ink with the substrate. In this context it is important to define in advance the requirements for the conductor structures and the preferred substrate material. This is essential in order to produce ultrafine structures, for example with conductor widths of 50 μm and 50 μm conductor spacing, of good quality in terms of conductance, adhesion, and reliability.

Curing Process

Curing is the process in which the liquid carriers are separated from the solid, generally metallic, constituents. At the same time, a supply of energy melts the nanoparticles to form a cohesive layer. Both these processes have to be considered in the light of exposure of the substrate to thermal loading. There are several ways of curing the printed aerosol structures.

It is standard practice to cure the printed circuits in a drying oven or reflow-soldering oven. There is a direct relationship between dwell time and curing temperature. Low temperatures necessitate long curing times to permit the ink to cure. This is the method of choice for substrates with low thermal durability. The dwell times for silver inks, for example, can be anything from 15 minutes up to 2 hours. The relationship between dwell time and temperature is also reflected in electrical conductivity, which tends to increase with sinter temperature.

Laser sintering and light soldering are selective curing processes. Infrared is used for laser soldering, whereas light soldering offers a choice of bundled, highly focused light and UV light. Ultrabright blue LED light can be bundled in parallel and focused through a glass lens for light soldering. The challenge in both cases lies in optimizing the power of the system. Wavelength and intensity are two additional factors that have to be taken into account for successful UV light soldering.

3.1.1.2.2 Inkjet Printing

Used for printed radio frequency identification (RFID) transponder tags, antennas, keypads, displays, sensors, and smart packages, the inkjet process is also a good choice for applications involving high-power electronics and organic LEDs. The major advantages are suitability for large conductor structures, high-precision metering of the ink, maskless printing of complex structures, and the wide choice of systems technology suppliers.

Printing Process

Inkjet printers can be operated in two different modes. One is continuous and the other is known as drop on demand (DoD). In continuous operation, connected structures are produced by a uniformly applied jet of ink. This mode is used primarily for high-speed printing. DoD, in contrast, produces a sequence of individual droplets of ink. Printing speed is limited, but precision is much better than in continuous printing.

Continuous-flow inkjet printing can produce structures 2 µm thick in the conductor width range from 50 to 150 µm. Arraying multiple jets in parallel increases printing speed. Thicker conductors can be produced by overprinting, running multiple passes over the same contour. Waste is a drawback of this printing process; the flow of ink is continuous, which means that the ink has to be electrically deflected at the places that should not be printed.

There are two ways of triggering ejection of ink droplets from the jet in DoD printing. The bubble-jet print head uses a thermal trigger, whereas a piezo print head is electrical. Bubble-jet print heads work on the principle of evaporation. A heating element positioned in very close proximity to the jet permits local, precision-metered heating of the ink. This causes flash evaporation, bubbling a defined quantity of ink and expelling it from the jet on to the substrate. A piezo print head uses the contraction of the piezoelectric crystal to produce a flow of ink. The action of the crystal produces deformation in a membrane that is attached to the ink tank, ejecting microfine drops of ink from the nozzle (Fig. 3.14). The higher the voltage applied, the larger the droplets of ink [19, 61].

Both of these methods require a constant gap of 2 mm to be maintained between nozzle and substrate in order to produce high-quality conductor structures. This can complicate process control to the point of impracticality, particularly when three-dimensional structures of considerable complexity are concerned.

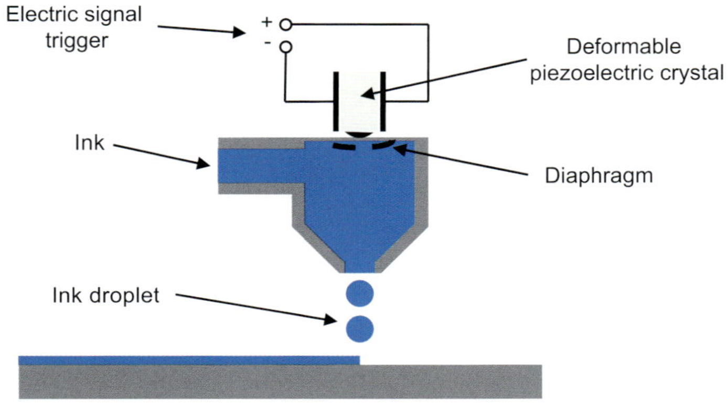

FIGURE 3.14 Schematic of the inkjet process [65]

Inks and Substrates

The viscosity and surface tension of the ink are the most important factors for inkjet printing. Viscosity should not exceed 20 cP for easy, optimum ejection from the nozzle. Surface tension is responsible for the globular shape of the ink droplets applied to the substrate. It is typically in the 28 to 350 mN/m range. There are enough different inks available on the market to render the inkjet process highly versatile. For example, conductive, dielectric, and semiconductive inks are all available for use [24, 28, 110].

Electrically conductive inks must contain nanosized solid particles to minimize clogging of the printing nozzles. A particle size of 50 nm is the norm, partly because the nanosize effect is beneficial for sintering temperature. The concentration of atoms at the surface is higher, so less energy is required to detach them from the mesh of solids. The tendency of nanoscale solids to agglomerate is problematic. Surface energy is minimized on account of the particles' very large surface area. In order to prevent agglomeration, the particles receive an organic coating immediately after manufacture. One beneficial side effect of this coating is a lower density than that of the carrier medium, which means that the heavy metal particles do not settle to the bottom of the ink tank [83].

As in Aerosol-Jet® printing, the conductor structure achievable by inkjet printing depends on numerous process parameters. The ink itself, the size of the ink droplets, the substrate material, the rate of advance of the print head, the gap between nozzle and substrate, and ink viscosity are all crucial.

Curing Process

Much as in Aerosol-Jet® printing, various different thermal curing processes can be applied to inkjet printing. The main objective is to remove the insulating coating and melt the stripped solid particles so that they form a cohesive track.

At this time, copper inks can be cured only by a high-energy light flash in the visible light spectrum. Depending on how powerful the flash is, it sinters the ink within a millisecond or 100 µs. This precludes the possibility of the copper oxidating. The power of the flash is in the 5 to 16 MW range. This method can also be used to cure other metallic inks. Thermal input to the substrate is low, so this method can be used in many different applications [83, 117].

Other ways of sintering the printed structures include thermal curing in a circulated-air oven, induction heating, laser, plasma, and microwave radiation. As in Aerosol-Jet® printing, circulated-air ovens and lasers are used. Plasma sintering is generally employed in combination with microwave radiation. Plasma energy for presintering is low at 150 to 300 W and is followed by microwave radiation to finish the process, much like the high-energy light flash used to cure copper inks. Microwave radiation at 1 W is applied for a few seconds. This process requires inks that favor thermal induction of microwaves. Excessive exposure to microwave radiation can result in hotspots that would ruin the substrate. The big advantage of this two-stage curing process is good conductivity, particularly with silver inks [83, 127].

3.1.1.2.3 Hot Embossing

Hot embossing is a fast, clean, and economical fully additive structuring process that has the added benefits of a limited number of process steps and low investment costs. A distinction can be drawn between stroke and roll versions of the process (Fig. 3.15). The stroke version is the more common in MID technology, simply because the parts are three-dimensional.

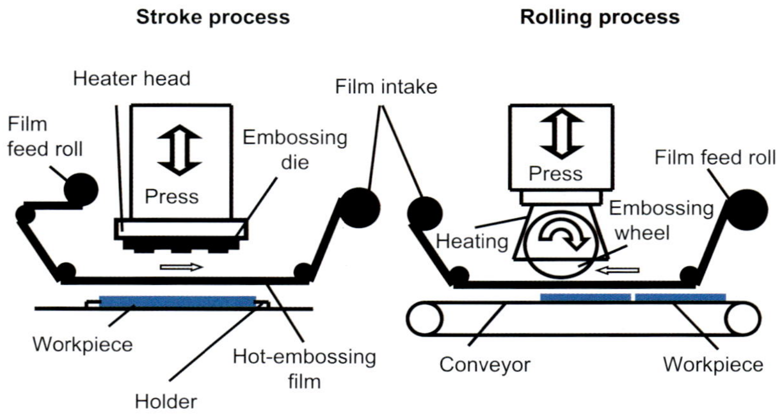

FIGURE 3.15 Schematic of the stroke and roll process variants

Versions of the Hot-Embossing Process

In stroke embossing, a heated die executes a vertical stroke to lift the hot film up and press it firmly against the blank. Press action is intermittent, and the embossing force depends on the type of drive used. Most presses used for MID have electromechanical or hydraulic drives designed for the high forces required. Presses designed for embossing forces from 70 to 90 N/mm^2 are needed, depending on the thickness of the copper film and the substrate. Sturdiness, reflected in the precision action of the die, is essential in a press designed to ensure the reproducibility of the conductor track contours. The system also has to be capable of precision temperature control up to 300 °C and embossing time up to 3 seconds [48].

The tool used for roll embossing is a wheel or roller with built-in heating. The plastic molding follows a linear track as it passes underneath the tool on a conveyor belt, for example. By contrast with stroke embossing, the force is applied only locally, mapping the line of contact between roller with film and the blank. The ability to produce large, uninterrupted surfaces without air inclusions is one of the big advantages of this process [48].

Hot-Embossing Process

In the hot-embossing process a heated die carrying the negative of the conductor layout presses a specially coated copper film on to a thermoplastic substrate, applying thermal loading and mechanical pressure. The die cuts out the film, forming a positive bond to the locally melted plastic close to the surface of the blank. Figure 3.16 shows the process chain in simplified, schematic form.

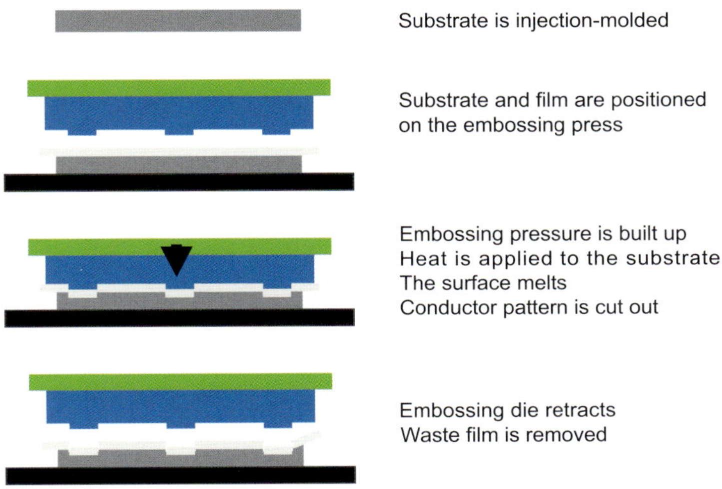

FIGURE 3.16 Process chain for hot embossing

The plastic molding requires no postembossing chemical treatment whatsoever, and this is one of the major advantages of the process. The plastic is exposed to less process load and is suitable for decorative finishes. The savings in process time and costs are also considerable. Hot embossing is a process that can be used for many different thermoplastic compounds.

Three-dimensional design freedom is restricted in the case of hot embossing. Vertical interconnects necessitate the use of additional processes. This entails inserting the film into preshaped holes in the substrate. Pins then press in enough conductive paste to fill the holes.

At this time, the sizing of conductor structures produced in this way is generally about 500 μm for conductor width and 500 μm for conductor spacing. In theory, even finer structures are possible, depending on the plastic substrate and the thickness of the film. This would entail using roll film with the lower limit for conductor structures within the adjustment accuracy of laser precutting and die embossing structure [142]. Table 3.1 shows possible conductor widths and spacing against film thickness.

TABLE 3.1 Conductor Width and Spacing in Relation to Thickness of Copper Film [48]

Film thickness [µm]	Min. conductor width [µm]	Min. conductor spacing [µm]	Distance from edge [µm]
12	300	400	200
18	500	500	300
35	800	800	300
70	1000	1000	500
100	1200	1200	500

Ideally, MID components are produced in medium- to high-volume runs with single-layer conductor tracks on planar surfaces. Inclined planes can be embossed, subject to certain restrictions, but it is virtually impossible to emboss cavities. All in all, the requirements that apply to the embossing tool are high. The conductor pattern to be embossed should measure $< 100\ cm^2$ and the length-to-width ratio has to be less than 4:1.

The conductor layout has to be designed to avoid possible problems. For example, transitions should always be rounded and neither conductors nor pads should have sharp corners or right angles. Conductor tracks should not taper. Thermo-mechanical stresses and strains can cause cracking, which would damage the film. This can be avoided by "rounding" the features of the circuit layout. Excess film should be easy to remove after embossing, so the structures should not form enclosures around excess film.

Hot-Embossing Films

Embossing films with copper coatings in the range from 12 to 100 µm for electronics applications have been around since the 1980s. Films with thin, vapor-deposited metal coatings were being used for embossing processes as long ago as the 1930s. At that time it was mostly gold and aluminum coatings that were being used for decorative purposes. [48]

The film can be fixed to the plastic substrate either by a glue or by mechanical adhesion. In the first instance there is a thermally activated glue coating the underside of the film. Hot embossing activates this glue, and when the film cools it adheres to the plastic. A very rough film surface can mechanically key the film to the plastic. The cauliflower-shaped roughnesses on the film are pressed into the softened plastic in the hot-embossing process, and when the surfaces cool a mechanical key has formed (Fig. 3.17).

The makeup of embossing films for MID components is the same for both versions of the process. Most films have three layers. The conductive copper film has a layer with adhesive properties on its underside. The top layer is surface metallization to

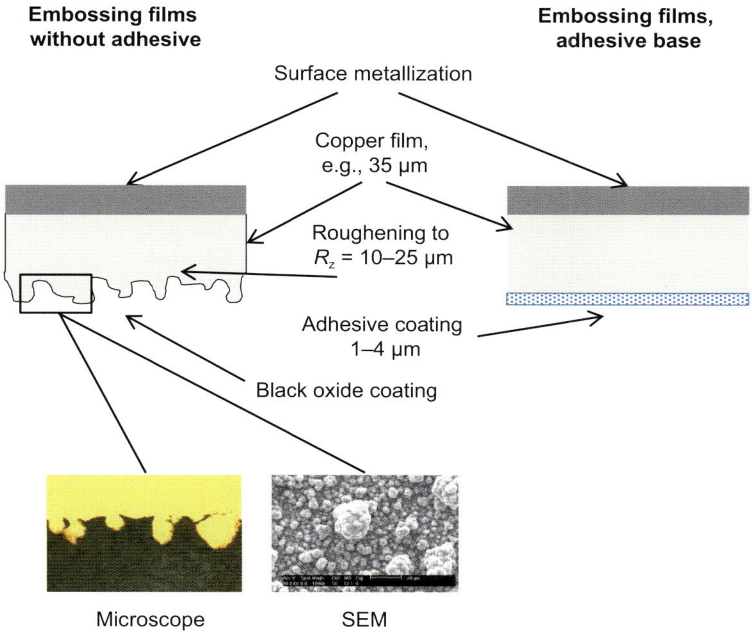

FIGURE 3.17 Schematic view of the makeup of embossing films for MID, with mechanical keying on the left and glue adhesion on the right [48]

prevent oxidation and improve solderability and contactability. The most common thicknesses for the layer of copper are 18 µm, 25 µm, and 35 µm. The surface metallization is tin, nickel, or gold, or some combination of the three [48, 142].

The manufacturing process is galvanic, from copper electrolytes. Copper is deposited directly on to a rotating roller made of titanium. The copper layer's low shear strength is brought about by a special process in film deposition that produces oriented crystal growth normal to the surface of the film [48].

The thermally activated glue applied in a thickness between 1 and 4 µm to the underside of the copper has to be matched to the embossing temperature for the substrate material. Heat and pressure are carefully controlled during the embossing process to soften the adhesive and key the copper film to the substrate (Fig. 3.17, right). The adhesive component of nonglue embossing films consists of surface roughening with a black oxide coating (Fig. 3.17, left). This roughening is effected when the copper film is deposited, and the peak-to-valley height R_Z is generally from 10 to 25 µm. The black oxide coating is applied after galvanic deposition of the coating in an alkaline, strongly oxidating solution in which a fine crystalline coating of copper oxide grows from a mixture of CuO_2 and CuO and adheres firmly to the surface of the copper. Local melting of the plastic in the process permits backflow of the material and mechanical anchoring around these rough structures [48].

Embossing Die

The cost of hot-embossed MID depends very much on the number of units that can be produced with a given toolset. The longer a die can operate and the higher the number of cycles it can withstand without maintenance, the lower the unit cost per MID. Conventional embossing dies are made of temperable steel such as 1.2767 and vanadium 6. Dies made of brass can be used for low-volume runs. Precision milling is used to pattern the die. After about 500 to 1000 cycles, tin deposits or carbonaceous pyrolysis products from the thermoplastic will have dirtied the surface of the die to the extent that mechanical or chemical cleaning is necessary. This entails interrupting production so that the die can be changed. [184]

The latest research indicates that application of a suitable coating can render dies considerably more durable. The most promising candidates appear to be a -C:H or DLC (diamond-like carbon) metal-free amorphous carbon coatings. They provide a very poor key for tin oxide deposits and alkaline cleaning is easy. In [184], cleaning after 3000 to 4000 cycles was sufficient. The die did not have to be remachined until it had performed 15,000 cycles. Industrial trials with these coated dies is in progress at this time [184].

3.1.2 Two-Shot Molding

The two-shot molding process chain is very short. As Fig. 3.18 shows, the interconnect device is produced in a sequence of two molding cycles prior to metallization of the electrically conductive structures. In this process the conductor layout is defined in the molding stage.

The two molding cycles differ as regards the type of plastic used. The plastic for the first shot is not plateable, whereas a plateable plastic is required for the second shot. Consequently, the cavities produced in the first shot are filled by the second shot to form the conductors. Alternatively, a plateable plastic can be molded in the first shot and then jacketed in nonplateable plastic in the second shot. As in the first case, the plateable conductors are exposed at the end of the injection molding process.

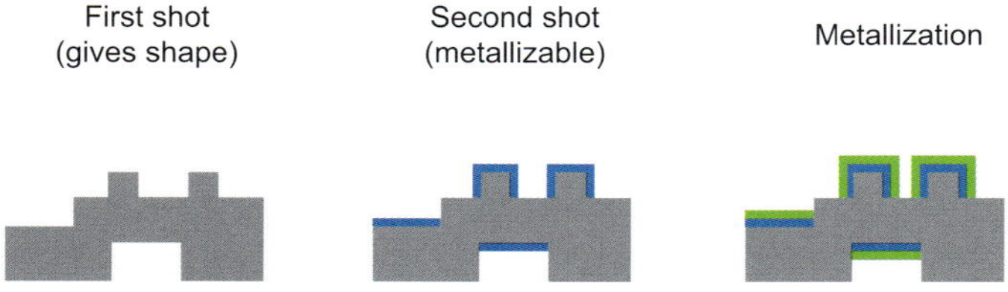

FIGURE 3.18 Process steps in two-shot molding [50]

Broadly speaking, there are two basic processes for two-shot molding. They are the PCK process (printed circuit board Kollmorgen), in which the two shots are molded directly one after the other, and the SKW (Sankyo Kasei wiring board) process, in which the plateable component is chemically catalyzed before the second shot is molded [48, 126].

Resolutions achieved with this process typically have conductor widths and spacing in the 200 to 250 µm range. The technological limit is currently 150 µm, although efforts are being made to achieve further refinement. Crucial in this respect are the flow properties of the plastic in combination with the design of the injection molding tool and the lengths of the tool's runner channels [83, 151].

The plastic combinations used must have good composite strength. This is the property that determines the quality of adhesion between the substrates molded in the two shots. Another option is to establish mechanical interlinks between the two plastics. This requirement limits the choice of plastics that can be combined. The issue is explored in greater detail in Chapter 2.

The main advantages of two-shot molding are high geometric design freedom and suitability for high-volume production with uniform precision and low costs per unit. A combination of electric circuitry and mechanical functionality is the typical requirements list for a two-shot MID such as a snap hook that enables the electrical part to connect to other components.

3.1.3 Insert Molding

Insert molding, or in-mold processing as it is sometimes known, is characterized by the combination of a printed or structured film with a thermoplastic substrate material. MID technology makes use of this property in a variety of process combinations.

3.1.3.1 Film Insert Molding

FIM is a method of producing MID that are combinations of rigid and flexible circuit boards and also integrate mechanical functions. This can be an ideal way of utilizing package space. The plastic is injected into a mold containing a flexible circuit-layout film, which can consist of one or more layers and is produced in a separate process. Prior to the molding process the film is structured and plated by conventional PCB techniques—an important advantage of this technology. Established, well-proven techniques can be used to produce the electrical circuit layout. Another strength is that the process consists of no more than a limited number of steps (Figs. 3.19 to 3.21). Other benefits include the versatility of using films plated on one side only or on both sides and suitability for producing decorative surface finishes [48, 181].

Film insert molding is challenging for the molding company, both in terms of the injection molding process itself and as regards tool and die making. In-mold decoration and in-mold labeling are established technologies that have been used for decades in the manufacture of decorative moldings. Empirical values from this pool of conventional process know-how can be transferred with little change to the MID environment [48]. MID manufacture can utilize the techniques of thermoplastic foam molding, injection compression molding, and press insert molding.

3.1.3.1.1 Thermoplastic Foam Molding

Thermoplastic foam molding combines mechanical strength with lightness. The parts have a cellular, lightweight internal structure and a smooth, closed-pore outer skin.

As can be seen in Fig. 3.19, the film is first placed in the dip-edge tool. A standard injection molding machine then injects the metered molding material, which contains a propellant, into the closed tool. The tool is opened slightly at the end of the cycle. The molding material foams because the chemical propellant heats past its decay temperature and generates gas. The high back pressure in the plastification cylinder prevents foaming before the material is inside the tool. The shot fills the tool cavity at a pressure of about 100 bar. Expansion of the foam is facilitated because the tool is slightly open, and this is also why almost 100% foaming can take place. The dwell-pressure phase common to most injection-molding processes is not necessary on account of the gassing. The mold cavities are filled even though melting viscosity is low by comparison with conventional injection-molding processes. Low pressure combined with viability for long runners and a low level of internal strain is another strong point of this process [48].

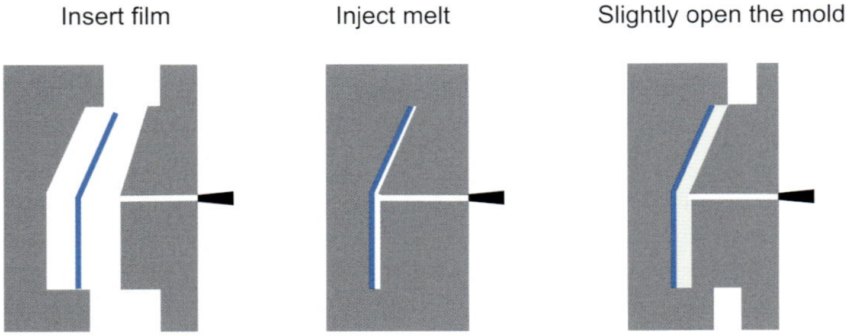

FIGURE 3.19 Process chain, thermoplastic foam molding [48]

3.1.3.1.2 Injection Compression Molding

Both thermoplastics and thermoset materials can be injection compression molded. Tool patterning is particularly accurate on thermoplastics. It is also simple to produce very large parts using this method. Figure 3.20 shows the usual process flow.

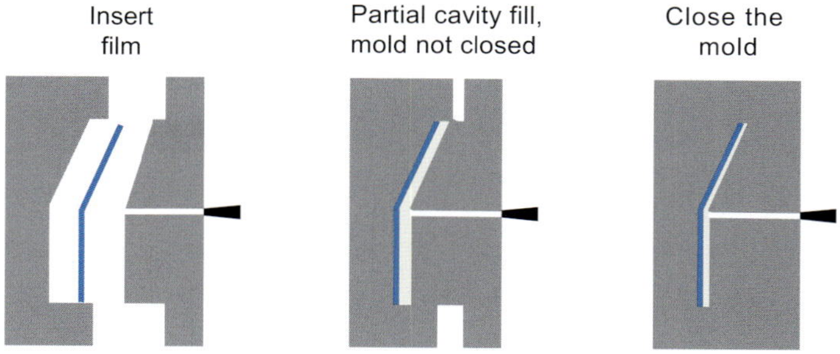

FIGURE 3.20 Process chain, injection compression molding [48]

Once the film has been placed in the open tool, the shot is injected with the tool not fully closed. The tool can be open from 1 to 20 mm, depending on the molding, and in isolated cases the opening can be as wide as 50 mm. The tool does not close until the molten plastic has started to cure. The closing pressure applied in this way is responsible for final shaping of the molding. Pressure is evenly distributed, so the parts are largely free of internal strains. Alternatively, the mold cavities can be filled with the tool closed. The tool is then opened slightly in response to the increase in internal pressure. Closing force is increased to exert counter-pressure on the shot and produce relatively strain-free moldings [48].

3.1.3.1.3 Press Insert Molding

Very high-strength keying between the plastic substrate and the film is characteristic of press insert molding. A broad distinction can be drawn between melt-strand and laminar-flow processes. Melt-strand press insert molding is very similar to injection compression molding as described above. An extruded flow of melt is injected into the partly opened tool, which then closes. The advantages are the short cycle time, low shear, low decoration loading, and the ability to mold either thermoplastics or thermosets. It is also possible to produce parts that do not have a weld line. Drawbacks are the limits on part size, the need for an injector that can be operator-manipulated, and the limited range of suitable plastics [48, 84].

In the laminar-flow version of press insert molding, the melt is injected into a vertical tool that is open at least 30 mm, as shown in Fig. 3.21. By contrast with injection molding, the film and the melt do not come into contact until the tool closes. This helps minimize thermal strain on the film. Relatively large parts can be produced using this method. Free component geometry and the wide range of usable materials are further advantages. Weld lines, however, are unavoidable, and this is a drawback vis-à-vis melt-strand molding. Long deformation runs, high melt temperatures, and high tool costs further restrict the practical reach of this process [48, 84].

The principle for both these methods is final shaping by pressing in the molding tool. The film and the melt are inserted prior to production of the molded part.

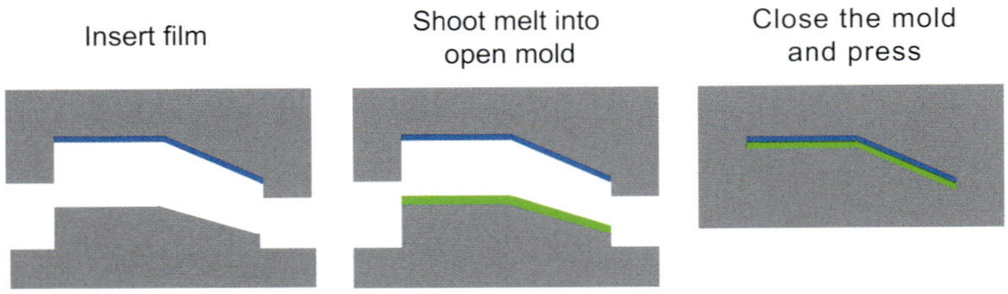

FIGURE 3.21 Process chain, press insert molding [48]

3.1.3.1.4 Other Varieties of Film Insert Molding

In addition to the three processes described above, **backfilling** conductor films is another way of creating an MID from a flexible, structured, and plated film material. For the sake of completeness, some of the possibilities are outlined here (Fig. 3.22).

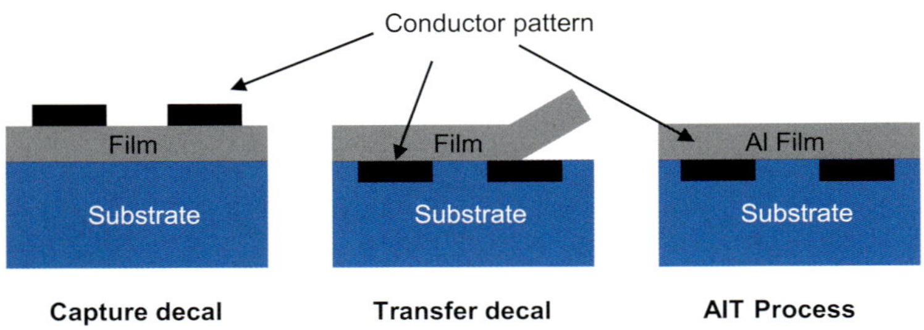

FIGURE 3.22 Diagrammatic view of the process chain for variants of film insert molding (FIM) [48]

The **capture decal process** is very similar to in-mold decoration (IMD). Plastic parts with decorative surfaces are the primary interest. The choice of backfill film and backfill melt for the combination is crucial in terms of strength of adhesion [48].

A method known as the **transfer decal process** can be used to transfer the conductor layout from a film to a plastic body. The conductor side of the film is backfilled and then the film is peeled off the finished MID component. This process requires low adhesion between backing film and backfill material. A low-strength bond between conductor tracks and backing film is also essential. The conductor tracks can be surface-treated to ensure good adhesion to the backfill material [48].

Yet another option is the **advanced interconnect technology (AIT) process**. It uses a multi-layer aluminum film carrying structured copper conductor patterns. Once the copper-structured side of the film has been backfilled, as in the transfer decal process, the aluminum is selectively etched away. The aluminum film and copper conductors are unaffected by high thermal loading, so high-melting plastics can be used to produce MID parts that are resistant to high temperatures [48].

FIM offers a wide choice of possible methods, each of which has specific advantages and disadvantages. Common to all is the need for ensuring maximum permissible formability of the plastic films. The relatively high fracture strain of the copper conductors is the deciding factor in this context. Consequently, certain restrictions apply to three-dimensionality and have to be taken into consideration on a case-to-case basis.

Thermoforming, vacuum, or compressed-air deep-draw forming of films are ways of countering these limits with regard to freedom of design. In this respect it is advisable to apply the conductor layout using the primer method, which is characterized by high elongation. This process is discussed in more detail in Section 3.1.4 [48].

3.1.4 Alternative Structuring Processes

At this time, the processes described above are the core methods for structuring and metallizing MID components. It is standard practice to choose from this pool when considering a method of production in line with application specifics and requirements. There are also other processes currently in the development phase. Older processes, on the other hand, have virtually disappeared from the scene, but for the sake of completeness they are mentioned here.

Primer and tampon printing are alternative structuring methods currently used in MID production along with the plasma technologies, which are discussed in more detail in Section 3.3 below.

3.1.4.1 Primer Technology

The term "primer" is used primarily in painting. It generally means an agent that can be applied in order to improve adhesion, or in other words, to form a strong key between two materials. In MID technology, materials with paint-like properties are classified as primers. They are applied to plastic substrates and then metallized or plated. The primer acts as an adhesive coating between substrate and metal plating.

Virtually all primers used in MID technology are metallizable. They are applied in structured form either by silk-screen or by tampon printing, thermally activated and then additively plated. There are also some primer systems that can be applied as full-surface coverings, then plated and structured using conventional masking and etching techniques. Selective activation of allover primer systems is also possible [48].

The usable primer systems consist of a polymer binding agent matrix, a mixture of functional fillers, an organometallic palladium compound as catalyst for subsequent copper plating, diverse fillers for rheological adaptation to the application technique used, and a suitable solvent mixture [48].

Figure 3.23 shows the process chain involved. Once applied, the primer has to be thermally conditioned to key it to the plastic substrate (Fig. 3.24) and also to activate the metallization catalysts it contains. At the surface of the primer the catalyst forms an effective base for the copper plating produced by chemically reductive deposition in the final step of the process. The conductor patterns can be temporarily contacted to facilitate galvanic thickening of the layers or surface finishing, if necessary.

Primer technology offers a raft of options for structuring three-dimensional interconnect devices. One possibility is to shape silk-screen printed films and then stabilize them by film insert molding (Section 3.1.3.1). Another option is to use tampon printing to transfer a circuit layout directly to a three-dimensional plastic body.

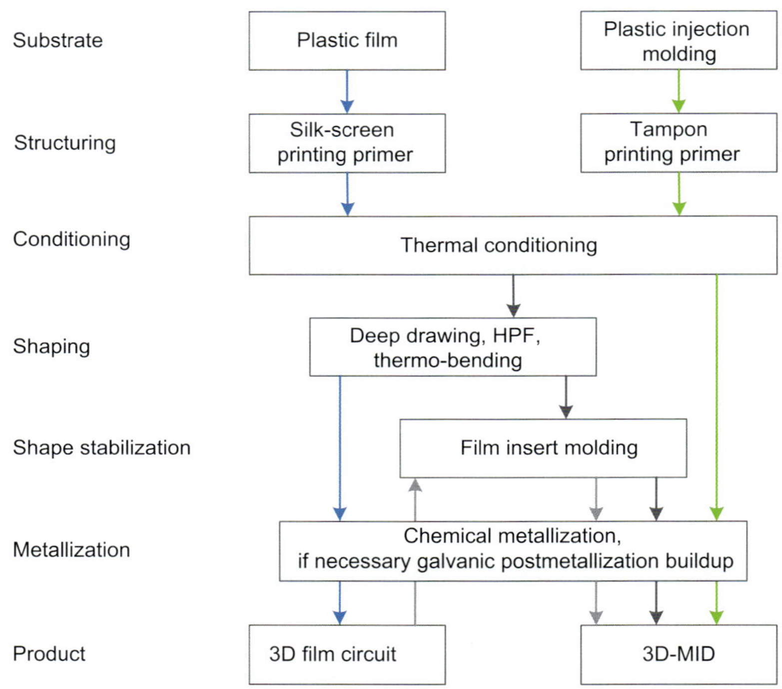

FIGURE 3.23 Process chain of primer technology [48]

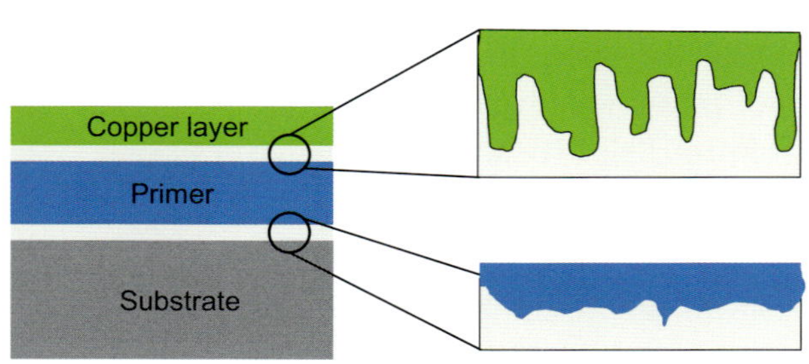

FIGURE 3.24 Adhesion mechanisms, plastic – primer – copper [48]

3.1.4.2 Tampon Printing

Tampon printing is indirect, using a resilient silicone tampon to transfer printing ink from a block to a substrate. In MID engineering the process is used to print insulating surfaces, solder stoplines, or the primer coats described above.

Structure sizes in silk-screen printing stop at a conductor width of about 130 μm. The corresponding figure for tampon printing is in the region of 100 μm. Tampon printing has the advantage of three-dimensionality over silk-screen printing, however, because the flexible tampon adapts well to different surface curvatures. This ability is further enhanced by variable guidance of the tampon.

3.1.4.3 Plasma Technologies

As yet, plasma technology has not established itself as a method of simultaneously structuring and metallizing MID components. In other areas of industrial engineering it is widely used, for example for activating, cleaning, etching, and coating surfaces. These areas of application in industry at large are primarily metal-engineering processes, so the thermal loads involved are generally negligible, but this does not hold true in the case of the thermoplastic materials used for MID.

The high-energy jet of gas, termed a plasma, is used to melt a nanoscale metal powder. In this way metal structures such as copper can be deposited directly on to the substrate surface. Plasma deposition remains at the fringe of structuring and metallization at this time. The technology is currently in the research phase. There is certainly potential for industrial application, but considerable research is still needed. The biggest advantages are the very short process chain (Fig. 3.25), simultaneous structuring and metallization, and the possibility of building thick layers and applying them quickly. So the technology has a great deal to offer, for example, when high currents have to be carried by the conductors. The automotive industry, the energy sector, and power electronics are all likely areas of application.

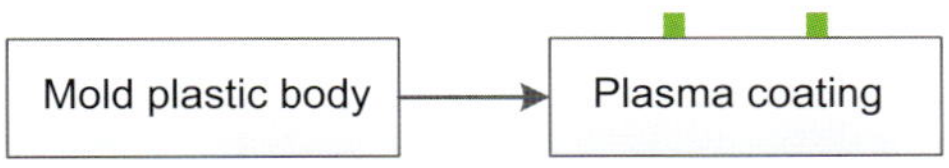

FIGURE 3.25 Process chain for plasma coating

3.1.4.3.1 Flamecon®

Flamecon® is an automated, thermo-kinetic metallization process. The technology enables fine metallic structures for electrical functions to be applied to a very wide range of carrier materials. Plastics, wood, ceramic, and even metal can all be metallized.

Figure 3.25 shows the process flow. Before being coated, the substrate is abraded by a laser, a jet of sand, or a pulverized metal abrasive to structure the surface. In the Flamecon® process as such, metal particles of the preferred coating material are injected into a hot chamber by a high-precision metering unit. The particles are melted and jetted under computer control in a stream of carrier gas to the nozzle head, which applies the plasma to the substrate. Both processes can be integrated in a single step. Width and thickness of the conductive structures can be parameterized [170].

The biggest advantage of this additive process is that it takes a relatively short time to produce conductive structures with high current-carrying capabilities. Conductor thicknesses between 20 and 100 μm can be laid down in each coating pass. Iterative coating over a given structure can produce thicknesses of more than 1000 μm. On account of porosities and other factors, the conductivity of the structures is about 50% of the material in the solid [170].

The adhesion of the structures to the substrate is influenced by the diameter and the material properties of the metal particles and also by the particles' thermal energy and velocity on impact. Adhesion, therefore, is a combination of mechanical keying and chemical bonding. Other advantages of this process include simple adaptation for new layouts by changing the robot's path program and the low level of thermal impact on the substrates.

The technology is still in the development phase at this time. The Flamecon® process is attracting considerable interest in many areas of application, particularly in the automotive and lightweight engineering sectors.

3.1.4.3.2 Plasmadust®

The Plasmadust® process uses a cold plasma to apply fully functional structures to a wide range of carrier surfaces in the single process step. No volatile organic compounds are required, and since the process is VOC-free it dispenses with environmentally suspect solvents and surface-activation, cleaning, and drying processes. Plasmadust® therefore has significant advantages—including extra value-adding potential—over conventional processes as a means of structuring MID quickly and cost-effectively.

The Plasmadust® process is patented by Reinhausen Plasma GmbH. It is a generative direct, noncontacting, solvent-free coating process for building structures from powder-form materials such as metals or polymers on different substrates (e.g., paper and textiles). It can build conductor structures at high speed and in-line. The central feature is the use of a cold plasma which, because its working temperature is low, results in a correspondingly low temperature at the surface of the substrate. The working range is from 10 to 150 °C, so the process can be applied to temperature-sensitive substrates such as paper or card. It has considerably less effect on the base material than competitor processes such as PVD (physical vapor deposition), CVD (chemical vapor deposition), and thermal spraying [39, 140].

The layers are deposited by a powder/gas mixture. In principle, this means mixing a material in powder form into a stream of carrier gas. The fine-grained powders with grain sizes in the range from 100 nm to 20 µm can be based on metals (Cu, Sn, Zn, Ag) or nonmetals (e.g., PTFE) and have to be fed into the module. As many as five powder cartridges or powder hoppers can be connected in a feeder array. The powder is channeled to an emulsifier that meters the supply and maintains a continuous feed to the nozzle. Precision powder metering is induced by vibrator oscillations that keep the powder from agglomerating and permit the module to lay down reproducible and uniform layers. The metered supply of powder is transported by a stream of carrier gas from the emulsifier to the plasma generator, or more accurately to the plasma nozzle [39].

As Fig. 3.26 shows, inside the nozzle the carrier gas is introduced at both sides by feed hoses and travels through a small channel directly to the center of the plasma nozzle. High-frequency ignition between cathode and anode generates the plasma. The metallic coating head is designed with its shape-defining hollow cylinder as the anode. The cathode is a probe-like hollow core structure that directs a process gas (e.g., nitrogen) toward the surface of the substrate. By application of a high-frequency current, on the inverter principle, a high-energy, pulsed arc is generated between cathode and anode, ionizing the process gas in its passage by energy transfer and transforming it into what is referred to as a low-temperature, unstable plasma [11, 139, 165, 171].

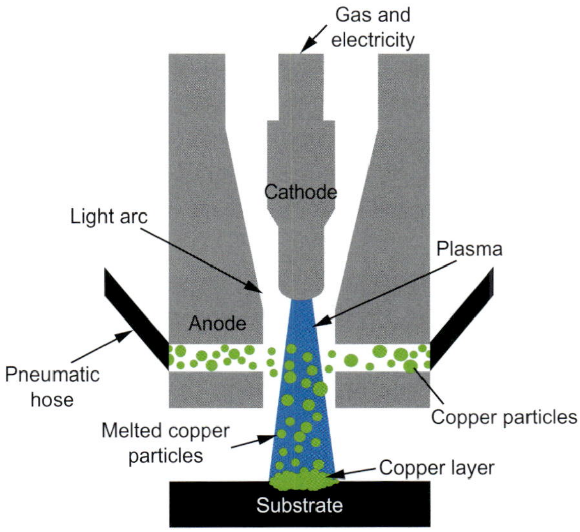

FIGURE 3.26 Principle of the Plasmadust® process

Unlike stable plasmas, which are in equilibrium, the temperature and energy of the electronics in the unstable flow are several orders higher than in the neutral gas particles and ions. These energy-rich particles strive toward achieving as low an energy level as possible, so they release their energy in pulse form to their immediate surroundings. This discharge is possible only to heavy particles such as copper atoms, so light particles such as gas molecules become only slightly hotter. It is on account of this relatively low gas temperature up to 150 °C that this kind of plasma is referred to as low-temperature plasma.

Gas pressure impels the copper particles melted by energy transfer against the substrate material. Roughened by the plasma, the surface is ripe for positive bonding, without excessively high thermal loading of the substrate (Fig. 3.26). The grains of powder are minute, so temperature-sensitive materials can be coated, because before they reach the surface the particles have cooled to temperatures at which they cannot cause damage [11, 139, 165, 171].

The technology has now evolved to where structures with conductor widths of 2 to 5 mm can be produced. Masking brings the size limit down to about 200 μm. However, this entails sacrificing the advantage of a very short process chain (Fig. 3.25). Structure height is in the range from 1 to 100 μm, depending on the number of passes. Maximum coating speed is currently 50 m/min.

The process has a relatively large overspray, which is a drawback. Overspray is the unwanted proportion of sprayed material deposited in droplet form on the component along the edges of the printed structures. Unmelted copper particles settle as dust on the substrate during deposition. This dust is very easy to remove with alcohol afterwards, but this is an extra work step. Reducing overspray would also cut down on overall consumption of copper powder.

The Plasmadust® process is suitable for a wide variety of materials, and this opens up tremendous potential. The prerequisites are that the material has to be available in powder form and it has to melt inside the temperature window of the Plasmadust® process. Along with the mainstream metals (copper, aluminum, gold, tin, bismuth, tellurium) and polymers (PTFE, PE, ABS, PP), glass, ceramic materials, semiconductors such as CIGS (copper, indium, gallium, sulfur, selenium), and even minerals like salt can be deposited on substrates such as metal, plastic, glass, ceramic, paper, card, and textiles. Another advantage is that the process can also handle alloys and material combinations such as carbon and metal [42, 139, 140, 179].

The important structuring processes and their specific properties are set out in Table 3.2. There are obvious differences between the individual processes, and clearly no one process has the ideal combination of properties for every sector. As stated several times in this chapter, the choice of structuring process depends very much on the application, and this too is clear from the comparison.

TABLE 3.2 Comparison of the Important MID Structuring Processes on the Basis of Specific Properties

	Min. conductor width	3D design freedom	Layout change	Material diversity	No. of units	Material costs	Tool costs	Investment
LPKF-LDS®	100 μm	high	simple	medium	high	high	very low	medium – high
ADDIMID	No info	high	simple	high	high	medium	very low	medium – high
MIPTEC	50 μm	high	simple	very low	high	high	very low	medium – high
Resist	30 μm	medium	moderate	very high	high	low	low – medium	medium
Aerosol-Jet®	10 μm	high	simple	very high	low	medium	medium	medium – high
Inkjet	50 μm	low	simple	very high	high	medium	very low	high
Two-shot	150 μm	very high	very high	low	very high	low – medium	very high	very high
Film insert molding	100 μm	medium	simple	very high	high	low – medium	high	very high
Hot embossing	300 μm	low	moderate	very high	high	low	medium	low
Flamecon®	No info	medium	simple	high	medium	low	very low	medium – high
Plasmadust®	200 μm *	medium	simple	high	medium	low	very low	medium – high
Primer	100 μm	high	simple	very high	medium	low	medium	medium – high

* Masked. Without masking at least 2000 μm

It is equally important to see that there is a suitable structuring process available for every requirements profile. Fine-pitch structures, mass production with simple layouts, a high degree of design freedom with full 3D capability, for high-performance or stock plastics, and also for low investment outlay: MID technology now offers a suitable structuring and metallization process for every product.

3.2 Metallization

Chemical metallization is not a necessity for each and every structuring process. The Aerosol-Jet® and inkjet printing technologies, hot embossing, Flamecon®, and Plasmadust® all dispense with postprocess metallization. The other processes all require chemical and in some cases also galvanic metallization to plate metal layers on to the structured MID.

3.2.1 Cleaning the Substrate Surface

Cleaning is an essential step, particularly when the process used is LPKF-LDS® or laser subtractive structuring (LSS). For the former, cleaning prior to electroless metallization is just as essential as after copper plating in laser subtractive structuring. There are various methods of cleaning, ranging from wet-chemical through water-jet to CO_2 snow-jet cleaning. The cleaning processes can be distinguished in terms of cost efficiency or component handling but also with regard to the laser structuring process employed. For example, care has to be taken to ensure that plastic surfaces activated by the LPKF-LDS® process are not deactivated by cleaning. Unwanted metallization is a possible result of insufficient cleaning and can lead to shorting in the metallized structures [189].

A **wet-chemical cleaning** process generally takes place in aqueous wetting solutions with a cleaning agent backed by ultrasonics and temperature. Ultrasonic frequency and power and temperature often contribute more to the cleaning effect than the cleaning medium as such. A rise in temperature reduces the positive effect of ultrasonics on cleaning, so it is always best to maintain room temperature when cleaning in an ultrasonic bath. Research has also indicated that multifrequency ultrasonics (40, 80, and 120 kHz) are considerably more effective for cleaning than a standard ultrasonic frequency of 40 kHz. Cleaning time depends on temperature and the ultrasonic parameters. Broadly speaking, a lengthy process time makes wet-chemical cleaning more efficient [189].

Another distinction should be drawn between frame and drum cleaning. In frame cleaning the parts are suspended from a specially designed and manufactured frame. In drum cleaning, on the other hand, the parts are placed in a drum not unlike that of a household washing machine. Drum cleaning has the advantage of quantitative cleaning in a short time. But it requires more intensive wet-chemical cleaning than basket or frame cleaning, with their slightly more complex process chain. This is because the parts while in the drum tend to shield each other from the cleaning medium [189].

CO_2 snow-jet cleaning makes use of the following mechanisms: pulse transfer of the high-velocity CO_2 crystals, resulting thermo-electric effects, extreme volume increase when the crystals evaporate, and the associated flushing of the surface and change in aggregate status as the crystals strike the surface. The ultrasonic two-component concentric nozzle is the essential component of the cleaner. Liquid CO_2 exiting this nozzle expands to form a snow/gas mixture. This core jet is jacketed inside a ring jet of compressed air, which directs the CO_2 crystals and accelerates them to ultrasonic speed. The high kinetic energy of the CO_2 crystals combined with the high dissolving force, comparable to that of cyclohexane, enables the removal of numerous solid and film-like contaminating substances without the need for additives. The nozzle design permits highly accurate spot cleaning and cleaning of extremely awkward areas. Large surface areas can be cleaned in an environmentally compatible way, leaving no residues and without any need for solvents, by batteries of suitably arrayed nozzles [1, 189].

Compared to the two methods described above, **water-jet cleaning** is much simpler and more cost effective. The structured substrate is cleaned by a jet of water at up to 150 bar. The high pressure necessitates the use of workpiece holders, and the method is not suitable for every 3D geometry. Nor is it possible to use this method to clean miniaturized components [189].

CO_2 snow-jet cleaning is eminently suitable for 3D substrates with pitches $< 250\ \mu m$. The wet-chemical method has advantages in terms of cleaning LSS surfaces and in high-volume production runs. The process of water-jet cleaning is the most economical solution for pitches $> 300\ \mu m$, assuming that the geometries are suitable [189].

3.2.2 Metallization

Galvano-Chemical Metallization

Specific metallization processes are necessary for the selective deposition of metal coatings on structured thermoplastics. Wet-chemical metallization consists of pretreatment and activation of the plastic surface, followed by chemically reductive or electroless metallization. The systems of metal plating are similar to those employed in the manufacture of printed-circuit boards, but in MID technology the pretreatment process that precedes metal deposition is of crucial importance. Pretreatment has to ensure the production of microrough plastic surfaces so that the plating can adhere securely [30].

For semiadditive laser structuring or two-shot molding, this microrough surface is produced by wet-chemical pretreatment of the plastic substrate. Depending on the plastic substrate used, its surface can be dimpled by etching away the amorphous parts or by leaching out fillers close to the surface. This enables the metal plating to key itself mechanically to the substrate. Thorough rinsing then removes traces of the pretreatment mediums and partly detached particles of plastic. These rinsing steps also have considerable effect on the quality of the subsequent coating. Dwell time in an etching bath determines the roughening of the plastic surface, which is linear relative to the adhesive strength of the plating to be coated. It is important to bear in mind, however, that an overly rough surface can have a negative effect on placement of the electronic components. The trade-off is adequate adhesive strength against acceptable surface roughness [30].

Etching is followed by seeding with palladium. The palladium nuclei are inserted into the exposed cavities and are catalytically active in the follow-up chemically reductive metallization process. The plastics used in two-shot molding are generally core-catalytic and require no additional seeding or activation. Only the surface has to be etched open to permit access to the catalysts embedded in the plastic [30].

Lasering is used in the LPKF-LDS® process to create the microrough structures and excite catalytic activity (see Section 3.1.1.1.1). This necessitates cleaning of the structured plastic body to remove the residues of lasering (laser debris). This can be done in an aqueous ultrasonic bath with tenside additive, by water-jet cleaning, or by CO_2 snow-jet cleaning. Water-jet cleaning is subject to constraints in terms of component geometry. It is generally only when the requirements for conductor-surface and edge roughness are extra-high that resort is made to snow-jet cleaning [30].

Plastics are very good electrical insulators, and this hampers galvanic metallization in MID technology. The electrons required for the purpose cannot be provided by an external electrical source. Chemically reductive metallization baths, also known as electroless baths, are used instead. The electrons are provided by a component of the bath, namely the reductant. A bath of this nature consists of an aqueous metal salt solution, a reductant, and various additives such as chelating agents and stabilizers. In technical parlance, chemical baths of this nature are termed thermodynamically unstable and kinetically inhibited. The art of designing the bath is to make sure that metal deposition starts only at catalytically active areas (the tracks lasered by LDS or the places activated by selective palladium activation) and does not take place where no plating is deposited. This is accomplished by special stabilizers in the bath and by injection of finely distributed air [30].

The copper bath is always the starter for metallization. This is followed up with nickel and gold overplating, depending on the quality of the copper coating. The catalytic activity of core-catalyzed or laser-structured surfaces is generally lower than when surfaces are catalyzed with palladium. This can be remedied by increasing the activity of the copper bath [30].

The advantages of chemical, electroless metallization are the homogeneity of the plating, the ability to metallize superfine structures, and the freedom of the process from the complexity of component geometry. The coatings produced in this way can be used in many ways. Soldering and conductive-adhesive gluing of surface mount devices are just as viable as flip-clip placement and aluminum wire bonding.

MID parts can be either secured to a frame or metallized in a drum, depending on size. Drum metallization is a high-capacity process for highly miniaturized parts. The drum has to be in constant rotation in order for metallization to be uniform. Whenever plating quality is subject to high requirements, however, or if component geometry is complex, it is advisable to use frames for metallization [30].

The diagram in Fig. 3.27 represents one possible setup for metallization with a Cu-Ni-Au buildup. Metallization is generally followed by several cleaning steps. Extra process baths are needed for activation, pretreatment, and coating. Constant automatic circulation in each bath keeps the concentration uniform at the plastic surfaces to be coated. Filters connected to the baths remove deposited particles from the mediums. Programmable logic controllers (PLC) are a frequent choice for controlling the heating and ensuring a constant process temperature in each bath. Automatic air injection regulates the activity of the copper bath. Chemical analysis of the various baths and balancing of the chemicals can be manual or, more commonly, automatic [30].

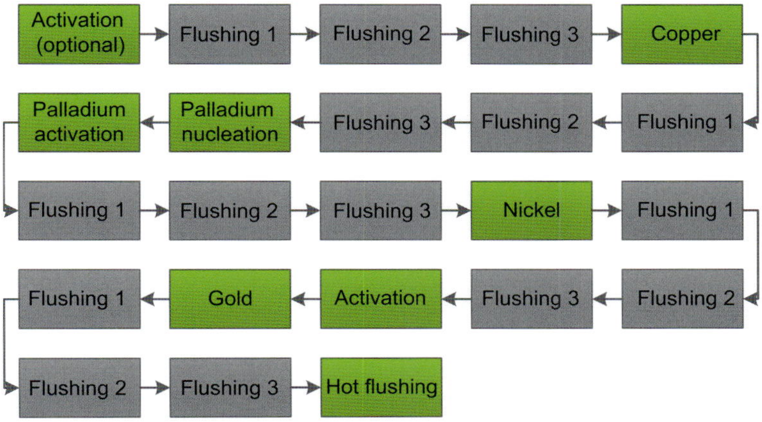

FIGURE 3.27 Possible configuration for MID metallization with copper, nickel, and gold plating

In much the same way as design rules have been set out for the various structuring processes, there are also design rules for metallization [30]:

- Avoid tracking conductors over sharp edges.
- Parts intended for drum metallization should be of a shape that prevents them from snagging on each other.
- Integrate the conductors in such a way as to prevent them from abrading on each other.
- Avoid tracking conductors across weld lines, because thermo-mechanical strains have negative effects on the metallized structure at the weld line.
- Position ejector pins and holders for a frame well away from metallization surfaces.
- Avoid cavities and blind holes that could entrain chemicals, or provide drillings in cavities to ensure that medium can drain off for each change from one bath to the next.
- Prevent contamination of the plastic surface end to end, from injection molding through the entire metallization process. Consequently, do not use release agents or machining additives for injection molding.

Galvanic buildup is an option beside electroless metallization of MID parts. This necessitates electrical contacting of all the conductor structures, and this has an effect on design freedom and conductor width. Once galvanic buildup has completed, the conductors have to separated again. Even so, this is certainly an option for high current-carrying capabilities.

Beaker

In terms of throughput, outlay, and costs, the process of metallization in glass beakers comes in midway between the conventional chemical processes discussed above and ProtoPlate LDS®, which is explained below. This is an electroless method of metallizing laser-structured MID, particularly after structuring by the LPKF-LDS® process, with copper, nickel, and gold. Metallization takes place in glass beakers with a capacity of 5 liters. Typical plating thicknesses are 4 to 8 µm for copper, 3 to 5 µm for nickel, and approximately 0.1 µm for gold.

Beaker metallization requires a grounding in chemistry, because the baths have to be analyzed and topped up before and after each metallization process. Analysis, moreover, requires complicated testing with relatively complex chemical consumption, measuring, and metering equipment.

Figure 3.28 shows the metallization workflow for an LPKF-LDS®-structured substrate. After thorough cleaning in the ultrasonic bath, the activated LDS plastic is immersed in the copper bath. Dwell time depends on the activity of the copper bath and the target thickness for copper plating. A short period of immersion in an

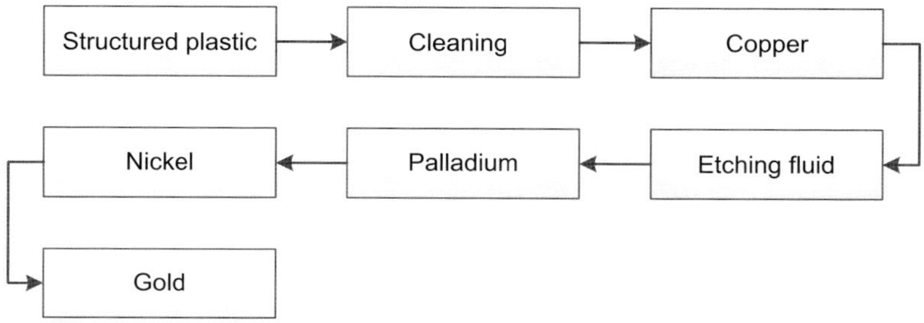

FIGURE 3.28 Flowchart for beaker metallization

etching solution is followed by seeding in a palladium bath. Seeding is required to start nickel/phosphorus metallization in the next step. Application of a very thin gold finish concludes metallization. Thorough cleaning with deionized water is necessary between all these steps in the process.

Beaker metallization is intended for short runs and for process evaluation. As many as ten parts can be metallized in each pass, although this depends on the size of the parts. All the baths have to be analyzed and corrected after each pass. Only then is the next metallization pass possible. Obviously, the number of parts that can be metallized is limited by the time needed to perform these analyses and also by the relatively long dwell times in each bath. It can take two to three hours to complete a pass, depending on how thick the plating has to be.

ProtoPlate LDS®

A joint development of LPKF Laser & Electronics AG and Enthone GmbH, this metallization process was intended specifically for prototyping. It is a bath metallization process for copper plating LPKF-LDS®-structured parts. No in-depth knowledge of chemistry is required, and the bath can be used for experimenting with metallization. The bath has three components based on Enthone LDS Cu 400 PC. Mixing all the components yields an activated copper bath that remains usable for a two-hour time window. Copper layers up to 15 µm thick can be plated in this time. The spent chemicals can be returned in the original packaging to the manufacturer for disposal [132].

This metallization bath is widely used in prototyping, particularly for testing different structuring parameters of the LPKF-LDS® process. It has a footprint of 50 × 40 cm and plugs into a standard socket outlet. The fact that it works only with copper is a drawback, as is the short life cycle [106].

Selective metallization by wet-chemical methods is currently the state of the art, although many complex processes call for special know-how and correspondingly costly equipment.

Various Metallization Coating Systems

Very widely differing connection technologies necessitate metallization layers of varying composition. Each process (soldering, conductive-adhesive gluing, wire bonding, pin contacts) has its own set of requirements for the surface finish. Cu-Ni-Au metallization is the conventional approach, but electroless Cu/Sn and Cu/Ag metallizations are other possibilities.

Excellent solderability, superb bondability with aluminum wires, and long storability are all characteristics of Cu-Ni-Au coatings. Multiple solderings are totally unproblematic. The only drawbacks are cost and the restricted suitability for bonding with gold wires [41].

Tin coatings are very solderable and the manufacturing costs are low, but they have weaknesses in terms of storability and bondability. Tin coatings, moreover, are repairable. The defective tinning can be removed to expose the copper, which can then be regenerated and retinned. A high level of copper depletion during soldering or in storage on account of diffusion effects and the problems associated with multiple soldering also constitute disadvantages [13, 41].

Silver, like tin, is eminently solderable and the coatings are relatively economical. Once again, problems such as storability and the extreme thinness of the coatings tend to complicate handling. As with tin, Cu diffusion takes place during the soldering process. Low mechanical loadability and high potential for electromigration are further drawbacks [41].

3.2.3 Thickness and Roughness of Coatings

MID applications are now widely used for signal transfer, electrical circuitry in sensor applications, and for antennas. A low power requirement is common to all these fields. This settling of past and present MID applications at the low end of the power range has a great deal to do with possible plating thicknesses associated with chemical metallization.

Possible Thicknesses in Relation to Metallization System

Electroless metallization is characterized by deposition rates of only a few μm per hour. This obviously means a certain dwell time in each bath, and this in turn is partly responsible for the costs of metallization. Copper plating thicknesses of well over 10 μm are theoretically possible with conventional industrial metallization equipment. However, the question of economy arises at this point. Standard practice in applications calling for thicknesses in excess of 10 μm is to resort to galvanic buildup, applying copper at deposition rates of around 1 μm per minute.

Beaker metallization is limited to about 10 μm of copper per bath pass, because the bath has to be analyzed and rebalanced as soon as a defined time has elapsed. A ballpark figure for growth rate is approximately 1 μm every ten minutes. Hypothetically speaking, copper can be plated in thicker layers even by beaker deposition. Repassing the copper-plated plastic part in the analyzed and rebalanced copper bath makes it possible to build copper layers thicker than 10 μm. The cost effectiveness of the process should not be the primary variable. The thickness of the copper buildup with the ProtoPlate LDS® process is limited by the life of the bath. The maximum plating thickness is 8 to 15 μm.

Much the same applies for the other metals that can be deposited on a conductive underlayer of copper. The deposition rate for nickel in chemically reductive metallization baths is roughly 6 to 12 μm per hour. It takes about 20 to 30 minutes to deposit a gold finish with a maximum thickness of 0.1 μm. Because the coatings are thinner, beaker metallization is much more closely comparable to other metallization configurations.

Smoothing

Since it is capable of producing conductor structures measuring less than 100 μm, LPKF-LDS® technology in particular has potential for bare-chip assembly. The surface roughnesses of the metallized coatings currently hamper contact by bonding. The plastic surfaces roughened by lasering are reflected in subsequent metallization (unlike galvanic baths, chemical metallization baths cannot be used for smoothing), so the surfaces have to be smoothed after being lasered. The possible methods of smoothing with dies and CO_2 snow jetting are outlined below [185].

Laser-structured plastic surfaces can be die-embossed using purely mechanical pressure, a combination of mechanical and thermal energy, or a combination of mechanical pressure and energy in the form of ultrasonic vibration. Figure 3.29 illustrates how mechanical force is applied in die embossing. The relatively soft laser-structured surface is flattened by the die. This is followed by conventional chemically reductive metallization [185].

In theory, die embossing with purely mechanical pressure loading can be done with soft, elastically deforming dies or hard dies. The advantage of elastic dies is that they do not have to be matched to each conductor structure, so one die can be used for very different geometries. By the same token, conductor tracks can be smoothed all the way to the edges. However, research has revealed that the smoothing effect that elastically deformable dies can apply to laser-structured surfaces is at best no more than slight. Hard dies, on the other hand, can achieve good smoothing results. Even so, the differences in quality are considerable. If die and substrate surface are off parallel the result is an offset in the embossed plastic (Fig. 3.30) [185].

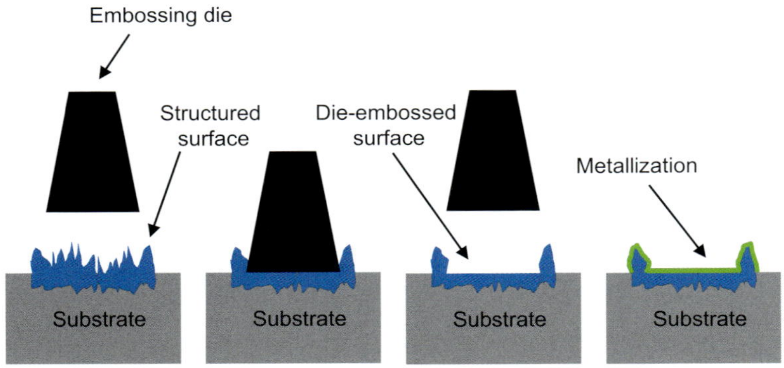

FIGURE 3.29 Diagrammatic illustration of smoothing by die embossing as in [185]

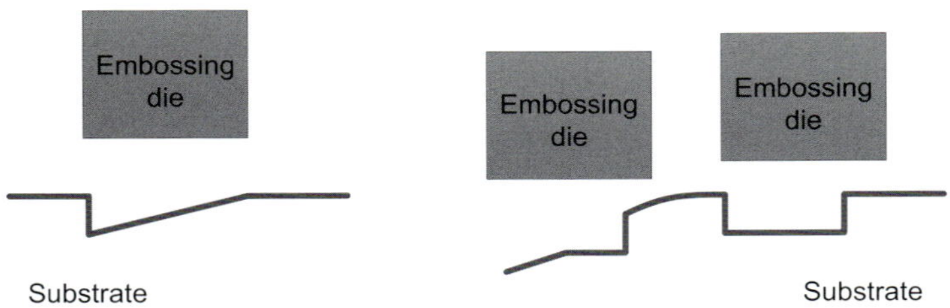

FIGURE 3.30 Possible faults due to lack of parallelism as in [185]

A circular die geometry in combination with a sequence of overlapping embossing strokes promises the most success in purely mechanical embossing. Reduction of around 50% in roughness has been achieved in research. Reliable wire bonding, however, is still not sufficiently viable [185].

One way of further reducing laser-structured roughness is to combine mechanical die-embossing pressure with ultrasonic energy transfer. Suitable die geometries transfer the ultrasonic vibrations selectively to the surface. Research has shown that there is no appreciable benefit to be obtained by employing ultrasonics. Differences in the smoothed surface were due entirely to die pressure, not ultrasonic power [185].

A combination of mechanical and thermal energy is another option for die-emboss smoothing. The die is heated. With the heater suitably calibrated the temperature at the die tip can be precision-set. This method returns a considerable reduction in surface roughness compared to purely mechanical die-embossing cycles. The temperature is crucial. No additional smoothing effect is evident if the temperature is either too high or too low. Die temperature depends on the specific melting temperature and has to be set for each plastic. With optimized process parameters, roughness can be minimized to the extent that wire bonding is possible. However, increasing smoothing gives rise to problems with edge metallization.

Force and temperature combined press the metallization so far into the surface of the plastic that a continuous structure is no longer detectable [185].

In the CO_2 snow-jet process, solid CO_2 ice crystals bombard the surface of the plastic at ultrasonic speed. A combination of mechanical, thermal, and chemical properties means that CO_2 snow can be used for cleaning and for smoothing. Figure 3.31 shows in diagrammatic form how the method is employed for smoothing. The liquid CO_2 coming from the nozzle produces a mixture of CO_2 gas and CO_2 snow that is accelerated to ultrasonic speed by a jacketing ring jet of compressed air [1].

This method can be used to smooth both laser-structured and metallized surfaces. By comparison with simple die embossing, the process is characterized by numerous process parameters. Traversing speed of the CO_2 nozzle, nozzle-to-surface distance, and angle and width of jet are only some among many. These parameters have been studied in detail by researchers. The slower the traversing speed, the smoother the finished surface. No relationship could be established between nozzle-to-surface distance or nozzle angle and surface roughness. CO_2 snow-jetting can bring peak-to-valley height R_Z down to below 10 μm, which is enough for bondability [185].

Smoothing laser-structured flats is possible, but it involves an additional process step and requires the appropriate equipment. Another point to bear in mind is that laser parameters such as frequency and power are responsible for the roughness of the plastic surface. Consequently, if smoothing is a possibility it is also important to look to optimization of LPKF-LDS® laser structuring [185].

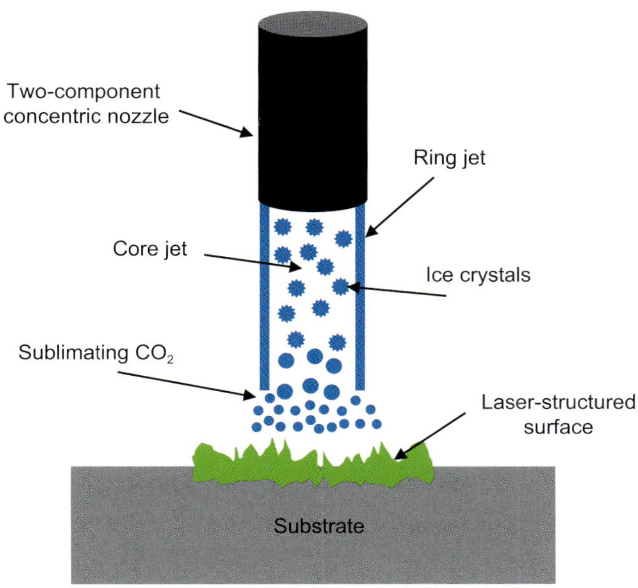

FIGURE 3.31 Diagrammatic view of CO_2 snow-jet smoothing as in [1]

Bonding

Aluminum wire bonding on die-embossed and CO_2 snow-jetted substrates was studied in the framework of a research project. Characterization was in accordance with DVS standard 2811 issued by the German Welding Society. When bonding to die-embossed plastic surfaces it is important to ensure that the distance between bond point and unembossed surface is sufficient. The step structure can lead to nonsticks and bad bonds if this is ignored. Numerous tests with three LDS plastic substrates showed that the requirements for tear-off load, standard deviation, lift-off, and nonstick as set out in the standard can be met [1].

The properties of the CO_2 snow-jetted substrates were similar, but LCP was the only substrate that proved capable of meeting the requirements of DVS standard 2811. It is also important to choose suitable laser parameters. With the other materials tested there were local roughness peaks that resulted in a reduction in bond surface and, as a result, wire pull-off.

3.2.4 Current-Carrying Capability

The factors influencing an MID can be many and varied. Conductor structure, electronic components, structural design, molding technique, thermal stability of the substrate, or the type of metallization have a certain effect on the MID's electrical and thermal capabilities. These factors all have to be taken into account in development, in order to prevent failures at a later juncture. At this time, current-carrying capability always has to be determined in relation to structural design and the best-fit layout identified. Work in this field is still at the development stage. The following is a brief overview of current research work into and results on the issue of current-carrying capability [54].

The IPC 2221 design guideline for conductor widths based on an epoxy-resin base material (e.g., FR4) plated with a copper film 35 µm thick is still in use, despite the fact that it suffices only for an initial estimate of the design of the electric circuitry. Considering current-carrying capability in depth entails factoring in the effects of numerous boundary conditions (test layout, combination of materials, sample geometry, ambient conditions). Determining heat rise in the conductor as a function of current applied also requires a knowledge of total power loss, which is affected by material-specific and environmental constants. Estimates of heat rise under electrification are given below for conductors deposited on thermoplastic substrates using different metallization processes [54].

Research work has been carried out into heat rise in conductors as a function of different load intensities. Conductor temperature was measured up to a constant value under current loading with discrete currents, and a short-term load (t = 1 s)

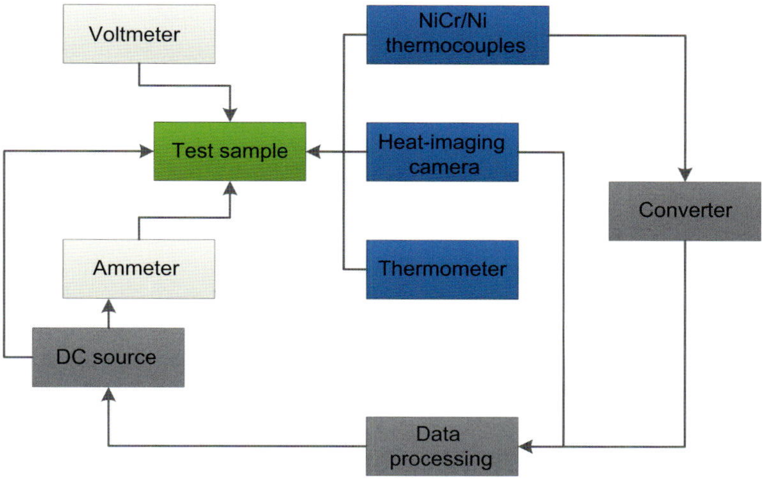

FIGURE 3.32 Schematic representation of the measuring layout in [54]

with discrete voltages of 12 and 24 V DC was also considered. Temperatures were measured with NiCr/Ni thermocouples and a heat-imaging camera (Fig. 3.32). In this way it was possible to determine the change in temperature in sync with current load. Once the conductivity of the copper plating had been ascertained it was possible to draw conclusions about the different metallization processes [54].

According to [54] the figures for specific conductivity for these metallization processes are as stated in Table 3.3.

Table 3.3 clearly shows that the values for the galvanically assisted and hot-emboss metallization processes are very good. The literature gives a value of 58.8 [m/Ωmm^2] for bulk copper at 20 °C [54]. The conductivity of chemically reductive metallization on LDS-activated surfaces is about 60%, which is a consequence of the lower crystallite density in chemical plating. The wide scatter is due to the roughness discussed in Section 3.2.3 and produced by the LDS process. The poor conductivity for Flamecon® is due to the process as such. Molten metal particles are sprayed on to the surface, and the plating forms as individual particles intermesh. The outside layer of the individual particles of copper hampers electron flow, and electric conductivity diminishes accordingly [54].

TABLE 3.3 Specific Conductivity for Different MID Metallization Processes [54]

Process	Chemical-reductive metallization	Galvanically assisted PVD metallization	Hot embossing	Flamecon®
[m/Ωmm^2]	35.7 ± 19.4	56.9 ± 4.9	56.7 ± 6.3	19.2 ± 4.2

There is a clear relationship between heat rise and current load of a conductor for the processes listed in Table 3.3. A power law is applied for calculation. Thermographic measurement is preferable to measurement with thermocouples for ascertaining conductor temperature. However, it is important to allow for the specific degree of emission of the surface in this respect. This can cause fluctuations in measurement. Blackening the conductors can minimize this effect and is recommended [54].

Regarding printing techniques and inkjet printing in particular, initial studies have been undertaken with fine-grained silver inks and a conductor cross-section of 0.0019 mm^2. Again, thermography was used to ascertain heat rise in the blackened conductors. Specific current-carrying capability was calculated at 160 A/mm^2. The value for 18 µm copper plating is about 135 A/mm^2. The higher current-carrying capability could be due to the better thermal conductivity of silver compared with copper, combined with the thin and wide printed conductor structures, which allow heat to dissipate more rapidly. The absolute current-carrying capability for the thin inkjet structures, however, is well below 1 A [83].

Other factors (e.g., power transferred, plating makeup, thermal conductivity of the materials, environment, load duration, conductor pitch) will have to be taken into account in MID technology for more in-depth consideration and computation of current-carrying capability. In the approaches used to date, moreover, no attention is given to the thermoplastic substrate materials and the different metallization processes. The latter in particular, however, lead to differences in conductivity due to microstructure, fluctuations in plating thickness, and differences in roughness. Future studies will provide more results in this area, and this in turn will lead to a more accurate assessment of current-carrying capability [54].

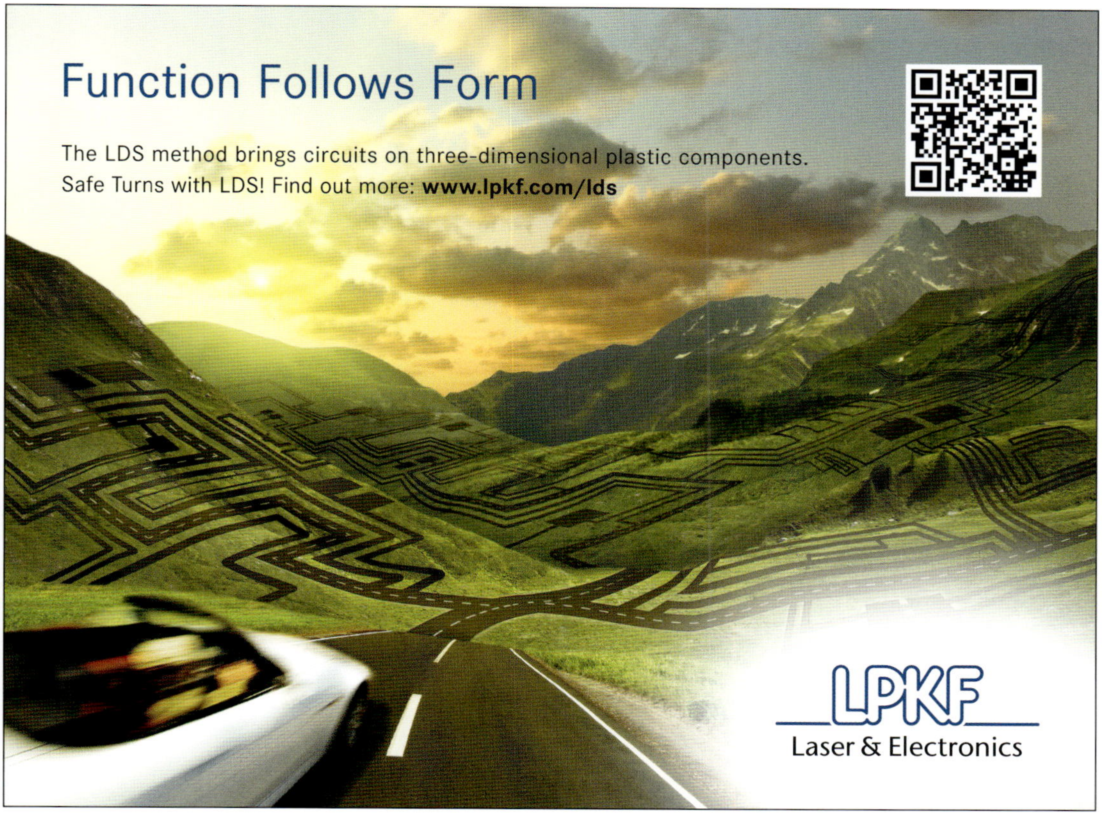

4 Assembly Technology for 3D-MID

M. Pfeffer

This chapter opens with an explanation of the underlying principle and the challenges of assembling three-dimensional bodies. It then goes on to study various methods of automatic component placement on three-dimensional circuit carriers and assess their suitability for MID production. MID-related changes that have to be made in the individual process steps are also explained.

4.1 Process Chain

The processes generally employed to fix electronic components on three-dimensional circuit carriers are reflow soldering and electrically conductive-adhesive gluing. The first step in reflow soldering is to print solder paste on to the pads of the circuit trace (Fig. 4.1). The electronic components are subsequently placed in these solder depots. The module, complete with its electronic components, is then heated in a reflow oven, the solder particles of the paste melt, and the molten solder flows over

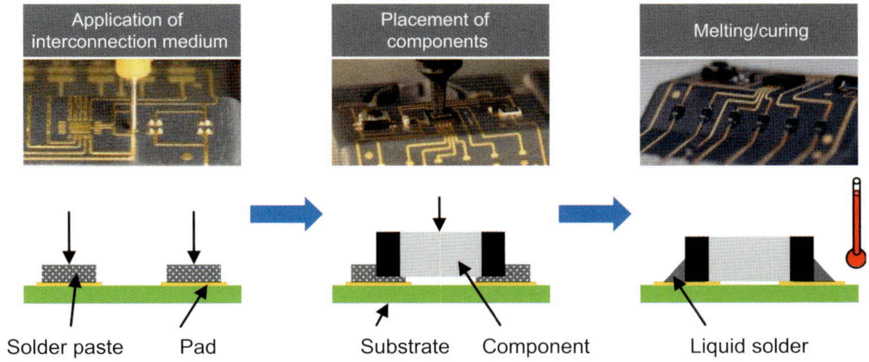

FIGURE 4.1 Process chain of reflow soldering or electrically conductive-adhesive gluing processes with application of the interconnection medium, placement of electronic components, and remelting of the solder or curing of the conductive adhesive

the pads of component and circuit trace. The module cools and a conductive connection forms between interconnect device and electronic component.

The steps in the conductive-adhesive gluing process are not unlike those in reflow soldering. Conductive adhesive is applied to the pads of the circuit carriers by a conventional printing process. Placement of the electronic components is followed by curing of the adhesive. The three-dimensionality of interconnect devices entails elevated requirements for medium application and for component placement, particularly in terms of process automation.

4.2 Assembly-Related Challenges

The automated assembly of MID is heavily influenced by the geometric structure and the complexity of the circuit layout. MID with a complex three-dimensional structure are the most interesting in terms of function integration and miniaturization of the module as a whole, but the requirements with regard to the assembly process are correspondingly high. Consequently, robots of the kind generally employed to apply solder paste and for component placement in PCB assembly lines are either unsuitable or only conditionally suitable for MID production. The designers of production systems have to be aware of the challenges specific to MID assembly in order to come up with solutions that are both suitable for and reliable in 3D-MID production.

4.2.1 Effect of Structure

3D-MID can be subdivided into different geometric types by the three-dimensional structure of the basic body and the layout of the circuit trace (see Section 1.1.2). The bodies have one or more surfaces carrying the circuit traces, and the interconnection medium has to be printed and the electronic components placed on these surfaces (Fig. 4.2). These surfaces are known as process surfaces.

Conventional printed circuit boards are flat, so all the process surfaces on which the connection medium is printed and components placed are planar. In 2½D configurations and higher, the circuit carriers have one flat face or multiple plane-parallel process surfaces plus structural elements vectoring in the *z* direction. Class 1A is characterized by one flat process surface with 3D elements on the reverse. This layout has no definitive effect on the printing and placement processes. Class 1B and higher MID, by contrast, affect the assembly processes. Class 1B has geometric elements on the process surface and class 1C has multiple plane-parallel process surfaces.

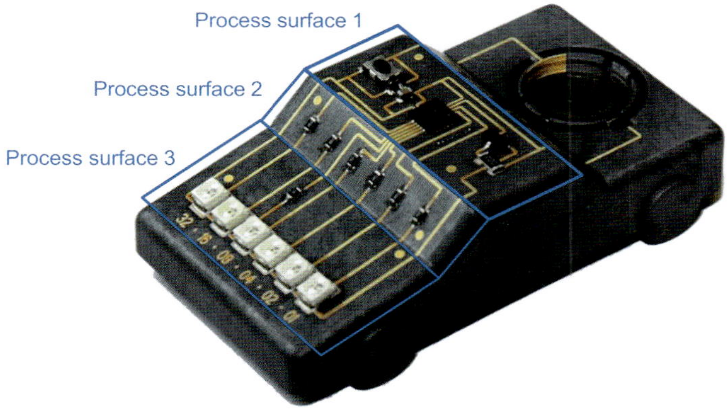

FIGURE 4.2 Ideal arrangement of the process surfaces

The geometries of n × 2D and higher interconnect devices have the biggest effect on the assembly process, consisting as they do of multiple flat process surfaces generally of varying orientation relative to each other.

4.2.2 Component Placement on Three-Dimensional Bodies

In the automated placement of electronic components in surface mount technology (SMT), a gripping tool—usually a vacuum pipette—picks up components from a feeder. These components are then carried by a gantry system to their destinations and placed on the interconnect device's pads, which are coated beforehand with solder paste or a conductive adhesive. As the components are set down, the process partners always have to be aligned such that the joining direction of the component is normal to the process surface (Fig. 4.3) [16]. If the joining direction is not perpendicular to the process surface as the component is placed, there is a good chance that the component will tilt and be misfitted. Off-square joining can also expel some of the solder paste from the depot on the pad, and this can result in a substandard soldered join.

When conventional machines are used to place surface mount devices (SMD) on printed-circuit boards, the joining direction is always normal to the process surface. Consequently, placement entails a four-axis joining movement consisting of three linear and one rotational degree of freedom. On the other hand, 3D-MID generally have multiple process surfaces with different orientations, necessitating a joining action with six degrees of freedom. Consequently, mounting components on MID in three dimensions require more degrees of freedom for aligning the two partners than are necessary for SMT.

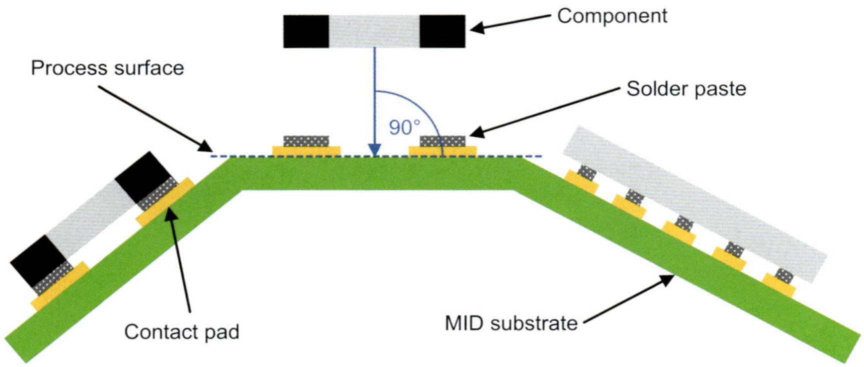

FIGURE 4.3 Perpendicular placement of the electronic components onto the process surfaces

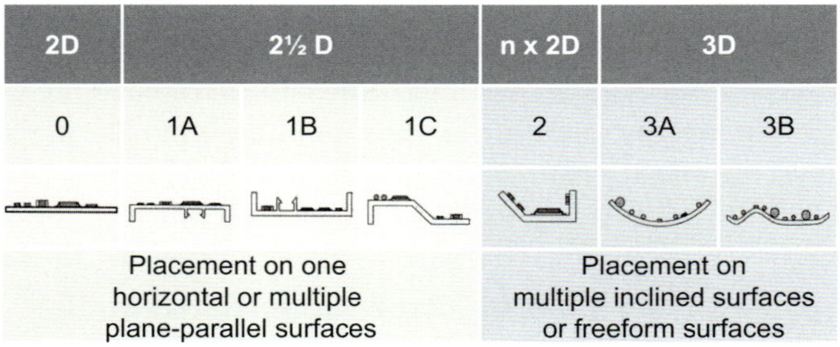

FIGURE 4.4 Required 3D capability of the assembly robots for dispensing the interconnection medium and the placement of electronic components

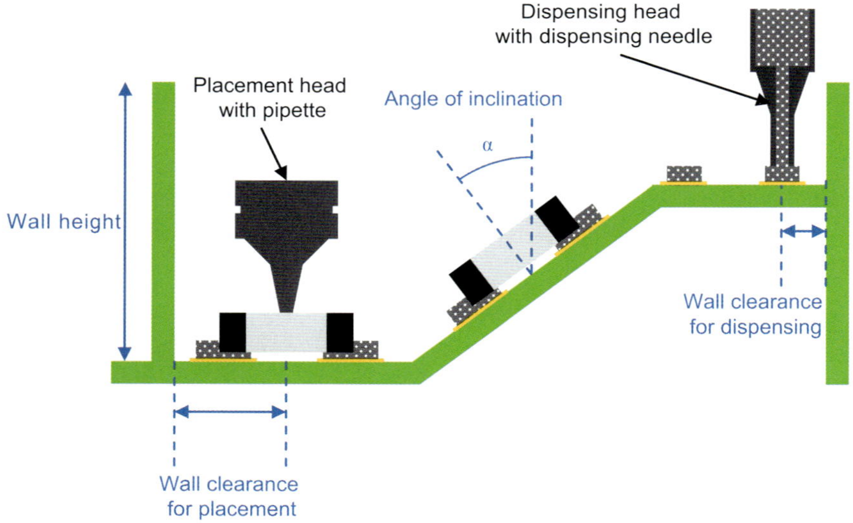

FIGURE 4.5 Accessibility of the electronic component positions and dispensing points (after [16])

As illustrated in Fig. 4.4, there are two categories for this 3D capability of the assembly robots for dispensing interconnection medium and placing components: 2D and 2½D geometries entail placement on one horizontal face or on multiple plane-parallel surfaces. The geometrically complex n × 2D and 3D-MID bodies require the placement of components on multiple process surfaces that are generally on inclined planes relative to each other, or on regular or freeform surfaces. Precision placement of the components on the MID requires accurate position detection of the interconnect device. Geometries in the circuit trace (e.g., corners of a pad) can be used for this purpose, as can extra reference marks (sometimes called fiducials), which are commonly in the form of a circle or a cross. These marks are located by camera imaging and used to determine the position of the process surface. Ideally, MID position detection is in 3D, which means that height can be measured along with lateral offset in the x and y directions.

The interconnect device has to be designed to ensure the accessibility of the dispensing positions by the medium dispensers and of the mounting positions by the component positioning tools (Fig. 4.5). The center of the component is generally defined as the mounting position. The positions of the dispensing points are set inside the component's pad. A crucial factor in automated production is ensuring that collisions between the module and the dispensing and positioning tools are reliably prevented. This, in turn, entails allowing sufficient component-to-wall clearance all around to ensure that the tool (dispensing/positioning head) does not come into contact with the wall. Maximum permissible wall height also has to be taken into consideration in this context. Wall height and component height determine the tool's z travel.

The mechanical strength of the bond holding the component in place is another issue. Components placed in wet solder paste or adhesive are not initially secured by a bond of adequate strength. The solder paste or adhesive has a certain adhesive strength when wet, but large, heavy components in particular are not held firmly in the medium, especially if the component-to-medium contact area of the pad is small. It is not until the adhesive has cured or the solder melted and solidified that a strong, material-bond connection is formed. Only then are the components held securely in position. The module is subjected to external loads during the assembly process. Rotating the modules, for example, produces accelerations that can cause components to shift position or even come loose. Other forces acting on the components are weight when the process surfaces are angled off the horizontal and vibrations occurring not only during transport but in the reflow-soldering or adhesive-curing process. Minor differences in position can be tolerated; misplaced components are unacceptable.

Gluing is one way of keeping components in place on MID bodies. The process known as *in situ* fixation entails curing the adhesive, applied beforehand, while the component is still held in position by the pipette. This reliably locates the component directly after it has been positioned in the solder paste. Adhesive systems used for this purpose must have the ability to cure very quickly so as not to prolong the production cycle. An alternative is to use electrically conductive adhesives that mechanically locate the component and also establish the necessary electrical connections. Chapter 5 provides an overview of possible connection mediums and processes particularly suitable for contacting MID.

4.3 Automated Assembly

4.3.1 Requirements

The requirements set out in the preceding section have to be satisfied by the individual process steps for the automated placement of electronic components on 3D-MID. The subsections below describe different methods for the individual process steps in reflow soldering of MID and discuss their suitability for MID. Necessary modifications to these processes are also explained.

The general requirements that apply to automated production systems for economical and reliable assembly of MID are close to those for printed-circuit board production:

- high-speed component placement and medium dispensing for high productivity rates and short process cycles
- sufficient placement accuracy of the placement machines
- sufficient dispensing accuracy (in terms of both position and quantity) for application of the connection medium
- ability to handle wide range of differing components (component sizes, contact grids, size of component contacts)
- in-line capability of the production machines

Broadly speaking, MID process times are longer than those for conventional printed-circuit boards, because it is usually necessary to allow for positioning of the process surfaces.

4.3.2 Application of Connection Medium

Reflow soldering is the most common assembly process. In the first step a defined quantity of the interconnection medium is applied to the pads of the circuit carrier.

In MID assembly this is generally done by stencil printing or dispensing. The electronic components are set into the printed solder or conductive-adhesive depots in the next step. Jet and pin transfer are two alternative printing processes.

Stencil Printing

In printed-circuit board production, stencil printing is the most common process for applying solder paste. A metal stencil is positioned on the blank circuit carrier. The holes in the stencil correspond to the spots where the solder depots will be applied to the board. A squeegee blade passes over the stencil, filling the holes with solder paste. When the stencil is lifted clear, the solder depots are left on the board (Fig. 4.6). All solder-paste depots on the board are applied in a single process step, permitting a high rate of productivity.

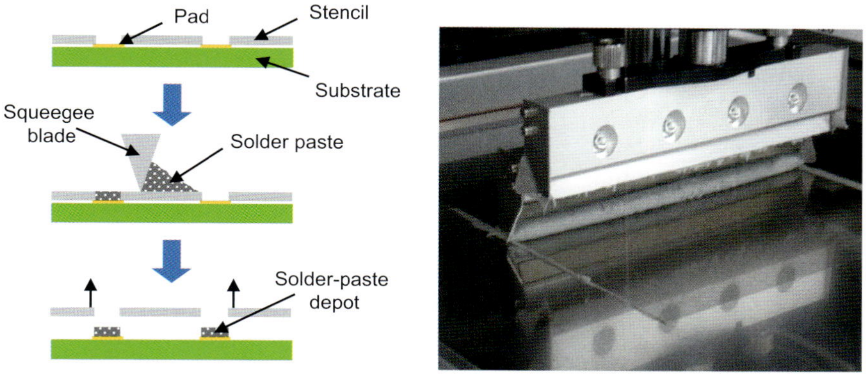

FIGURE 4.6 Stencil printing for the application of solder on planar circuit carriers

Dispensing

Dispensing is a widely used and versatile method of applying solder paste, conductive adhesive, SMD adhesive, and other mediums. The requisite quantity of the medium is applied sequentially to the blank with the aid of a special dispensing valve. The medium itself is kept in a cartridge, precision-metered by a dispensing valve and precision-applied at the appropriate point by a needle (Fig. 4.7). The flexibility to deal with changes in circuit layout is an important characteristic of the dispensing process. High-precision dispensing with ultralow-volume medium shots is necessary on account of progress in the miniaturization of the electronic connection structures.

As explained in [16], quality-influencing parameters for dispensing are needle geometry, needle inside diameter, dispensing time, waiting time, temperature, and dispensing angle. Keeping the tip of the needle accurately positioned at a precise distance from the surface of the substrate is crucial for an accurate print pattern, as is precision control of the dispensed quantity of medium. The viscosity, composition,

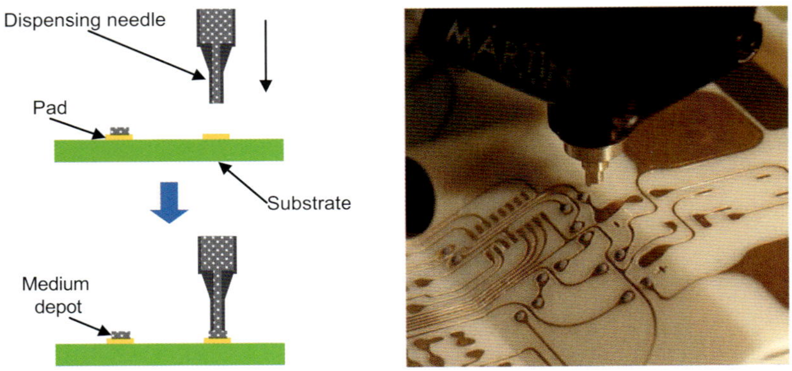

FIGURE 4.7 Dispensing system for the sequential application of interconnection media onto MID

and temperature stability of the medium are all properties that have an important effect on the process. The dispensing angle should be normal to the target surface, but when the medium is solder paste, virtually any angle relative to the process surface is possible given the thixotropic properties of the paste.

Two commonly used types of dispensing valve are the compressed-air dispenser and the screw dispenser. The two parameters pressure and dispensing time define the quantity of medium dispensed by a compressed-air dispenser. Quantity can be controlled by changing the needle inside diameter, varying dispensing time, or changing dispensing pressure. Simplicity of design and easy cleaning are two properties characteristic of compressed-air dispensers. A screw dispenser defines the quantity of medium dispensed by the turn of the thread on a leadscrew. The rotation of this leadscrew forces the print medium out of the valve. The quantity of medium dispensed in this way depends on the angle of rotation of the leadscrew. Consequently, quantity can be controlled with considerably more accuracy than is achievable with compressed-air dispensers, particularly when the dispensing quantities involved are small.

Jet Printing

Jet printing is a noncontacting process that can be used to apply the connection medium (Fig. 4.8). Dispensing is metered by the action of a plunger expelling the medium from a chamber. Timed advance and retraction of the plunger controls quantity dispensed per pulse with a high drop-to-drop precision. The number of pulses defines the total quantity of medium dispensed. Clogging of the jets can be a problem with mediums that contain additives. For example, solder particles can cause clogging when the medium is solder paste. The composition of solder pastes gives rise to the possibility of solder particles precipitating out of the paste and compacting sinde the jet. This can add significantly to the outlay for cleaning and maintenance.

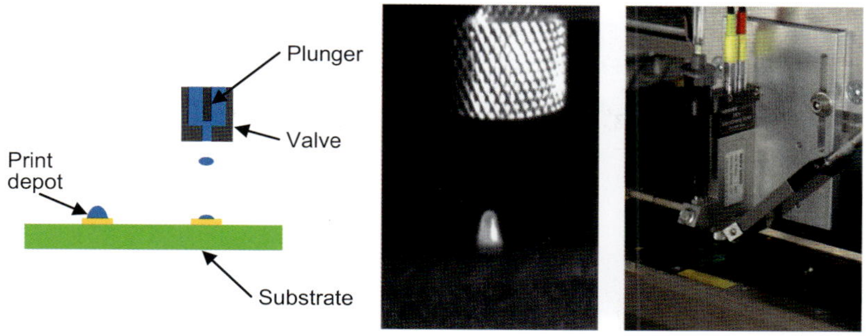

FIGURE 4.8 Noncontact application of the interconnection medium by jet printing

Pin Transfer

The medium is applied by a pin that is alternately dipped into a coating on a medium carrier and then touched to the surface of the substrate (Fig. 4.9). A cavity in the medium carrier accommodates the medium in a thin, homogeneous layer. The quantity transferred depends on the viscosity and composition of the medium and also on the shape and diameter of the pin, the depth of insertion into the coating, and the dwell times of the pin for picking up the medium and for transferring it to the substrate [92]. This method is viable only for low-viscosity mediums with small particles. A certain amount of the medium always adheres to the pin, which means that these residues have to be removed in a cleaning step repeated at defined intervals.

Pin transfer is a simple method that is also suitable for uneven substrate surfaces. The thickness of the coating on the medium carrier has to be monitored and the pin cleaned regularly in order to ensure the reproducibility of the quantity of medium transferred.

The process known as dipping components is a variant of pin transfer. In a defined action, the component's contacts are dipped into flux or connection mediums, producing a defined depot on the contacts. Squeegee blades sweep the cavity at regular intervals to maintain the thickness of the medium coating. This process is generally used for large, fine-pitch components such as ball grid array (BGA) and flip-chip components.

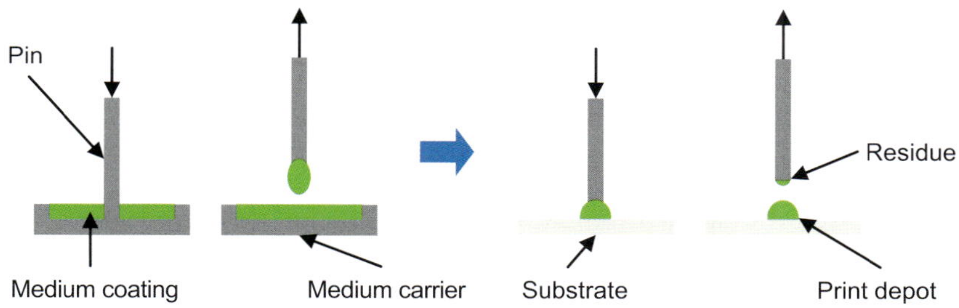

FIGURE 4.9 Transfer of the interconnection medium by pin transfer

Suitability of the Methods for MID

The structure of the 3D interconnect device is important when it comes to selecting a method of medium application for MID, because the geometrical properties of the MID impose severe restrictions on the printing technique. Figure 4.10 compares printing processes in terms of their suitability for the various geometric types.

	2D	2½ D			n x 2D	3D	
Geometry type	0	1A	1B	1C	2	3A	3B

Stencil printing	+	+	0	0	–	–	–
Dispensing	+	+	0	+	0	0	0
Pin transfer	+	+	+	+	0	–	–
Jet printing	+	+	+	+	0	0	0

+ = suitable, 0 = conditionally suitable, – = less suitable

FIGURE 4.10 Suitability of the different printing techniques for the various MID geometry classes

Stencil printing is only conditionally suitable for MID technology because the method requires a printing surface that is planar, on which the stencil can lie flat. Geometric elements are therefore inadmissible on this side of the board. Stenciling is suitable for type 1A geometries only. Class 1B has planar process surfaces, but 3D elements are present on the process surface and the stencil cannot be laid flat. Class 1C and higher MID have several plane-parallel or planar, angled process surfaces, so it is not possible to lay stencils flat for printing. Under certain circumstances, classes 1B and 1C can be stenciled by employing a split stencil and squeegee blade, for example. Class 3 geometries are generally unsuitable on account of the regular and freeform surfaces.

In principle, dispensing is suitable for all classes of MID. Dispensing solder paste on to flat surfaces easily accessible from above is a straightforward procedure (classes 1A and 1C). Class 1B interconnect devices can be printed in this way if the dispensing head has good access to the individual dispensing positions. The other geometrical types can be printed with standard dispensers if the angle of inclination of the process surfaces is slight [16]. Steep angles of inclination necessitate additional degrees of freedom for positioning the tool or the process surfaces of the MID, ideally so that the target process surface will be at right angles to the dispenser needle.

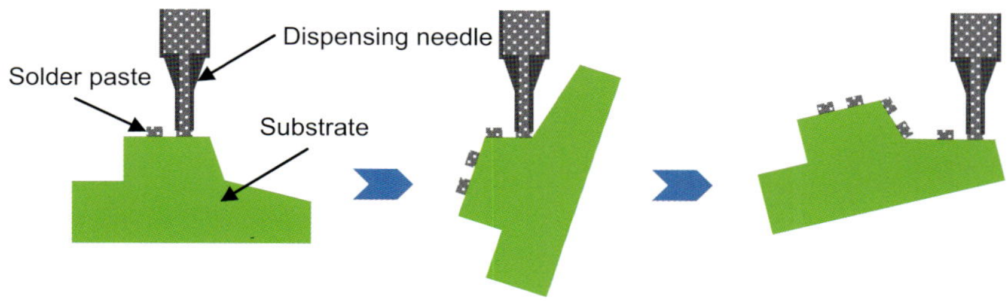

FIGURE 4.11 Dispensing of solder onto MID with horizontal alignment of the process surfaces

By way of example, Fig. 4.11 shows how solder paste is dispensed onto an MID with three process surfaces that can be brought through additional degrees of freedom to the horizontal and consequently positioned normal to the dispenser tool.

Jet printing is noncontacting, so in principle it is suitable for geometries of all classes. Another advantage is the variable working gap between jet tip and substrate with virtually no change in the print result. Consequently, it is ideal for plane-parallel process surfaces, n × 2D, and regular and freeform surfaces.

In terms of its suitability for MID, pin transfer is comparable with dispensing. However, the orientation of process surface to pin has to be very precise in order to ensure that the correct quantity of medium is transferred. Consequently, it is only conditionally suitable for 3D. Both pin transfer and jet printing require the provision of a handling unit for positioning the process surfaces of n × 2D and higher geometries.

Dispensing is sequential, so throughput depends on the number of points for medium application and on the tool's positioning traverses in the cycle. Typical MID applications generally have only a limited number of components, however, so the dispensing time necessary for the entire MID is relatively short.

4.3.3 Placement of Components

Once the interconnection medium has been applied, the next step is to place the electronic components. Current advances in MID technology and the growing number of series-production applications have given the market a variety of production systems for automated 3D-MID assembly and placement. In terms of implementing 3D assembly, they all satisfy the primary requirement: the component's joining direction must be normal to the process surface. The systems achieve this by integrating additional degrees of freedom and drives in various kinematic variants (Fig. 4.12). Broad distinctions between these kinematic variants can be drawn as follows:

- Throughout the entire placement process, the circuit carrier is held immobile and the components are rotated, positioned, and placed.
- Or the interconnect device is rotated and positioned and the components are placed.

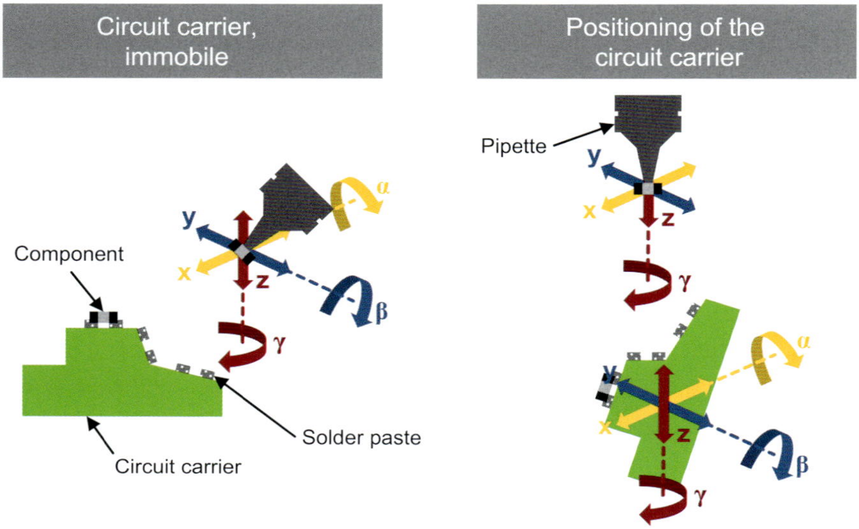

FIGURE 4.12 Kinematic variants with immobile or mobile circuit carriers

By way of example, Fig. 4.13 shows how components are placed in the kinematic variant with a moved circuit carrier. The process surface is brought to the position at which the components can be placed vertically on the process surface. Once all of the components have been placed on one process surface, the MID is repositioned and the components are placed on the next process surface.

The choice of assembly solution depends primarily on the product and the MID's three-dimensional structure. The number, arrangement, and type of degrees of freedom are defined by the number and orientation of the process surfaces and the number of electronic components for placement.

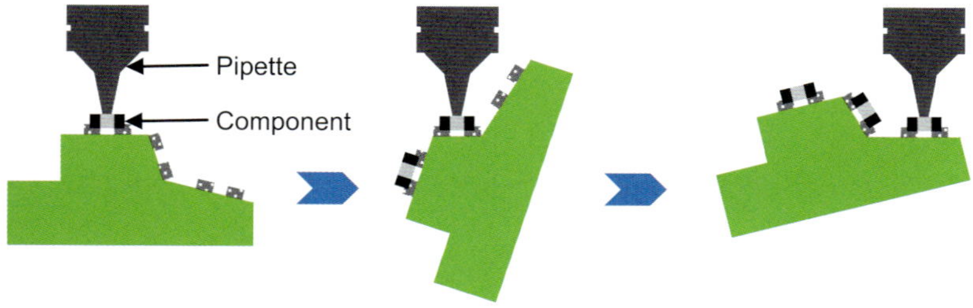

FIGURE 4.13 Placement of electronic components onto an MID with horizontal positioning of the process surfaces

This in turn means that the following distinctions can be drawn between machines:

- type of kinematics: industrial robot, Cartesian gantry system, or individual kinematics
- number of degrees of freedom
- application of the degrees of freedom, to interconnect device and/or component for placement
- type of degrees of freedom: translational or rotational axes
- MID classes handled
- circuit carrier maximum sizes
- maximum range of travel of the axes, positioning accuracy

Broadly speaking, the machines have much the same periphery as those for SMD placement. They can be coupled to component feeders for the common types of SMD package or for special types (e.g., bare die). There are gripping tools for all the various types of package. Most of these tools are variations on the vacuum-pipette principle. The robots have the requisite interfaces and conveyor belts for integration into production lines.

The next section affords an overview of the possibilities and solutions for automated 3D component placement on MID. Interest focuses primarily on the different kinematic types of solution. This is not intended as a comparative assessment of available system technologies.

Conventional SMD Pick & Place Machine

Conventional SMD pick & place machines are optimized for manufacturing planar printed-circuit boards. High throughput, high precision, and the versatility to cope with a wide range of different components are possible characteristics, depending on the type of machine. Components are picked up and held by component-specific vacuum pipettes arranged on the placement head. A Cartesian gantry system is used to position the head on the *x-y* plane. The head picks up one or more components from the component feed units, moves to each placement position in turn, and sets down the components one after the other. As it is set down, each component is turned to the correct angle relative to the board. In order to perform these actions, the placement machine has two linear axes x and y plus one linear axis z and a rotational axis about z in the placement head (Fig. 4.14).

These kinematics are only conditionally suitable for placing components on three-dimensional circuit carrier. Only class 2D and 2½D MID can be handled. Classes 1B and 1C are suitable, but only if free of geometric restrictions and only if the placement positions are accessible. It is particularly important to maintain adequate clearances between the tool and structural elements (walls, lands) on the component

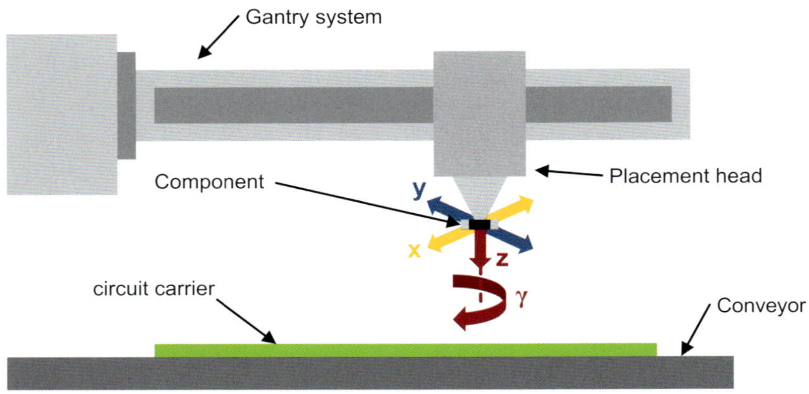

FIGURE 4.14 Cartesian gantry system of an SMD pick & place machine

TABLE 4.1 Properties of the Kinematics of an SMD Pick & Place Machine

Kinematics, component side	Cartesian gantry system with ▪ three linear axes: *x*, *y*, *z* ▪ one rotational axis: γ about *z*
Kinematics, MID side	None. Circuit carrier immobile
MID classes handled	2D and 2½D

side of the circuit carrier. The difference in height between plane-parallel process surfaces cannot be more than the maximum *z* travel of the placement head. It is also absolutely essential to ensure that there are no collisions between the three-dimensional circuit carrier and the placement head or pipette. Consequently, it is not permissible for elements of the circuit carrier to project into the head's working space. A safety margin is invariably required for overpassing walls or other structural elements.

It may well be necessary to design a workpiece carrier for station-to-station transport of the circuit carrier along the production line and also for loading the devices into the individual stations. Depending on circumstances, it can be very difficult to convey small MID parts and complex 3D bodies dependably on the PCB conveyor belt. Reusable carriers are one way of shortening cycle times, with each carrier accommodating multiple interconnect devices.

SMD Pick & Place Machine with Manually Operated Workpiece Carrier

One way of extending the 3D capability of an SMD pick & place machine is to use a workpiece carrier with defined angular settings for the individual process surfaces of the MID (Fig. 4.15). These settings are presets deriving from the design of the

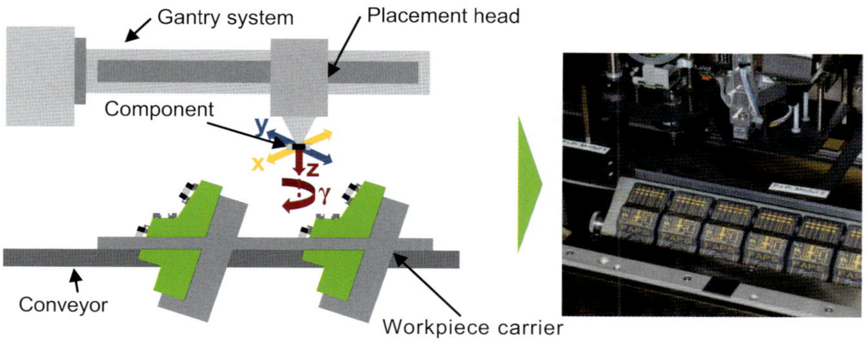

FIGURE 4.15 Manually operated workpiece carrier in an SMD pick & place machine

TABLE 4.2 Properties of the Kinematics of an SMD Pick & Place Machine with Manually Operated Workpiece Carrier

Kinematics, component side	Cartesian gantry system with ▪ three linear axes: x, y, z ▪ rotational axes: γ about z
Kinematics, MID side	Workpiece carrier with manual rotation α about x and/or β about y
MID classes handled	2D, 2½D, and n × 2D

workpiece carrier. An MID with multiple process surfaces could be worked on a single workpiece carrier with the appropriate number of angular presets, or on the appropriate number of workpiece carriers, each of which has a single angular preset. Once all the components have been placed on one process surface, the workpiece carrier withdraws from the robot, the angle settings are changed, and the carrier moves back into position in the robot. Changing angles or workpiece carriers, however, necessitates additional handling processes that have to be carried out either manually or by an industrial robot. Depending on circumstances, it might also be necessary to change placement programs. An alternative is to design a workpiece carrier with several manually selectable angular positions. Once all the components have been placed on one process surface, the carrier is moved by hand to the next angular position.

Using workpiece carriers is a straightforward, low-investment solution for MID component placement. The major drawback, however, is low productivity. Changing angles, repeatedly withdrawing and inserting the workpiece carrier, and changing the placement program are all very time-consuming steps in the process. Consequently, this solution is viable only for low-volume runs and for manufacturing prototypes.

Cartesian Gantry System with Extended Kinematics

Placement machines based on a Cartesian gantry system like that of an SMD pick & place machine generally have three linear axes *x*, *y*, and *z* and one rotational axis, usually about the *z* axis. The precision of these axes and of the sensors is such that Cartesian gantry systems are capable of high positioning accuracy. This translates into high placement precision on the interconnect device. Class 2 and higher MID require kinematics extended beyond those of the Cartesian gantry system, in order for the machine to place components on all process surfaces. In most instances, this extension is affected on the circuit carrier side. A suitable array of actuators positions the target process surfaces of the circuit carrier horizontally in the work zone of the gantry system, and the components then require only vertical movement for placement.

The alternative approach is to extend the kinematics at the component side. This could entail integrating additional degrees of freedom into the placement head, for example, for tilting and turning the component into position for setdown. Piezoelectric drives have the outstanding dynamics in combination with excellent positioning precision that make them eminently suitable for this purpose. Moreover, piezoelectric drives are very small, so they have great potential for integrating additional degrees of freedom into the placement head.

The concept of a Cartesian gantry system with extended kinematics on the circuit carrier side is implemented by the following two industrial systems.

The chip and 3D microassembly machine from Häcker Automation unites a high-precision component placement machine with a highly accurate dispenser (Fig. 4.16). The integrated rotational and tilt axes bring the process surfaces to the

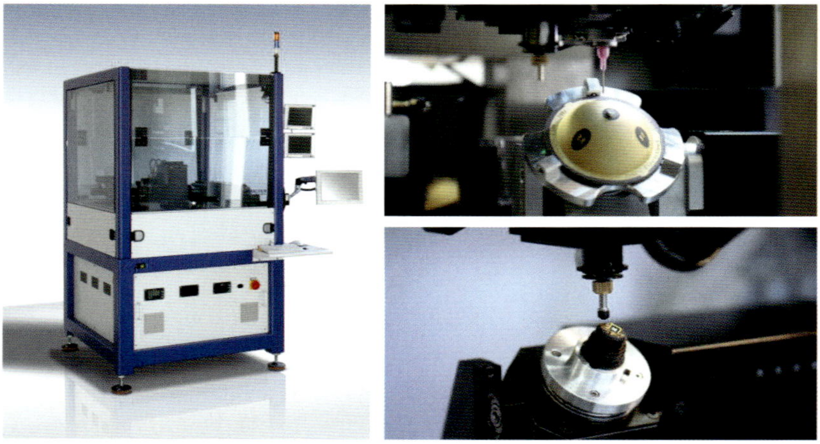

FIGURE 4.16 Microassembly machine with integrated 3D substrate pickup for MID [Häcker Automation]

TABLE 4.3 Properties of the Kinematics of the Microassembly Machine [69]

Kinematics, component side	Cartesian gantry system with ▪ three linear axes: *x*, *y*, *z* ▪ rotational axes: γ about *z*
Kinematics, MID side	3D substrate pickup in machine bed with rotational axes α and β
MID classes handled	Up to 3D
Optional functions	▪ Microdispensing unit ▪ 20-axis manipulator ▪ Laser soldering system ▪ 3D camera system ▪ Wafer station ▪ 3D placement head

horizontal [154]. The placement head has 150 mm of travel in *z* for an extended access range to the placement positions. A 3D camera system measures the position of the circuit carrier in *x*, *y*, and *z*, so height measurement is also possible. The system is modular, so retooling is quick and straightforward.

The Space 400 3D precision assembly module from Xenon is a versatile in-line system with a four-axis gantry system and a component manipulator for 3D applications (Fig. 4.17). The conveyor brings carrier-mounted MID to the component manipulator. The manipulator grips the workpiece carrier and can then position the interconnect device's process surfaces about two rotary axes and along one linear axis. There are two parallel process axes for dispensing and component placement [5].

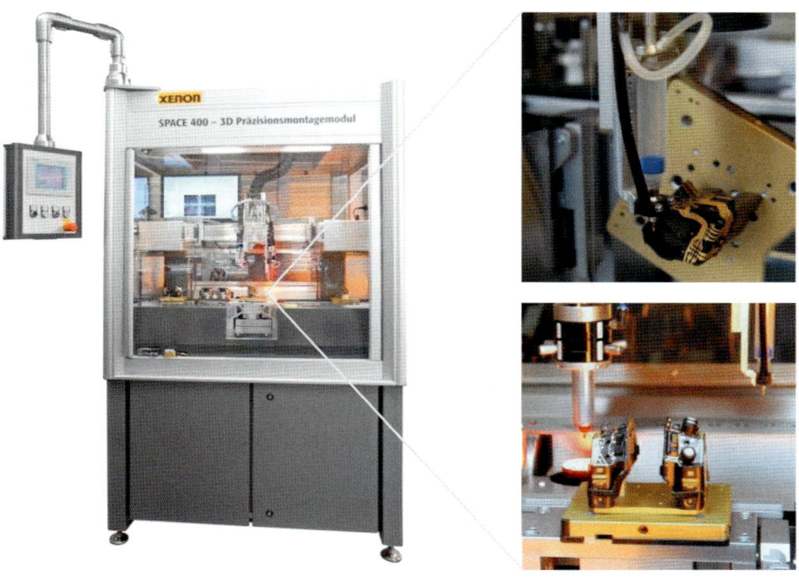

FIGURE 4.17 3D Precision assembly module [Xenon]

TABLE 4.4 Properties of the Kinematics of the 3D Precision Assembly Module [173]

Kinematics, component side	Cartesian gantry system with ▪ three linear axes: *x*, *y*, *z* ▪ rotational axes: γ about *z*
Kinematics, MID side	3D component manipulator in the machine base with ▪ rotational axes α and β ▪ linear axis *z*
MID classes handled	Up to 3D
Optional functions	▪ Two parallel process axes for dispensing and placement ▪ Microdispensing system ▪ 3D component metrology

Cartesian Gantry System with Integrated Six-Axis Multiple-Articulation Robot

Another industrial solution based on a Cartesian gantry system takes the form of a 2D placement machine with in-line capability with an integrated six-axis multiple-articulation robot: the Hydra 3D-MID dispensing and placing system from Essemtec (Fig. 4.18). The robot located in the base of the placement machine and grips the MID-loaded workpiece carrier as it arrives on the printed-circuit board feed conveyor. The MID's process surfaces are then brought to the horizontal position in the placement zone so that the dispensing head can dispense interconnection medium and the placement head can set down the electronic components [53]. The machine can be programmed to work in 3D mode for MID or in 2D mode for planar printed-

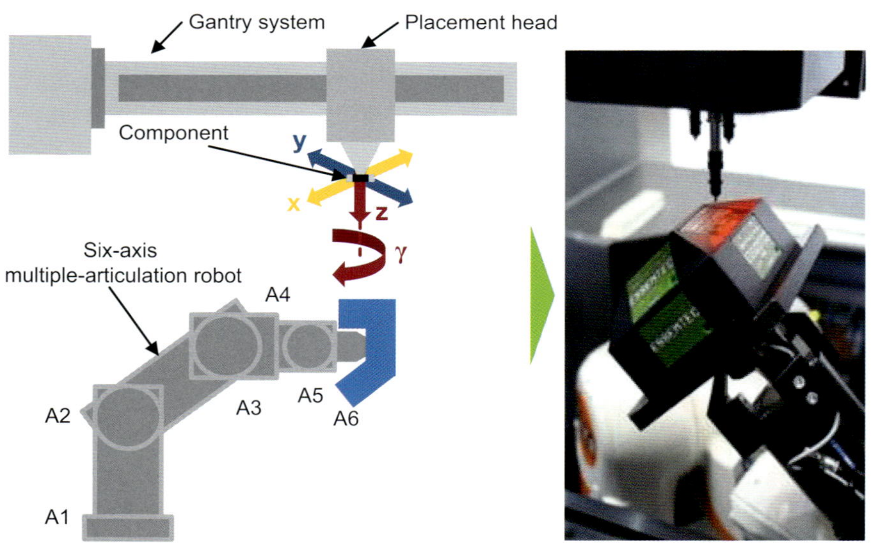

FIGURE 4.18 SMD pick & place machine with integrated six-axis multiple-articulation robot for MID positioning [Essemtec]

TABLE 4.5 Properties of the Kinematics of the Placement Machine with Integrated Six-Axis Multiple-Articulation Robot [43]

Kinematics, component side	Cartesian gantry system with ▪ three linear axes: *x*, *y*, *z* ▪ rotational axes: α about *z*
Kinematics, MID side	Six-axis multiple-articulation robot in the machine base with six rotational axes
MID classes handled	Up to 3D
Optional functions	▪ Quadruple placement head and additional dispensing head

circuit boards. The extended *z* axis enables the head to lift vertically to pass over obstructions up to 50 mm high or to lower components into deep cavities.

Robot Solution

Given their versatility for positioning and the high level of their 3D capability, industrial robots are suitable for automated placement of electronic components on three-dimensional interconnect devices. Six-axis multiple-articulation robots are eminently suitable for placing components on complex, class 3 MID structures with three-dimensional regular or freeform process surfaces.

The 3D-MID placing and assembling system from Xenon employs a six-axis multiple-articulation robot for assembly of a steering-wheel switch (see Chapter 9). Integrated into one of the cells in the production line, the robot dispenses solder paste and places the electronic components (Fig. 4.19). The robot was extended by integration of a gripper for SMT components and a tool changer. The cell also includes the component feeders and a camera system for identifying the positions of the components on the process surfaces of the interconnect device. The interconnect device is held in one position in the placement process, while the robot positions the components on the target process surfaces.

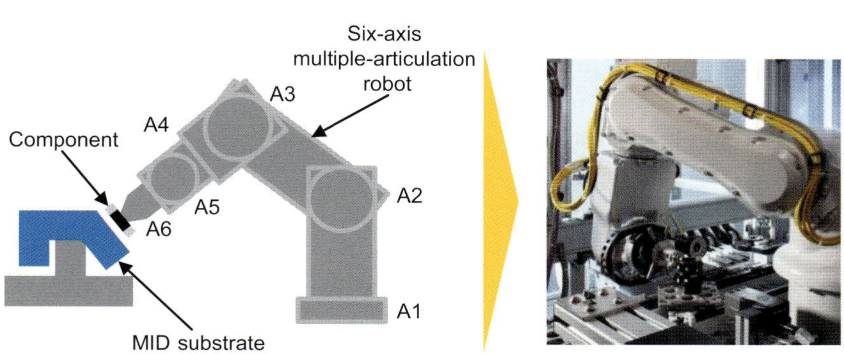

FIGURE 4.19 Six-axis multiple-articulation robot for the flexible placement of 3D-MID [Xenon]

TABLE 4.6 Properties of the Kinematics of the Assembly Machine [174]

Kinematics, component side	Six-axis multiple-articulation robot with six rotational axes
Kinematics, MID side	None; interconnect device is held in fixed position
MID classes handled	Up to 3D
Optional functions	▪ Pressing in of contact pins ▪ Solder paste dispensing ▪ Automatic Optical Inspection (AOI) of the solder points ▪ Protective lacquering ▪ Electrical and mechanical testing

Automated Workpiece Carrier

Another possibility for placing components on three-dimensional interconnect devices is to use an automated workpiece carrier to extend the kinematics of an SMD pick & place machine (Fig. 4.20). The drives integrated into the workpiece carrier orient the process surfaces of the MID horizontally on the machine's placement zone [128]. The multi-MID carrier enables batch production with placement of all the components on the carrier's MID bodies in a single operation. The conveyor for printed-circuit boards conveys the workpiece carrier into the placement machine (Fig. 4.21). Positioning by two rotational axes (90° tilt axis and 360° rotational axis) enables component positioning around an entire hemisphere. A *z* axis compensates for height difference and positions the process surfaces in the machine's placement zone. As with planar printed-circuit boards, the exact position of the process surface is determined using reference marks imaged in 2D by the machine's vision system. Fixations secure the circuit carriers relative to the axes of the workpiece carrier. The fixations have to be designed to match MID size, and in some cases the same might apply to the workpiece carriers. In general, and assuming that the placement positions are reachable, all MID classes can be handled by a machine with this kinematic arrangement.

Automated production requires a workpiece carrier that can be used all the way along the production line. Consequently, the design is such that the carrier can be conveyed by the printed-circuit board conveyor to all the machines in the process chain. The appropriate number of workpiece carriers is required if production is to be continuous. An interface for energy supply and communication between the automatic machines and the workpiece carrier is necessary, particularly for the dispensing unit and the automatic placement machines. The electronics built into the workpiece carrier render it unsuitable for reflow soldering, so a handling step has to be included so that the MID can be lifted on to a separate, passive workpiece carrier for this process. This transfer can be manual or by means of an industrial robot such as the Scara robot.

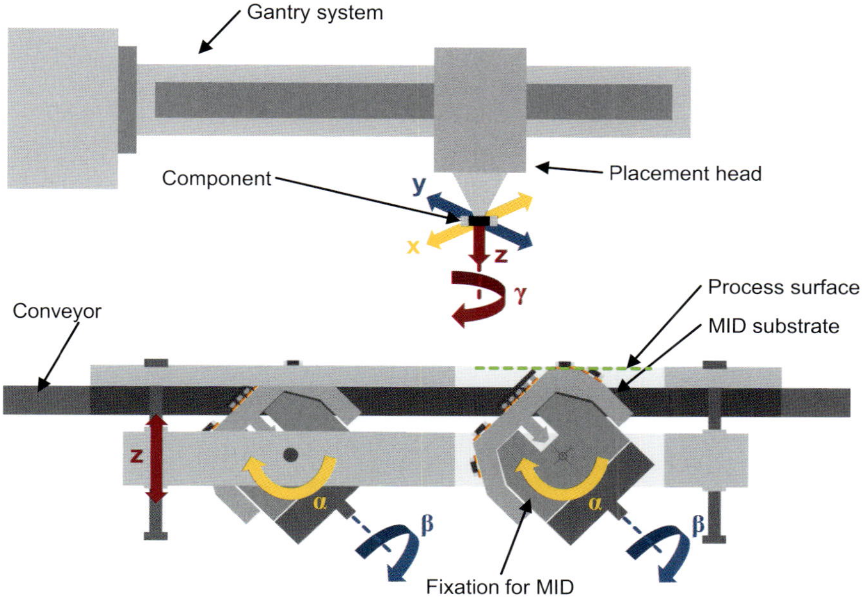

FIGURE 4.20 Automated workpiece carrier in an SMD pick & place machine

TABLE 4.7 Properties of the Kinematics of the Automated Workpiece Carrier

Kinematics, component side	Cartesian gantry system with ▪ three linear axes: x, y, z ▪ one rotational axis: γ about z
Kinematics, MID side	Automated workpiece carrier with ▪ rotation α about x and β about y ▪ additional linear axis z
MID classes handled	Up to 3D

An alternative to progressing down a production line is to integrate stationary workpiece carriers into the placement machines. In this arrangement the MID incoming from the printed-circuit board conveyor are readied for component placement by being transferred by the placement machine to the workpiece carriers.

The workpiece carrier can be used in an existing SMD placement machine and the machine itself can still be used for assembling planar printed-circuit boards, and this versatility affords manufacturers two-fold benefit from having an efficient, cost-effective solution for MID production.

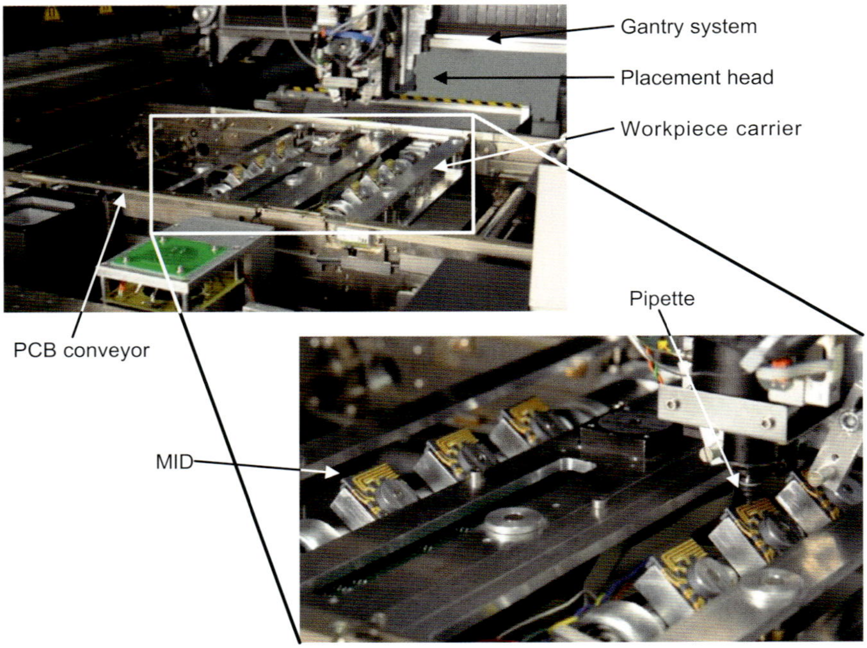

FIGURE 4.21 Automated workpiece carrier in SMD pick & place machine Siplace HF

4.3.4 Reflow Soldering

The process of vapor-phase soldering of MID parts is subject to the constraints deriving from the geometrical properties. The structural elements on the component side of class 1B MID can obstruct flow of the vaporized medium, which in turn can have a detrimental effect on heat transfer. Accumulation of the liquefied medium on the circuit carrier can also increase the loss of the medium. MID with localized cavities can experience the same problems. In general, tilting the circuit carrier slightly on the workpiece carrier suffices to help the medium flow properly. The overall dimensions of the MID body can introduce further constraints. If the module were higher than the thickness of the vapor blanket there would be no transfer of vaporization energy and vertical temperature distribution would not be homogeneous.

Carrier thermal mass or, more accurately, capacity becomes important when workpiece carriers (Fig. 4.22) are used along the entire production line. The soldering process in particular is very much affected by this thermal capacity. The temperature profile has to be adapted to ensure dependable electrical and mechanical interconnecting that will also result in the requisite reliability. If the choice is going to be in favor of an automated workpiece carrier, it is important to bear in mind that the integrated drives and the electronics are not suitable for the soldering process.

FIGURE 4.22 Use of workpiece carriers in vapor-phase soldering of MID parts [graphic left courtesy of 2E mechatronic]

Consequently, prior to soldering the MID will have to be transferred to a passive workpiece carrier, for example.

Selective soldering by laser beam, for example, requires the geometric structure of the MID to be taken into account along with the dimensional positioning of the process surfaces, in order to ensure that the laser beam coincides with the connection point at the correct, defined angle. A laser beam deflector or an extra handling unit could be necessary to position the process surfaces of the MID so that the laser beam coincides with the connection point at the correct angle. As with the up-line process steps, the accessibility of the connection points is essential for laser-beam soldering.

4.3.5 Optical Inspection

The optical inspection of postplacement MID requires an inspection system capable of scanning the entire device [10]. This calls for corresponding depth of focus or appropriately positioned sensors. The inspection points also have to be accessible to the sensor array. Shadowing or obstruction by geometric elements has to be avoided by using a variety of viewing angles and directions and also lighting angles for the optical sensors.

It may well be necessary to adapt an optical inspection system to accommodate the three-dimensional layout of MID process surfaces. Class 2½D MID can be inspected without additional kinematics and adaptation of the optical inspection system. This holds true only if the process surfaces are plane-parallel in the inspection plane of the sensor array. Classes n × 2D and 3D call for changes to the AOI system if the electronic components are widely spaced. This could well mean integrating an extra handling and positioning unit for MID, for example an automated workpiece carrier.

5 Interconnection Technology

C. Goth

Broadly speaking, the connection mediums and techniques familiar from conventional printed-circuit board technology can be used for mechanically locating and electrically contacting electronic components on MID. Soldering, conductive-adhesive gluing, bonding, and flip-chip placement are all in successful use under series-production conditions.

5.1 Specifics and Challenges

On account of the process chains described in Chapter 3, the specifics of connecting an electronic component and conductor have to be taken into consideration in terms of the connection processes, the mediums, and the parameters for production of 3D-MID (Fig. 5.1).

Depending on the manufacturing process used, the plating can be very thin and in some cases also very rough, and this imposes certain constraints on the subsequent connection processes. Owing to anisotropic and high coefficients of

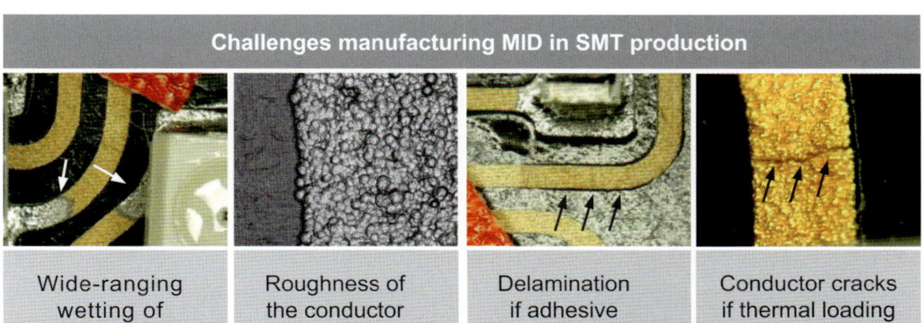

FIGURE 5.1 The manufacture of 3D-MID in the standard SMT process chain involves special challenges in terms of connection technology [63]

TABLE 5.1 Thermo-physical Properties of Thermoplastic Materials Compared with Standard Materials in Electronics Production, as in [178]

	Standard materials in electronic production				MID substrate materials for laser direct structuring		
	Silicon	Copper	Al_2O_3	FR4	LCP	PA 6/6T	PET+PBT
CTE (*x*/*y*) [ppm/K]	2.5–3	16.5	6–8	15–18	12/30	30/50	36/56
T_g in °C	–	–	–	135	–	90–95	60
max. *T* [°C]	> 260 °C	> 260 °C	> 260 °C	> 260 °C	> 260 °C	> 260 °C	< 255 °C

thermal expansion (CTE), high thermal loading can in some cases produce thermo-mechanical stresses at the boundary surfaces. This in turn can cause delamination of the conductors or cracking in the structures. The thermo-physical properties of thermoplastic materials are compared with those of the standard materials for electronics production in Table 5.1.

The comparison shows that the thermo-mechanical mismatch, characterized by the CTE, is significantly higher for MID than for printed-circuit boards. This means that higher compression/tension strains than in FR4, for example, can occur at the substrate/conductor, conductor/connection medium, and connection-medium/component boundaries (Fig. 5.2). On account of the differences between the individual materials, differences in long-term reliability can be expected, depending on the conditions of use. [178]

The differences between MID-compatible plastics themselves are also considerable. Silicon has a coefficient of thermal expansion between 2.5 and 3 ppm/K, and the figure for FR4 is approximately 15 to 18 ppm/K. The CTEs of the relevant MID substrate materials, by contrast, are approximately 36/56 ppm/K (parallel/normal to injection-molding direction) for PET+PBT, approximately 30/50 ppm/K for PA 6/6T, and approximately 12/30 ppm/K for LCP.

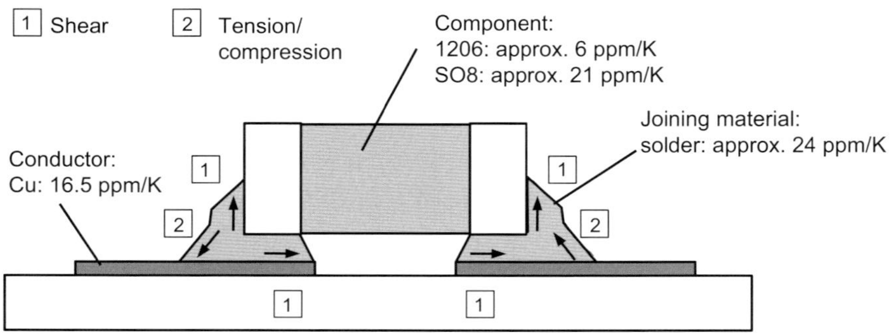

FIGURE 5.2 Deformations in SMT (surface mount technology) connections due to different coefficients of thermal expansion

As regards choice of components, it is important to bear in mind that the dimensions of the electronic components and the corresponding connection structures are defined by the minimum structure sizes. In MID technology, and depending on the manufacturing process, minimum conductor widths are in the region of 100 to 150 µm (technologically possible: 60 to 80 µm), whereas smaller structures are viable with printed-circuit board technology.

Generally speaking, there are two different assembly technologies for electronic components:

- through-hole technology (THT)
- surface mount technology (SMT)

This is why there are also different component shapes. Virtually the only components of interest for MID technology are those having metallized pads for connection that can be set directly on to plated lands on the surface of the substrate. These are surface mount devices, SMD for short. Through-hole devices, THD for short, are wired components and are not used for 3D-MID, so they are not discussed in more detail below. The range of SMD components can be subdivided as follows [109]:

- function (resistors, diodes, transistors, etc.)
- shape (chip, SOT, QFP, etc.)
- connector shape (metal caps, L-,I-,J-shaped legs, humps, etc.)
- number of contacts (two-terminal, multiterminal)

Other aspects include

- package material,
- metallization structure,
- geometry, and
- sizes.

Components classified by function can be further subdivided into active and passive components. The passive components include primarily capacitors and resistors. They are electrically contacted by a metallized pad. Passive components are usually block-like in shape, having two (or possibly more) sides metallized for the purposes of electrical contacting. Table 5.2 lists the sizes in which most two-terminal components are produced.

On account of structure sizes, components of the 01005 and 0201 size codes cannot be used in MID technology at this time.

Active components integrate a chip, usually inside a package, and can be classified by the arrangement of their connections. Table 5.3 provides an overview. The individual component shapes also differ in the size and grid sizing of the connection structures, and the higher the level of miniaturization the higher the requirements in terms of

TABLE 5.2 Dimensions of Common Sizes of Chip Capacitors and Chip Resistors

Shape	Length [mm]	Width [mm]	Length [inch]	Width [inch]
01005	0.4	0.2	0.016	0.008
0201	0.6	0.3	0.024	0.012
0402	1.00	0.50	0.040	0.020
0603	1.60	0.80	0.063	0.031
0805	2.00	1.25	0.079	0.050
1206	3.20	1.60	0.126	0.063

TABLE 5.3 Classification of Active Surface Mount Devices (SMD) by Arrangement of their Terminals

Terminals on two sides of the package	Terminals on four sides of the package	Terminals underneath the package
▪ SOIC (small outline integrated circuit) ▪ SOP (small outline package) ▪ SOT (small outline transistor)	▪ PLCC (plastic leaded chip carrier) ▪ QFP (quad flat package)	peripheral: ▪ QFN (quad flat no leads package) all over: ▪ LGA (land grid array) ▪ BGA (ball grid array) ▪ CSP (chip scale package)

connection. Highly integrated semiconductor components such as microprocessors have a multiplicity of connections, often forming a grid array on the underside of the component. A chip scale package is an evolved form of ball grid array in terms of smaller dimensions and grid sizing (400 to 800 μm). Naked silicon chips can also be wire-bonded directly to the substrate or connected by various flip-chip technologies.

SMD brings MID technology further toward the goal of miniaturized, lightweight systems producible in high quality (by high-precision pick & place robots and systems for automatic optical inspection, AOI). It has to be borne in mind, however, that surface-mount devices are more easily damaged by mechanical and thermo-mechanical loads than wired, through-hole components. Another point is that the components are exposed to higher thermal loading in the manufacturing process. What is more, the nondestructive testing of mounted devices with connections on the underside necessitates the use of X-ray technology. These issues are dealt with in more detail in Chapter 6.

There is little to be gained by using relatively large, two-terminal devices on thermoplastic materials. Two-terminal devices such as chip resistors are based on ceramic bodies, and consequently they have extremely low coefficients of thermal expansion (Table 5.1). The difference in CTE vis-à-vis the thermoplastic substrate material is highest for this type of component. Given the absence of conditionally elastic connector legs, this type of device is extremely stiff and consequently unable to prevent deformation at the point of connection.

5.2 Connection Mediums

Suitable connection mediums matched to the process and to the substrate material are essential for the dependability of mechanical and electrical connection. Solder alloys are the most widely used in standard electronics production. Viable alternatives for MID technology along with solder pastes are conductive adhesives for low thermal loading during the connection process, for example. Nonconductive adhesives can also be a good choice in certain circumstances.

The soldering and curing temperatures of the connection mediums differ. The composition of the solder alloy is the main factor determining the temperatures that have to be achieved in the soldering process to ensure that the solder melts in the peak phase and wets the contacts. Consequently, when selecting a connection medium or, to be more precise, a solder alloy for MID, it is very important to take into consideration the maximum temperature capability or the temperature resistance of the plastic.

5.2.1 Solder Paste

Solder paste consists of an alloyed metal powder and a flux. The workability-relevant properties of a solder paste are determined by the individual components metal powder and flux, and the mixing ratio of the two.

The purpose of the flux is to remove layers of oxide during soldering and improve the wetting properties of the pads. This facilitates the formation of a dependable soldered joint. The metal powder consists of particles of a defined solder alloy. It melts, flows over the pads of the component and the attachment points in the conductor pattern, and forms the solder meniscus or the soldered joint. Solder pastes are typically distinguished by

- particle size of the metal powder,
- composition of the solder alloy, and
- flux.

Particle size has to be chosen to suit the pitch of the structure. The size of the particles in the solder paste for a fine-pitch application has to be correspondingly small in order to facilitate printing. The smaller the particles, however, the greater the tendency of the solder alloy to oxidize during soldering, and this can detract from the paste's wetting properties. Powder particle size is the feature used for classifying solder pastes. [144]

TABLE 5.4 Classification of Solder Pastes by Metal-Powder Particle Size (ANSI/J-STD-005) (DIN 32513-1)

Solder-paste type	Less than 10 weight % smaller than	At least 80(1) or 90(2) weight % between	Less than 1 weight % greater than	No particles larger than
3	20 µm	25 and 45 µm(1)	45 µm	50 µm
4	20 µm	20 and 38 µm(2)	38 µm	40 µm
5	15 µm	15 and 25 µm(2)	25 µm	30 µm
6	5 µm	5 and 15 µm(2)	15 µm	20 µm

TABLE 5.5 Comparison of Different Solder Alloys [90, 153]

Alloy	Melting range	Comment
SnAg4.0Cu0.5	217 to 224 °C	The melting point is generally given as 217 °C
SnAg3.5Cu0.7	217 to 218 °C	More brittle than SnPb alloys
SnAg3.0Cu0.5	217 to 220 °C	Poorer wetting than alloys containing lead
SnAg3.5	221 °C	In combination with constituents containing lead a phase forms with a melting point of 179 °C
SnBi58	138 °C	Bi lowers the melting point and is brittle; critical if used with alloys containing lead

By law, only lead-free solder pastes have been permitted since 2006. This has considerably increased the variety of solder alloys. Prior to this change SnPb was the most widely used alloy (with varying combinations of these constituents as percentages by weight). One major change accompanying the switch from lead-based to lead-free soldering alloys is an increase in convection-soldering peak temperature from between 200 and 220 °C to about 265 to 270 °C. This temperature range is not critical when thermoset plastics are used, but it most definitely has to be taken into consideration when working with thermoplastics. Depending on the substrate material, however, the mainstream solder materials from conventional printed-circuit board assembly can be used as connection mediums for MID technology. For example, an SnAgCu solder alloy with a liquidus temperature of 217 °C can be used for soldering the high-temperature thermoplastic LCP or the engineering plastics PA and PET+PBT. The low-melt alloy SnBi with its liquidus temperature of 138 °C can be used for plastics such as PC/ABS or PBT, which are less temperature-resistant. [178]

Homologous temperature is a characteristic variable for the behavior of metals. It expresses the actual working temperature as a decimal fraction of the metal's melting point on the Kelvin scale. It determines the working range of solder alloys, because the strength of a metal drops off sharply at a homologous temperature higher than 0.4 [148]. Consequently, soft-solder joins with a rated melting point of 220 °C are usable only up to 120 °C [117a]. True maximum working temperatures, however, are also governed by other factors such as substrate materials, component size, and the nature and frequency of temperature loading. New ways of increasing the temperature resistance of soldered joints are the subject of much research, because the requirements for temperature resistance are increasing all the time. Reactive solders hold out considerable promise in this context. [144, 153]

5.2.2 Conductive and Nonconductive Adhesives

The substrate material is exposed to relatively high localized temperature loading, particularly when SnAgCu solder is used. An alternative connection medium would be a conductive or nonconductive adhesive, which cures at a much lower temperature and does not call for a matched temperature profile. An adhesive can be made conductive by the admixture of electrically conductive metallic or metallized particles. These adhesives can work at steady-state temperatures in the range from 150 to 170 °C. Another advantage is the availability of hard and soft adhesives, so the choice can be matched to connection technology to afford optimum long-term reliability.

Adhesives are distinguished as

- isotropic (conductive paths are directional; ICA, isotropic conductive adhesives),
- anisotropic (conductive paths vectoring only in *z*; ACA, anisotropic conductive adhesives) conductive, and
- nonconductive adhesives (NCA, nonconductive adhesives).

5.2.2.1 Isotropic Conductive Adhesives

The electrical conductivity of isotropic conductive adhesives is established in all directions by metal particles, generally silver flakes. These silver flakes are embedded in a matrix consisting of a base polymer, generally epoxy resin, and hardeners such as amines or carboxylic acids. It is the hardeners that ensure the mechanical strength of the joint. The choice of these two components can influence reaction speed and curing temperature. Broadly speaking, any chemical reaction will be fast at a high temperature and slow at a low temperature.

Figure 5.3 shows that each isotropic conductive-adhesive system has its optimum range for the proportion of fillers. Below what is commonly referred to as the percolation threshold only short conductive chains form, surrounded by an insulating polymer matrix, because at best only a few filler particles are in contact with each other. Increasing the proportion of filler above the percolation threshold enables the first conductive paths to form. Specific conductivity increases dramatically up to a plateau. The critical concentration depends on different factors such as

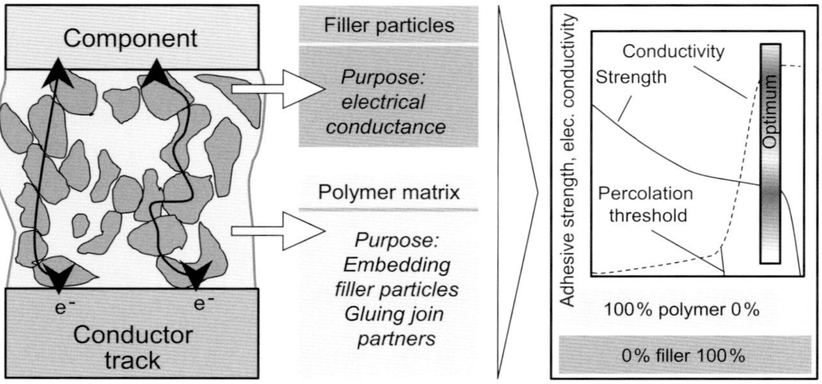

FIGURE 5.3 Principle of isotropic conductive adhesives [109]

filler material and particle geometry. Without significantly boosting conductivity, further increasing the proportion of filler produces a sharp reduction in mechanical strength. The resulting electrical conductivities are considerably lower than those of the pure conductor material. The proportion of silver particles in most of the mainstream adhesives available on the market is between 70 and 82 percent by weight, corresponding to 20 to 30 volume %. [109]

A study of the single-component, temperature-cured adhesives and the two-component adhesives available on the market reveals that the single-component systems predominate. Most two-component adhesives cure at room temperature. No external energy is required and the curing process starts when the reaction partners are mixed together.

Initial cost estimates of conductive adhesive/solder paste as alternative connection mediums indicate that the prices for conductive adhesives are considerably higher. A viable case-to-case estimate, however, has to take into account the joining material costs accruing for the production of a component joint:

$$\text{joining material cost} = \text{price of material} \times \text{density} \times \text{volume applied} \tag{5.1}$$

Given that conductive adhesives have a lower density than solder pastes and the fact that the application volume for conductive-adhesive joins can be lower, there is some compensation for the higher material prices. In practical terms and depending on the conductive adhesive and the components in question, joining material cost ratios between solder paste and conductive adhesive run between 1:6 and about 1:2.5. In the light of pricing in electronics production, it becomes apparent that proportionately speaking the significance of these joining materials costs is relatively minor. The connection material is generally reckoned to account for about 1% of aggregated costs for a module.

So substituting conductive adhesive for solder paste is unlikely to contribute much to direct cost savings. Since, however, switching to conductive-adhesive gluing for an MID application can permit the use of more cost-effective base materials on account of the lower process temperatures, certain indirect cost benefits could accrue. Conductive-adhesive technology opens up considerable cost benefits for applications characterized by large base-material volume combined with few conductive joins. [48]

5.2.2.2 Anisotropic Conductive Adhesives

In contrast to isotropic conductive-adhesive gluing, anisotropic conductive-adhesive gluing forms an electrical connection between component and interconnect device that is unidirectional, generally normal to the adhesive surface (in the z direction). The adhesive is electrically insulating in the two spatial directions (x-y plane). The matrix of the conductive adhesive consists of an epoxy resin admixed with a much lower proportion of conductive particles as filler. On account of this lower proportion (1 to 30 percent by weight) and uniform mixing of polymer and conductive pigments, the material in its native state is not conductive. The essential properties of the filler materials are stable electrical conductivity and a certain plasticity. This allows for a degree of noncoplanarity between the joining partners and differences in the size of the filler particles. The most common materials are gold-coated polymer or nickel balls or solder balls. [48]

5.2.2.3 Nonconductive Adhesives

If the adhesive contains no conductive filler, electrical conductivity relies on contact between the joining partners with adhesive-film thicknesses of the same order of magnitude as the surface roughnesses. Polyimides and epoxy resins are good materials for nonconductive adhesives, on account of their property profiles. Compared to adhesives admixed with conductive fillers, material costs are much lower on account of the lower price for the adhesive and the reduction in consumption.

5.2.3 Press-In Pins

A press-in connection is a solderless electrical contact made by pressing a pin into a plated hole in a printed-circuit board. The press-in technique is defined in EN 60352 and shown in Fig. 5.4. The press-in pin has a larger diagonal across its cross-section than the plated through-hole, or via, in the printed-circuit board. The overpressing caused by pressing in is absorbed by deformation of either the board or the pin.

The flexible type of press-in pin has become established on account of various technological advantages. Figure 5.5 compares the advantages and disadvantages specific to flexible and rigid press-in pins.

Press-in connection technique – DIN EN 60352 Part 5 definition

- A press-in connection is a solderless electrical connection established by pressing a pin into a plated hole in a printed-circuit board.
- The press-in pin (solid or flexible) has a specially shaped zone (press-in zone) of a design suitable for the establishment of a press-in connection.
- By means of a suitable press-in tool (machine), the press-in pin is pressed into the through-plated hole in the printed-circuit board.
- A gas-tight connection must be made between the press-in zone of the press-in pin and the metallization of the hole wall.

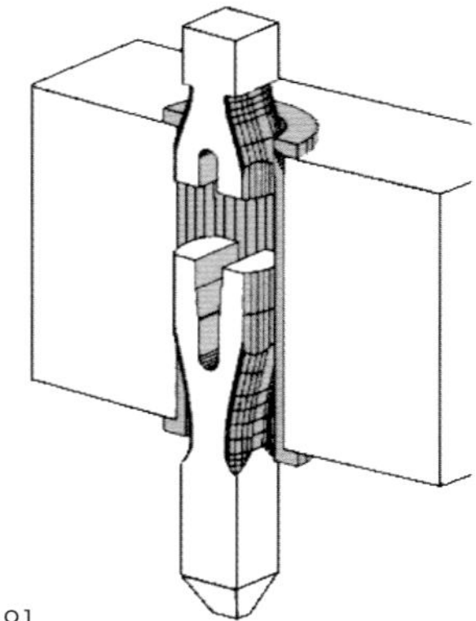

FIGURE 5.4 EN 60352 definition of a press-in connection [48]

Press-in pin designs

Rigid

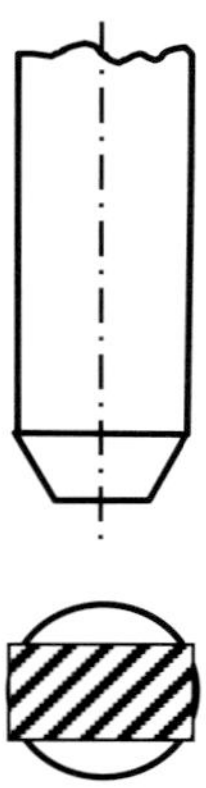

- Pin with solid cross-section
- Pressing in causes deformation of the through-hole
- Higher press-in forces
- Economical in production
- Tolerance requirements for solid press-in pins are higher
- Repair not possible because metallization is deformed
- Has not become established in the market

Flexible

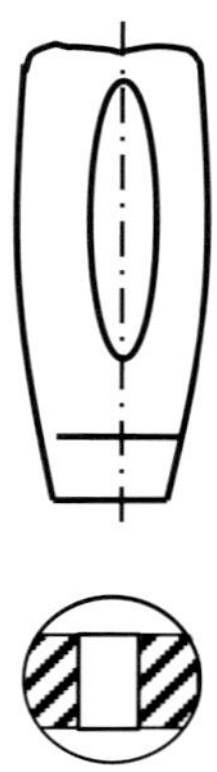

- Pin with flexible cross-section
- Pressing in causes primarily deformation of the pin
- Low press-in forces
- Production costs are higher
- Repair possible by replacement of faulty pins
- Tolerance requirements for flexible press-in pins are lower
- Elastic press-in zone has become accepted in the market

FIGURE 5.5 Technological comparison of rigid and flexible press-in pins [48]

The requirements for the press-in zone of a press-in pin are high. The connection between the plated via and the press-in pin must be gas-tight, without the plated hole being subjected to excessive strain. A pin must also test to a certain holding force and must not exceed a maximum permissible volume resistance. Even an elastic press-in pin will undergo some degree of plastic deformation when pressed into a through-hole of small diameter close to the bottom limit of tolerance, so even elastic pins can be pressed in only once. In a sleeved hole, by contrast, a new pin can be inserted in the event of repair or replacement. [37]

5.3 Connection Process

On account of its excellent electrical and mechanical properties, soft soldering is the primary connection technology for surface mount devices (SMD). Soft soldering is defined as a cohesive joining technique at temperatures up to 450 °C. Processes involving melting temperatures in excess of 450 °C are commonly termed hard soldering or brazing. Gluing with conductive and nonconductive adhesives are other cohesive joining techniques. A press-in connection, on the other hand, is a positive-lock connection with no connection medium.

Reflow soldering is used for more than 80% of the electronic assemblies produced worldwide, so it ranks as easily the most important connection technology in the industry. Any technique in which solder applied beforehand is melted to form the connection without the application of more solder is reflow soldering. In the various reflow processes a defined minimum quantity of solder that depends on the joint to be soldered is applied. This is the distinguishing feature in contrast to wave or selective soldering, in which a large quantity of solder material is present in the molten state and is applied as a wave to the point where the joint is to be made. Reflow soldering covers mass soldering processes and sequential soldering, both by automatic machines and by manual means.

A solder joint is characterized by two diffusion zones or intermetallic phases (one on the circuit board, one on the component) and the intermediate solder material (Fig. 5.6). The diffusion zones account for a volume of between 5 and 15% of the solder joint, with the formation of the zones depending on solder composition and also heavily on the level and duration of heat input during the solder process and subsequent use. In principle, the intermetallic phases are brittle, and these areas, therefore, are often responsible for the formation of cracks due to thermo-mechanical stresses. The standard Cu-Ni-Au plating system of MID technology, however, is comparatively uncritical with regard to the formation of intermetallic phases [153, 178].

In reflow soldering, heat is transferred by heat transport and can also be transferred by conduction, convection, condensation, or radiation. It is characterized by the coefficients of thermal transfer listed for the various heat-transfer mechanisms in Table 5.6.

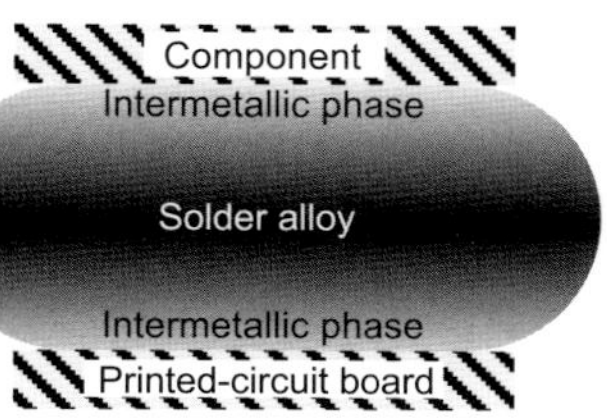

FIGURE 5.6 Basic makeup of a solder joint

TABLE 5.6 Coefficients of Thermal Transfer for the Various Heat-Transfer Mechanisms

Type of heat transfer	Coefficient of thermal transfer α [W/m^{2}K]	Comment
Convection	5 (still air) 20 to 100 (forced convection)	Solid body intercepts a flow of warmer fluid (or vice versa)
Conduction	4000	Transfer of kinetic energy between neighboring atoms or molecules without material transport
Condensation	100 to 400	Heat from the transition of a substance from gaseous to fluid state
Radiation	20 to 30 (warm-up) 60 peak	Energy is transported (primarily) by infrared waves

The coefficient of thermal transfer depends heavily on the type of flow, the surrounding fluid, geometric relationships, and surface conditions. Unlike thermal conductivity, the coefficient of thermal transfer is not a material constant. In a mass soldering process the entire component is heated, whereas in a selective soldering process only selected parts are heated. Reflow soldering units differ particularly as regards homogeneity of heating on account of process specifics. Consequently there are differences in thermal loading in the individual zones, and this is all the more significant with three-dimensional modules. [12]

MID technology generally dispenses with solder-stop paint. A consequence is widespread wetting, which can allow the solder to flow away from the intended joint. This can lead to inhomogeneity of the soldered joints and misshapen solder menisci, which under certain circumstances can detract from long-term reliability. Optimizing the geometric design of the pads, the quantity of solder, and the solder parameters can reduce this effect (Fig. 5.7). It is best to avoid sharp changes of conductor track direction and allow sufficiently large radii to reduce stresses induced by changes in temperature. The transitions from pad to conductor should be smooth, enabling the capillary effect to prevent the solder paste from flowing away. Rounding the corners of pads reduces the risk of the metallization detaching from the substrate when exposed to thermal loading. [104]

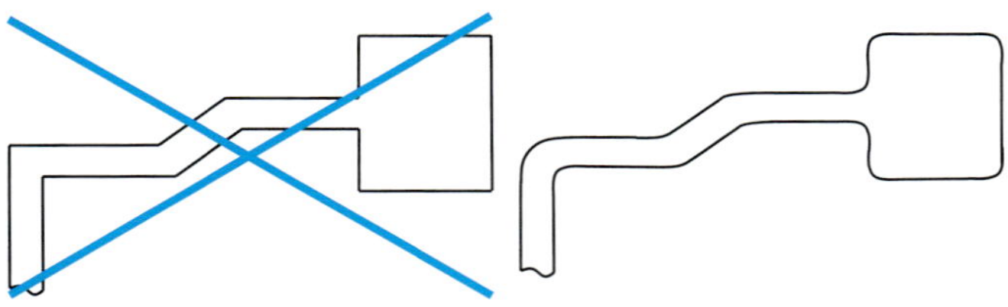

FIGURE 5.7 An optimized layout can compensate in part for the lack of a solder-stop mask [104]

5.3.1 Reflow Soldering Techniques

5.3.1.1 Infrared Soldering

The emitters in an infrared soldering oven emit electromagnetic waves in the medium- to long-wave section of the spectrum, between 1.5 μm and 3.0 μm and 4 μm to 6.24 μm. When these waves coincide with the material for soldering, they are partially reflected and partially absorbed and produce heat. However, the absorption of the different parts of an electronic module varies widely based on the various materials (substrate, electronic component, pad metallization, etc.) and surface specifics (e.g., color), so temperature distribution can be inhomogeneous across the module. This problem of inhomogeneity can be compounded by the shadows thrown between closely spaced components, or behind large components, and the walls of three-dimensional interconnect devices. Even under optimum process-transient conditions (soak zone as long as possible to homogenize temperature distribution), in infrared soldering temperature differentials up to 40 K can occur across a module. So the danger is that individual components or the substrate could experience thermal overload, while other components have not even reached a sufficiently high soldering temperature. Peak temperatures for infrared soldering have to be very high. Although IR soldering is one of the most widespread methods, medium-based techniques are becoming more common on account of the drawbacks inherent to heat-only ovens. [48]

5.3.1.2 Convection Soldering

In convection systems, heat is transferred by a turbulent flow of hot gas (air and/or nitrogen). The gas temperatures are only marginally above the required working temperature. The heated gaseous medium from tangential and axial fans is directed by special nozzle systems on to the module, transferring a certain amount of heat. Boosting flow rate, consequently, increases heat transfer. The advantage of these systems is that heat can be distributed very evenly over different modules. There is no energy-rich radiation, so there are no hotspots on the printed-circuit carrier. Temperature differentials in the peak zone are generally less than 20 K. Vis-à-vis infrared soldering, the more uniform transfer of heat means that the overall temperature level can be lowered considerably, reducing thermal loading for components and substrate alike. The coefficient of thermal transfer for forced convection is lower than for condensation, however, so in order to safeguard reliability in the soldering process the medium temperature has to be well above the liquidus temperature of the solder. Shadowing can be a problem, especially with complex 3D-MID.

5.3.1.3 Vapor-Phase Soldering

Condensation is another means of heat transfer. In vapor-phase soldering units, saturated inert vapor condenses on the cooler module, and the direct transfer of condensation heat causes rapid warm-up to vaporization temperature (Fig. 5.8). Thermally stable halogenated or heterocyclic hydrocarbons that cause no toxicological effects whatsoever are the vapor mediums. There is a range of mediums with different boiling points, so the process can be adapted to the solder materials (low-melt, high-melt) and the substrate materials. The volume of condensing vapor depends directly on the temperature differential between module and vapor. The result is that more vapor condenses on cooler surfaces (e.g., components with complicated shapes), so temperature distribution is extremely homogeneous regardless of geometrical complexity. The boiling temperature of the selected vapor medium determines the maximum possible solder temperature of the module. Under real-life conditions the temperature differentials on vapor-phase-soldered modules are only 0 K to 2 K. This process, consequently, offers the lowest thermal loading due to temperature peaks. [98]

The specific thermal differences of the individual mass reflow soldering techniques can be aggravated by the geometrical complexity and profiling of the 3D-MID.

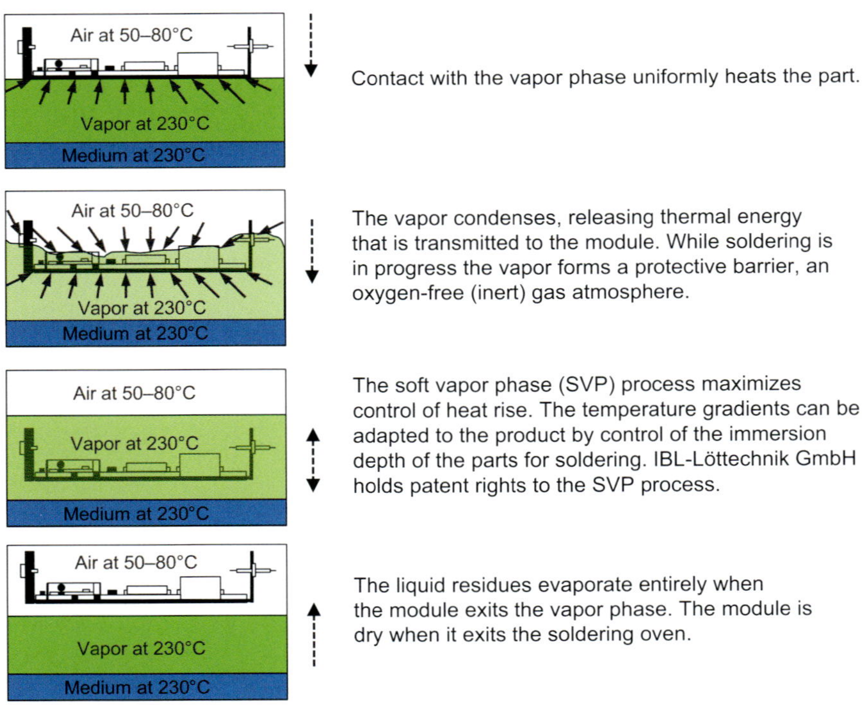

FIGURE 5.8 Process flow for vapor-phase soldering [98]

The differences in question apply to the soldering temperature level required and the temperature differential occurring in the peak zone. Figure 5.9 compares the pros and cons of the various mass soldering processes. IR is much less suitable than convection or vapor-phase soldering for 3D-MID. The geometry of the interconnect device tends to aggravate the process-related temperature differences in the peak zone. This is the result of shadowing and similar effects, and the net outcome is that a very high aggregate temperature level has to be achieved in order to ensure that soldering will be successful everywhere on the MID. So the choice of thermoplastics for this process is limited to high-temperature materials. Forced-convection soldering systems are much better overall. The turbulent flow means that the uniformity of temperature distribution is affected to a much lesser extent by walls or other 3D elements than is the case with infrared soldering. All in all, therefore, temperature levels are much lower. There is considerably less risk of damage to material or insufficient temperature at the solder depots than in IR soldering. The principle of condensation or vapor-phase soldering proves ideal for many MID geometries when uniformity of temperature distribution is applied as the criterion. Extremely uniform surface temperatures with fluctuations of only a few Kelvin are achieved, depending on the vapor medium used (boiling temperatures of maximum 230 °C for lead-free standard alloys, 155 °C or less for low-melt solders). Working conditions for thermoplastics and the alloy alike are outstandingly reproducible, with virtually no local differences. [48]

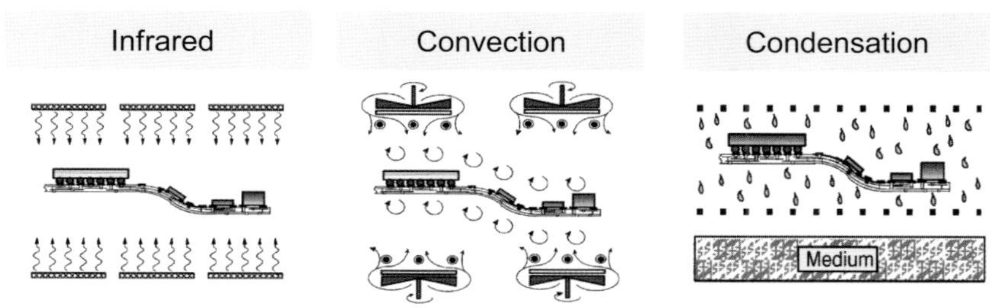

Infrared

- Heat transfer depends on coefficient of absorption, complicated profiling
- ΔT is large in the peak zone
- High thermal stress on board and components
- Shadowing effects, particularly with 3D-MID

Convection

- Uniform temperature distribution across different interconnect devices
- ΔT is low in the peak
- Reduced shadowing effects
- Less risk of thermal overload of the modules

Condensation

- Homogeneous temperature distribution (almost no ΔT in the peak zone)
- High reproducibility
- Simple profiling
- No risk of hotspots because working temperature is a given

FIGURE 5.9 Comparison of different soldering processes for MID applications

However, vapor-phase soldering is not suitable for class 1B MID with concave structures. If the condensate is unable to drain off, heat transfer is obstructed and irregularities in temperature distribution occur. Vapor-phase soldering accounts for only about 10% of ordinary electronics production, but it is the standard technique for MID.

In combination with matched solder materials, the techniques of convection and vapor-phase soldering widen the range of usable thermoplastic base materials. Using these processes, soldering on high-temperature thermoplastics (e. g., LCP) with standard solders (SAC 305, for example) is perfectly reliable, and with process control duly adapted the same applies to certain engineering thermoplastics. Assuming compliance with material-specific limits for moisture content (about 0.8 to 1.0% by weight), there are no difficulties in reflow soldering on glass-fiber-reinforced polyamides up to maximum temperatures of around 265 °C, which in any case are easily managed in vapor-phase soldering. If these temperatures are too high, solders with lower melting points (such as SnBi58) requiring process temperatures on the low side of 175 °C enable reliable reflow soldering on polyamides (maximum water absorption 10%). The reduced process temperatures possible with special alloys like these also permit reflow soldering on economical standard plastics like PC and PBT. Under certain circumstances, therefore, using matched solder materials for reflow processes can open up significant cost benefits for MID.

Special MID carriers are necessary for the soldering process. These carrier systems have to be matched to the geometrical boundary conditions of MID production and should be designed to carry multiple modules through the soldering stage. Workpiece carriers should be as skeletal as possible in order to minimize mass. It is also important to ensure that there are no pockets or cavities where the condensing solder medium can collect in vapor-phase soldering and be freighted out of the soldering unit. [192]

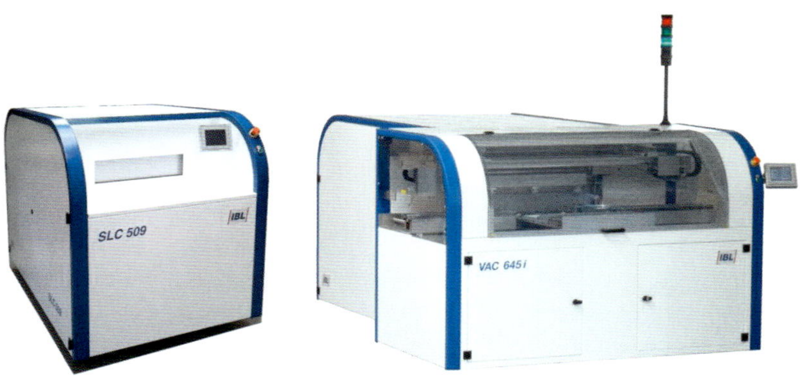

FIGURE 5.10 Batch- and inline-compatible vapor-phase soldering units (graphics courtesy of IBL-Löttechnik)

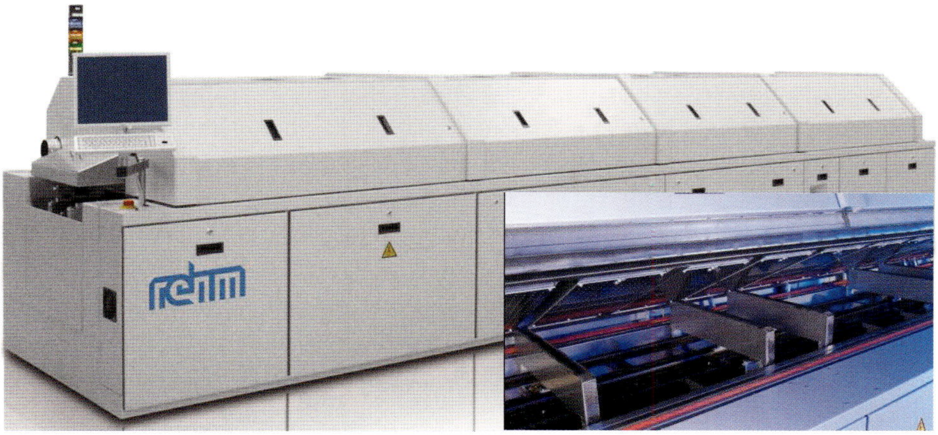

FIGURE 5.11 Convection soldering oven with view inside the process chamber (graphics courtesy of Rehm Thermal Systems)

5.3.2 Selective Soldering Processes

In selective soldering processes, temperature loading for electronic components and substrate material alike is not generally dependent on the process in question and is both localized and of limited duration. Only the connection points are heated to working temperature, and surrounding areas are exposed to only slight thermal strain, if any, so with carefully adjusted process control thermally sensitive components and thermoplastic substrates with limited dimensional stability when exposed to heat are easily workable. It depends on the application whether the filler material and the requisite energy are applied at the top surface or the underside of the module, or possibly with a time lag. In selective soldering with laser light, for example, filler material applied beforehand as solder paste is melted, whereas in fountain soldering as a selective version of wave soldering both filler material and energy are made available in the form of a permanently overflowing miniature solder wave.

When classified on the basis of contacting and noncontacting processes (meaning contact between energy source and solder stock), a distinction can be drawn between soldering-bit or soldering-iron, ultrasonic, and fountain soldering on the one hand, and hot-gas, infrared, and laser soldering on the other. The sequential nature of the technique is a drawback for any selective soldering process. This is why they are used primarily for jobs with only a limited number of solder points on the product, for soldering special components into position, or for soldering on injection-molded thermoplastic interconnect devices made of base materials with low levels of resistance for hot working.

Laser Beam Soldering

The process of laser beam soldering (Fig. 5.12) uses the energy from a laser source to melt the solder. Laser beam soldering, light beam soldering, and electron beam soldering are a group of processes that use radiation for energy transfer (DIN 8505). As in all other reflow processes, the solder is deposited prior to soldering in the form of solder paste, preforms, or solid solder depots. An alternative is provision of the solder in continuous-feed wire form while the laser beam is applied to the surfaces for joining.

The laser beam is focused directly on the surfaces to be joined and the solder depot. The localized energy input rapidly heats the partners and the solder, with ramps from 500 K/s to well in excess of 1000 K/s possible, depending on the masses to be heated and the power of the laser.

The solder melts and wets the partner components. As soon as the laser is switched off the solder cools and solidifies quickly, producing a fine-grained microstructure in the solder. A pyrometer can be used to measure the temperature in the soldered joint so that laser power can be adjusted to produce a desired temperature profile.

The influencing factors are numerous and include solder volume and volume of the partner components, but process times are between 0.2 s and several seconds. The spatial temperature gradient is large because energy input is localized and the process times are relatively short. So this process is also suitable for soldering in thermally sensitive conditions (e.g., temperature-sensitive components or thermoplastic interconnect devices). The main advantages of laser beam soldering are that the process is noncontacting and beam shape can be adapted to different joining arrangements. For example, all the connections of a component (such as a small outline package, for example) can be soldered in a single pass with an array of laser beams focused in a line.

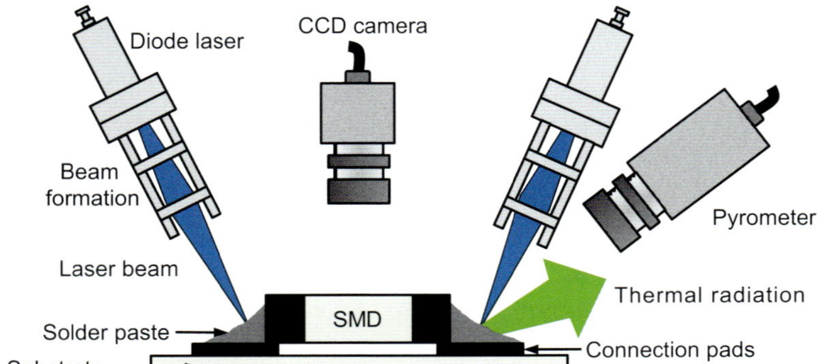

FIGURE 5.12 Schematic of simultaneous laser beam soldering of a two-terminal component

Consequently, like other selective soldering processes, laser beam soldering is used when

- mass soldering is not an option on account of technological or economic reasons, for example if there is only a very small number of solder points,
- if the materials/devices are temperature-sensitive,
- if convection is obstructed, or
- if the device has to be held down on the substrate.

5.3.3 Gluing

When devices are glued with electrically conductive or nonconductive adhesive, the mechanical strength of the bond is the result of thermal curing of the adhesive. Considerably lower than process temperatures for soldering, adhesive curing temperatures have to be chosen to suit the adhesive system, the substrate material, and required process time.

Adhesives do not have wetting properties. The self-centering effect familiar from soldering does not occur, and, consequently, positioning has to be considerably more accurate. An advantage from the MID point of view is that overspread wetting without solder-stop paint does not occur in gluing.

Conductive-Adhesive Gluing

The mechanical strength of conductive-adhesive gluing is achieved by thermal curing, while the electrical connection is established by the three-dimensional mesh of touching particles embedded in the polymer matrix (Fig. 5.13). Silver is the most common filler, on account of its good electrical conductivity and high thermal conductivity [109]. Curing temperatures are low (80 to 150 °C), so this process can be used on thermally sensitive materials.

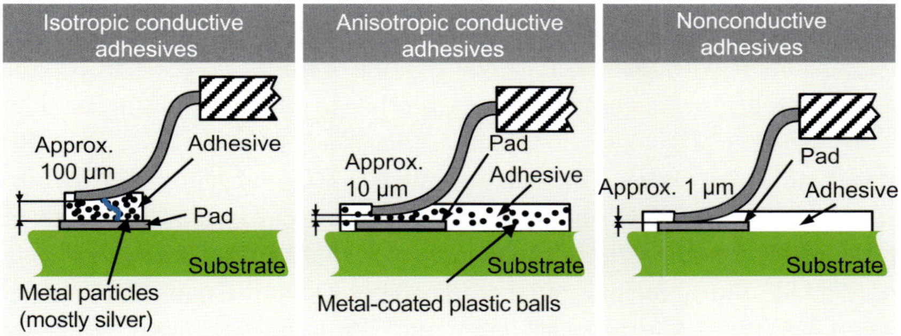

FIGURE 5.13 Principles of electrically conductive adhesives and adhesives without fillers [48]

The requirements applying to the surface properties of the two partners electronic component and substrate metallization are high for conductive-adhesive gluing. Surfaces with noble-metal plating such as gold, silver, or palladium are eminently suitable for making electrically and mechanically stable connections. By contrast, the tinned surfaces used for surface passivation in SMT production render specially adapted gluing processes necessary. Contaminants or layers of oxide present on the surface have a negative effect on electrical conductivity and the wetting properties of the adhesive.

Like solder pastes, conductive adhesives can be applied by stenciling/silk-screen printing or by dispensing. Pin transfer and die embossing are excellent possibilities with certain adhesives for MID. Jetting has tremendous potential for three-dimensional interconnect devices on account of the high speed. The basic viscosity of most adhesives is low, so there is more likelihood of bridges forming between the individual connections of small components. For this reason, connections with isotropic conductive adhesives are only conditionally suitable for highly miniaturized modules.

Anisotropic Conductive-Adhesive Gluing

The functional principle is not the same as that of conductive adhesives, and by analogy the processing characteristics differ as well. The properties of electrical insulation in the x and y directions means that the adhesive can be applied as a bead or as films cut to size. The risk of shorting such as when solder bridges form can be excluded down to grid sizes well below 100 µm [52]. So fine-pitch components can be contacted in this way, and less precision is required for application of the adhesive. A distinction has to be drawn between conductive paste and film adhesives in terms of application. Pastes are usually applied by stenciling or silk-screen printing or by dispensers, whereas film adhesives are tacked to the substrate by pressure and heat. The components are then positioned, pressed into the adhesive at comparatively high temperatures and pressures, and allowed to cool with pressure still applied.

Gluing with Adhesives Free of Fillers

Connections made with insulating adhesives free of fillers are electrically conductive if the partners are in contact with each other in adhesive no thicker than the surface roughness (< 10 µm). In principle the process is very similar to that of anisotropic adhesive gluing. The glue is applied by a dispenser or by print application of a paste. The electrically nonconductive adhesive can be applied allover across multiple connections. This implies low requirements for the process as such and therefore good affinity for fine-pitch applications. Once the partners to be joined have been positioned, the adhesive cures under pressure and temperature within a matter of seconds.

The process-related requirements applicable to the geometry of the component connections are high, so a considerable part of the SMD range is not suitable for this method of joining. Devices with nonplanar contact surfaces (e.g., BGA, PLCC) or with planar but not parallel pads (all small outline (SO) packages) pose problems in terms of electrical contacting, as in anisotropic gluing. The result is that filler-free adhesives are used in MID technology primarily for chip placement as described in Section 5.3.5.

5.3.4 Press-In Connection Technique

Another alternative to soldering without subjecting the module to thermal loading is to use the press-in connection technique. This solderless connection establishes electrical contact by pressing a pin into a plated through-hole (through-contacting) with a suitable press-in tool. The connection between press-in pin and plated hole must be gas-tight. On account of the thin plating applied by all-chemical metallization and the viscoelastic properties of thermoplastics, press-in connection is of only minor significance as a solderless connection technique for MID at this time. Scientifically speaking the subject has been investigated in depth and appropriate rules have been elaborated. [37]

The potential for automation, reliability, and quality all help make press-in connection the preferred technique for numerous applications. High resistance to shock and vibration loading are contributory factors along with good processing and EMC properties. Typical advantages vis-à-vis soldering are illustrated in Fig. 5.14.

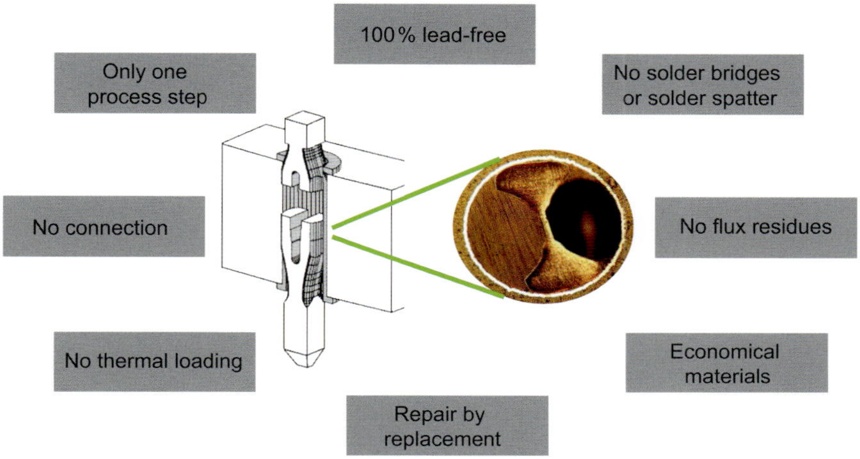

FIGURE 5.14 Technological advantages of press-in connection over conventional soldering [48]

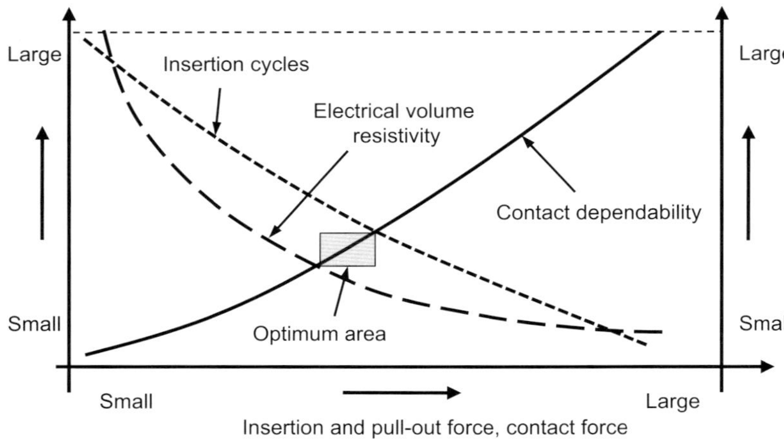

FIGURE 5.15 Insertion cycles, electrical volume resistance, and contact dependability in relation to insertion force, pull-out force, and contact force [118]

As regards the quality of a press-in connection, it is vitally important to have interconnect devices manufactured in line with the requirements inherent to press-in connection technology. Material and hole makeup have at least as much of an effect on the press-in connection as the press-in zone itself. Substrate material, through-plating, and press-in pin (press-in zone) all have to be perfectly matched in order to ensure a reliable connection. There is a direct relationship between the individual characteristics: lower insertion and pull-out forces or high insertion cycle rates are easily achieved at the cost of high resistance, but this often detracts from contact dependability. Figure 5.15 shows an optimum target area.

Applied to the press-in connection technique, this means that the holding force of the press-in pin in the through-plated hole in the MID has to satisfy precisely defined minimum requirements in order to ensure sufficient security of the connector with regard to insertion, pull-out, and contact forces. The tolerances applicable to production of the interconnect device are stated in Fig. 5.16 reference to a 1.00 mm hole diameter.

When these specifications are applied directly to thermoplastic interconnect devices, the following problems arise. Assuming that the hole diameter before plating (drill diameter) is the same as the specified diameter for injection molding an MID, in order to have a surface-finished (in other words plated) hole of the specified diameter it would be necessary to metallize with wall thicknesses of 42.5 μm and 92.5 μm (plating diameter, small: 1.125 mm to 2 × 92.5 μm plating thickness per wall gives the specified finished diameter of 0.94 mm; by analogy, the thickness of the plating for the large initial diameter is 42.5 μm). Galvanic overplating is the only way to achieve the thicknesses that MID needs to accommodate press-in pins.

Nominal diameter of hole	1.0 mm
Diameter of plated hole	0.94 - 1.09 mm
Diameter of hole before metallization	1.125 - 1.175 mm
Thickness of copper coating	≥ 25 µm
Finished surface	Must be stated by manufacturer

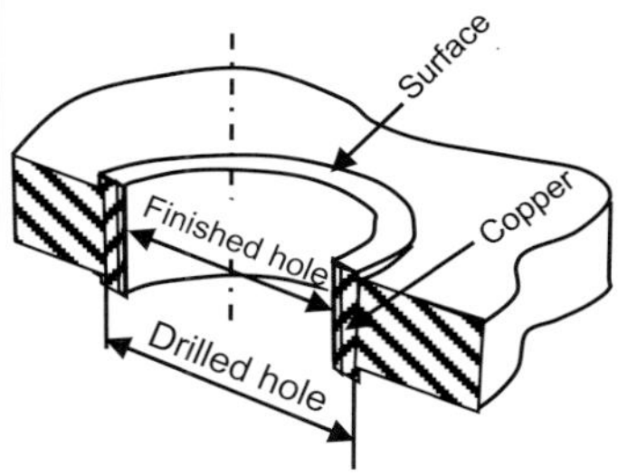

FIGURE 5.16 Tolerances for 1.00 mm hole diameter (DIN 60352)

Consequently, the standard is not directly applicable to injection-molded interconnect devices. The permissible metallized final diameter has to be used as the point of departure to calculate the plating diameter with a defined metallization thickness. The underlying difference is that thermoset circuit boards have to be drilled and the projecting glass fibers from the reinforcement are present inside the hole, reducing effective diameter and resulting in smaller final diameters after metallization. The finished surface has to be adapted to suit the requirements stated by the manufacturer of the press-in pins. Another factor is that although normally brittle (amorphous) or viscoelastic (partially crystalline), thermoplastics exhibit an elastic behavior as temperature increases. A further increase in temperature transforms these materials into a state of plastic flow. This has to be taken into account in the design, in order to prevent high temperatures from causing contact problems. [48]

Scientific studies show that laser-structured MID can be contacted reliably without soldering. The prerequisites are precision design of drilled hole geometry, compliance with relatively close tolerances, and close matching of the partners to be joined. This is particularly important if there is no additional means of securing pin contacts. Dependable press-in connection is also possible with hot-embossed MID, but the pins also have to be soldered to establish the electrical connection. Under these circumstances the pins are additionally secured, so tolerancing for drilled-hole geometry can be relaxed to some extent. C-section round contact pins are the best shape; closed round-contact or rectangular sections can easily result in damage to the hole plating or even the substrate plastic. The design of the holes in the substrate depends heavily on the structuring process. Laser direct structuring enables the hole geometries to be lasered, producing a slightly conical shape with a tapering diameter. Plating 10 to 15 μm in thickness means a 20 to 30 μm taper and correspondingly more when galvanic overplating builds up. [37, 194]

5.3.5 Chip Placement

Chip-placement techniques are very important in MID technology as elsewhere on account of developments in miniaturization, the associated reduction in space required, and the higher packing density. Mounting silicon chips on 3D-MID permits considerably higher integration density, opening up completely new areas of application. Chip-on-MID applications are particularly viable in sensor technology but also in other fields such as RFID tagging. Figure 5.17 shows a number of sensor concepts. The MID is a chip carrier and can also act as interface to the higher-order system (e.g., by mapping a connector layout). It can position sensors accurately at precisely defined angles and integrate mechanical connections (e.g., fluid ducts).

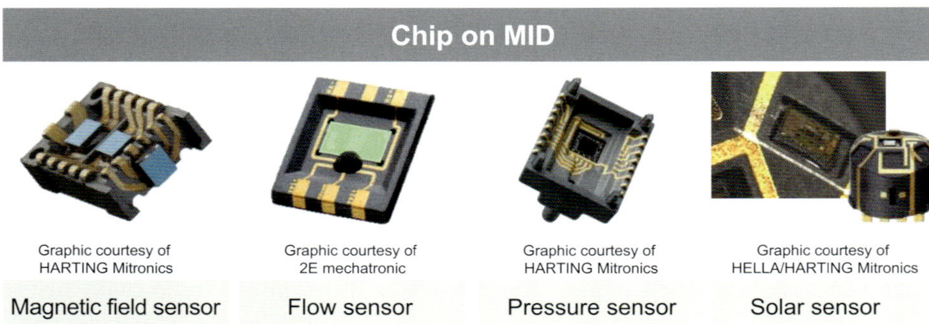

FIGURE 5.17 Sample silicon chip on MID applications [63]

Wire bonding and the flip-chip technique are suitable for mechanically and electrically connecting the silicon chip to the MID substrate with appropriate protection against environmental influences (Fig. 5.18). The technique of wire bonding

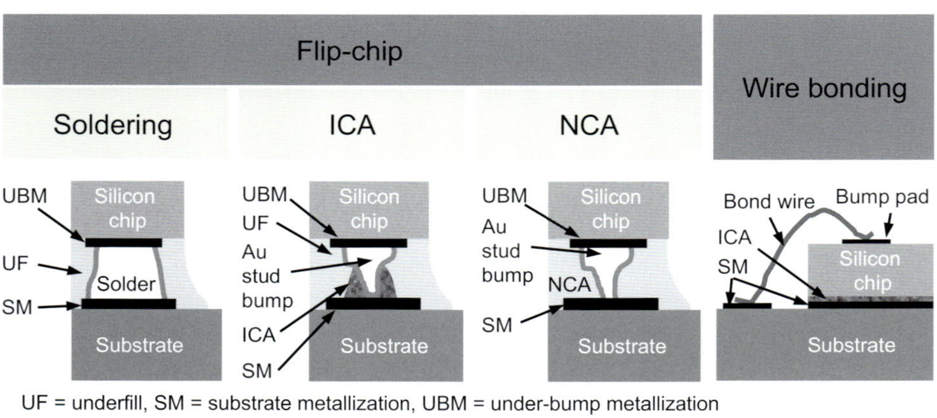

FIGURE 5.18 Schematic view of relevant chip placement technologies for MID [63, 191]

permits flexibility in substrate layouting because the positions and lengths of the bond wires are variable. The wire bonds have a certain amount of give, making electrical contacting all the more reliable. A flip chip has the functional side turned toward the substrate. The silicon chip's pads are directly connected to the substrate plating by soldering or gluing. One major advantage of this configuration is that flip chips need less space, and the entire surface of the chips can be utilized to increase packing density. By contrast with wire bonding, which is a sequential process, all the connections are contacted in a single step. [81]

Chip-on-MID reliability depends primarily on the chip placement technology used and also on the materials, chip layout (I/O), and chip dimensions. Under varying thermal loading, differences in coefficients of thermal expansion between substrate, connection medium, and silicon chip induce mechanical stresses and strains. This effect is aggravated in the MID scenario because the substrate materials are thermoplastics. At a temperature differential of 100 K, the CTE difference in the shot direction of PET+PBT is 33 ppm/K compared with a silicon chip. Over an edge length of 4.3 mm the difference in length at the edging is approximately 7 μm.

When a chip is mounted by soldering or by Au stud bumps and isotropic conductive adhesive, the space between the substrate and chip is filled with underfill as a means of extending durability. The underfill spreads internal stresses, reducing the local shear forces occurring at the points of connection. Figure 5.19 (left) shows stress loading over the solder balls for a silicon chip. The solder balls at the chip edges, for which DNP (distance to neutral point) at the center of the chip is greatest, are subjected to more severe loading. In MID materials this effect is aggravated at the chip edges that parallel the flow direction of the shot, on account of the anisotropic CTE [191]. Studies have shown that using underfill for chips on organic substrates can significantly prolong service life, assuming that the cavity underneath the chip is well underfilled with no voids. Underfills are distinguished as capillary-flow underfill (CFU) and no-flow underfill (NFU). NFU is a combination of underfill and flux and is dispensed in a defined pattern before the chip is placed. CFU is applied to the edges of the chip when it is in place and is drawn by capillary action into the space beneath the silicon chip. [23, 176]

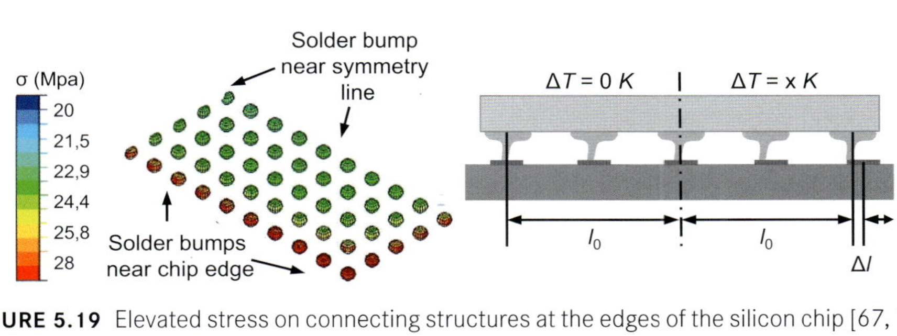

FIGURE 5.19 Elevated stress on connecting structures at the edges of the silicon chip [67, 138]

5.3.5.1 Wire Bonding

The most widespread process for electrically contacting bare silicon chips is wire bonding. This applies in particular to areas of application characterized by high climatic and mechanical requirements because of the flexibility of the wire bonds.

Wire bonds are made either entirely by ultrasonic vibration (ultrasonic wedge-wedge bonding) of the bonding tool or assisted by heat (thermo-sonic ball-wedge bonding). The advantage of wire bonding with ultrasonics (US) over thermo-sonic bonding (TS) is that the process can be carried out at room temperature. US bonding with Al wire (e.g., 33 μm diameter) is the most common way of electrically connecting the chip to the substrate of wire-bond MID. This is on account of the lower temperature loading of the substrate and the comparative thinness of the gold plating. The metallization is rough, particularly after laser structuring, so Au ball-wedge bonding is not used for MID.

The first step is to locate the silicon chip mechanically on the substrate with an adhesive. Adhesives with a conductive filler are often used, because their good thermal conductivity helps in removing process heat. The conductive adhesive has to be chosen firstly for high mechanical strength and secondly for temperature stability matched to the application. Application can be by dispensing, pin transfer, or stenciling, although this third option is only conditionally suitable for 3D-MID.

The partners are then welded together to establish electrical contact. The energy is provided by ultrasonics. A wedge contact is made by welding the wire to the silicon chip and the substrate by vibrations of the bonding tool parallel to the contact surface. The loops should be as short and shallow as possible, in order to minimize the space required by miniaturized MID systems. Loop lengths between 2 and 2.5 mm are typical, but loops can also be as short as 0.5 mm. The shorter the loop for given loop and chip heights, however, the more critical is cold deformation at the heel of the first bond as the wire is bent. So it is important for the MID designers to consider the possibility of positioning the chip in a recess in the substrate, as shown by way of example in Fig. 5.20.

The space needed by the bond head during the process is another aspect that has to be taken into account for US wire bonding on 3D-MID. Deep-access bond heads of some types can bond inside recesses. This permits a high level of 3D design freedom. On the horizontal plane, approximately 1.2 mm of clearance is typically required between the second bond and the next raised object (component, MID wall, etc.).

Ultrasonic wire bonding on MID has to allow for other factors influencing bondability that are of minor significance at best in the context of conventional printed-circuit boards or ceramic substrates. For example, the thickness of the plating, metallization surface roughness, and the type of substrate structuring are all of crucial importance. With two-shot MID the structural design and the mechanical

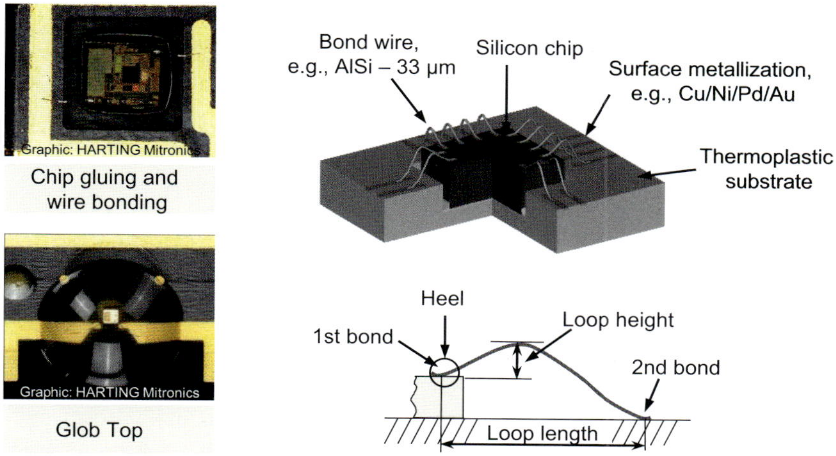

FIGURE 5.20 Die bonding in a cavity to reduce the distance between chip and substrate surface [48]

properties of the bonded materials also factor into these considerations. The 3D geometry, moreover, necessitates a special system of clamps because if the module is not held correctly at the bonding points vibrations can be excited that can prevent the formation of a durable weld. [182]

The thermoplastics used for MID are neither as rigid nor as strong as the glass-reinforced themosets routinely used for printed-circuit boards. The layer of metal deposited on the substrate is typically hard and compensates for the poor wire-bonding properties of the soft substrate material. This is why bondability is largely independent of the polymer used. It is important to bear in mind, however, that there is a minimum thickness specification for plating (Table 5.7). The thicker the metallization, the more dependable the wire-bonding process becomes. [182]

Regarding the surface of the substrate, moreover, the lower the peak-to-valley height of the surface, the better it is for the bonding process. Laser direct structuring in particular leaves the substrate surface badly roughened. It is important to parameterize the process to minimize surface roughness on the one hand while on the other ensuring sufficient adhesion of the plating to the plastic.

TABLE 5.7 Minimum Plating Thicknesses for Dependable Bonds on Various Thermoplastics [48, 182]

Material	Metallization	Cu [µm]	NiP [µm]	Au [µm]
LCP	Electroless	Approx. 4	Approx. 4	0.1
LCP	PVD and electroless	0.5 to 1	≥ 4	0.1
PA6	Electroless	–	≥ 10	0.1
PC/ABS	Electroless	–	≥ 7	0.1

TABLE 5.8 Maximum Permissible Roughness of the Plating for Reliable Bonds with Different Wire Diameters [48]

Wire diameter [µm]	Arithmetic peak-to-valley height R_a [µm]	Average peak-to-valley height R_z [µm]
25	$< 2 \pm 0.2$	$< 10 \pm 2$
33	$< 3 \pm 0.2$	$< 16 \pm 2$
50	$< 3 \pm 0.2$	$< 16 \pm 2$

High requirements therefore apply to selection of the laser parameters and for metallization to the chemical composition of the bath and to process control. For good bondability, the R_Z peak-to-valley height of an MID substrate surface should not exceed 10 µm and bud formation should be only minor. Nicks, holes, notches, discoloration, and contamination are all unacceptable for good bond surfaces. Bonding properties can be optimized by selecting the right finish, for example by application of NiP-Au. Galvanically overplating the copper is one way of producing smoother and, even more importantly, thicker conductors. [104]

Heat-assisted die embossing and CO_2 snow jetting are smoothing processes that can be applied to reduce roughness. Plating with average roughness $R_Z < 10$ µm can also be produced on laser direct structured MID. [185]

Shrinkage of the plastic shots in two-shot injection molding can result in a minute gap and a poor frictional lock between the metallized and nonmetallized polymers. Owing to the possibility of oscillation in the conductor structures this can have a negative effect on bondability, so it is always advisable to bond in the direction parallel to the conductor. If this is not possible, process-secure bonds can be achieved by design measures such as widening the conductors and integrating positive-lock structures and also by adapting the injection-molding parameters accordingly. [182]

5.3.5.2 Flip-Chip Technology

The primary joining techniques used in MID technology are soldering for chips with solder bumps and, for chips with stud bumps, gluing with isotropic conductive adhesive (ICA), anisotropic conductive adhesive (ACA), or nonconductive adhesive (NCA) [81].

In soldering and when an ICA is used, the electrical connection is first established by the solder or the conductive adhesive before the bond is mechanically stabilized by application of an underfill. Using ACA or NCA makes it possible to combine electrical connection and mechanical stabilization in one process step. Electrically conductive connections made with nonconductive adhesives are based on mutual contact between the joining partners. A constant contacting force therefore has to be applied until the adhesive has cured. Contacting flip chips with adhesive

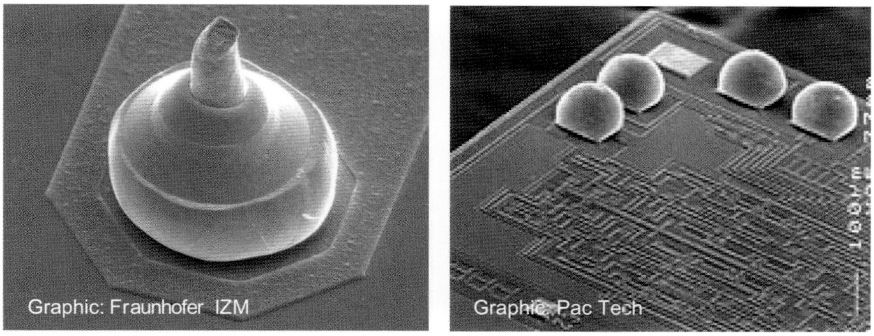

FIGURE 5.21 Stud bump (left) and solder bumps applied by stenciling printing (right)

requires very high positioning accuracy because the adhesive does not permit the chips to float into position in the way solder does. Vis-à-vis soldering, gluing allows the silicon chips to be worked at lower process temperatures. Stud bumps are produced by bonding gold wire to the chip pads and then tearing off the tails to leave an array of tiny bumps. The Au stud bumps are then flattened to a uniform height with a special tool. The process is known as coining, and it makes the chips easier to bond. [29, 81, 183]

The solder bumps can be applied to the chip dies by various wafer-bumping techniques that differ with regard to metallization and the characteristics of the bumps. Candidates as standard processes for producing solder bumps are electroplating, jetting, and stencil printing. Electroplating makes for high yield with very fine connection structures, but the complexity of the process makes it relatively expensive. Standard SMT machines can be used for stencil printing. A 60 μm pitch is possible given optimum stencil thickness, solder-powder class, and settings for the stencil printing process (e.g., printing and peel-off speeds). Jetting is sequential, so it is a flexible process that is also suitable for three-dimensional constructs. The wafer level solder sphere transfer process (WLSST) and the controlled collapse chip connection new process (C4NP) are two other possibilities. [137, 153]

Figure 5.22 shows a specimen process flow for flip-chip gluing with ICA. Conductive adhesive is stenciled on to the substrates. The devices are then placed on the chip, and the conductive adhesive is thermally cured. As in flip-chip soldering, underfill is applied as mechanical protection and cured.

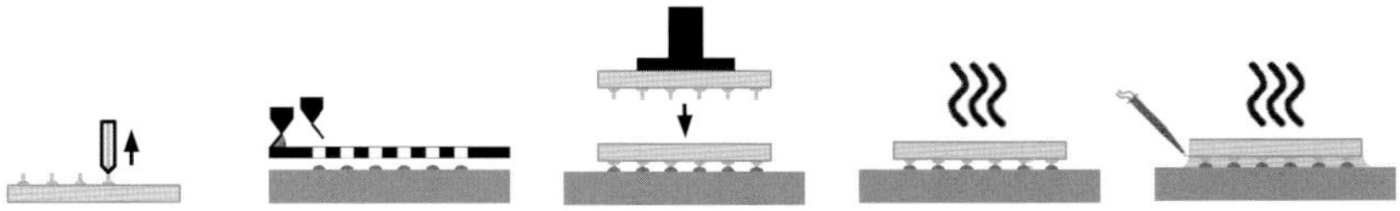

FIGURE 5.22 Specimen process flow for flip-chip gluing with ICA [63]

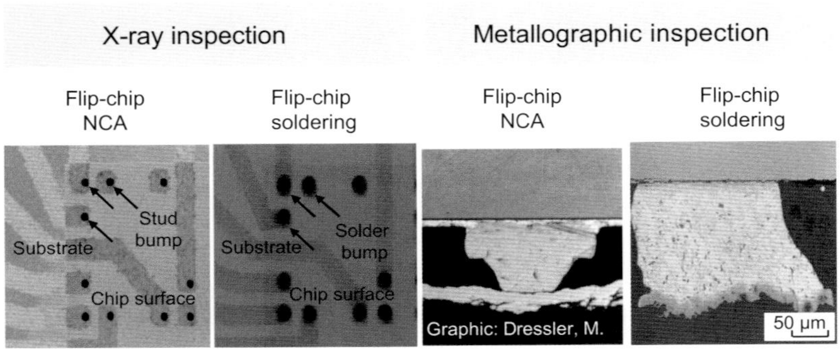

FIGURE 5.23 X-ray and metallographic images of flip chips mounted on MID substrates by gluing with NCA and soldering [63]

Figure 5.23 shows examples of X-ray and metallographic images for flip-chip gluing with NCA and flip-chip soldering on MID substrates. The graphics show that there is no offset between stud bumps and metallization. The X-ray image reveals slight solder outflow. There is no evidence of this effect on FR4. The solder can outflow because there is no solder-stop paint on the substrate. Laser-structured surfaces aggravate the effect; the deliberately roughened surface makes for good wetting. The higher stand-off when flip chips are soldered instead of glued with NCA is clearly visible. Isolated air voids cannot be entirely prevented, despite the good flow properties of the underfill and the NCA and defined temperature control on application. Under thermal loading, both these effects could have a negative effect on the durability of the modules.

5.3.5.3 Glob Top

Bare chips and wire-bond contacts are easily damaged by environmental influences, but they can be easily protected against corrosive mediums, moisture, and mechanical impact by application of casting compound in what is known as a glob top. There are various casting compounds available, each with its own set of properties.

Epoxy resins are very hard, so they provide good mechanical protection. However, they are only conditionally suitable for use in conditions on the high side of their glass transition temperature, because the coefficient of thermal expansion drops off very sharply above this threshold. The glob top can separate from the substrate, shearing the wire bonds. Practical areas of application, therefore, are consumer electronics, automation, and medical technology with operating temperatures up to a maximum of 85 °C.

Silicone gels afford only limited mechanical protection, but they ensure reliable passivation of the bond contacts. Their viscoelastic properties enable them to withstand even severe thermal cycling up to 125 °C and higher. This is a type of loading common in the automotive engineering sector.

5.4 Interfacing to Periphery

Interfaces for data and energy transmission are an important factor in the use of MID technology. A wiring harness, a printed-circuit board, a sensor, a leadframe, or wires could all be contact partners for the MID. With only a few exceptions (RFID being one), information is transferred by an electrical connection implemented in standard build technology by the provision of plug components. A standard plug-type connector or wires can of course be soldered on to an MID. This approach, however, falls short of system integration on the basis of injection-molded interconnect devices. Integrated separable and nonseparable module-to-periphery contacting for mechanical and electrical interfacing is very important in the implementation of an overall mechatronic system.

Solutions have been implemented in series production and are outlined in Chapter 9. Figure 5.24 shows some approaches in detail. One possibility is to map a connector layout with metallized plastic pins, which in turn can be connected to the contact partner (e.g., leadframe) by gluing. Contact pins can be pressed in, enabling the plug connector to be integrated directly into the package. The metal pins are electrically contacted by being soldered to the conductive structure. The mating piece can be a conventional plug. The design freedom of MID technology allows additional strain-relief elements to be integrated. Spring contacts are another possibility. This is the arrangement used to establish the connection between pressure sensor (see Chapter 9) and ESC control unit. Conductive adhesive gluing is the technique used to bond the sensor to the carrier. The same or similar means can be employed to contact a printed-circuit board. The MID is positioned like a surface mount device and contacted by metallized plastic pins, for example, to the printed-circuit board by soldering or conductive adhesive gluing. [150, 187]

As it regards to integrating elements for mechanical contacting into injection-molded functional modules, it is important to bear in mind that problems of relaxation over

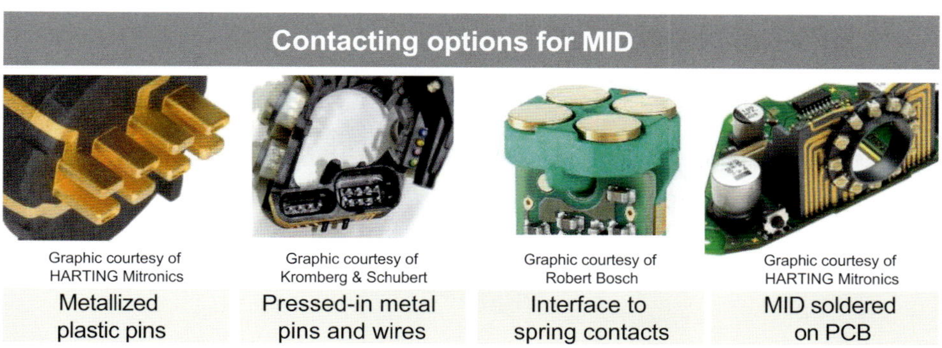

FIGURE 5.24 Possible solutions for contacting MID

time with associated mechanical fatigue and the temperature response of the material are classic terms of reference with plastics. There is also a risk of mechanical damage to the contact elements due to high stresses and strains.

5.5 Protecting Connections Against Environmental Influences

The requirements applicable to electronic modules both during production and in use have increased to a tremendous extent. Legislation is a contributory factor, as are new installed locations, the need for more reliability, and increasing integration density on the module and component levels. MID too are subjected to increasing loads, and the demand for modules that ensure dependable operation despite extreme environmental conditions is increasing enormously. Consequently, protection against mediums and against climatic and mechanical loading is essential in many different areas of application. The protective mechanisms can cover the module in its entirety or only the connections that are commonly relevant in terms of module failure.

Not infrequently, it is damage to the connections that results in nonavailability of the module. So suitable protective mechanisms have to be developed for MID in line with the applicable requirements. Points of departure include for example the provision of a screw-down cap or laser plastic welding to protect the electronic components. This latter approach requires a laser-transparent plastic and a laser-absorbing material.

An innovative and practical approach is encapsulation of the fully assembled MID to protect against extreme environmental influences. Figure 5.25 provides a graphic illustration. A suitable process chain is essential, since conditions differ significantly

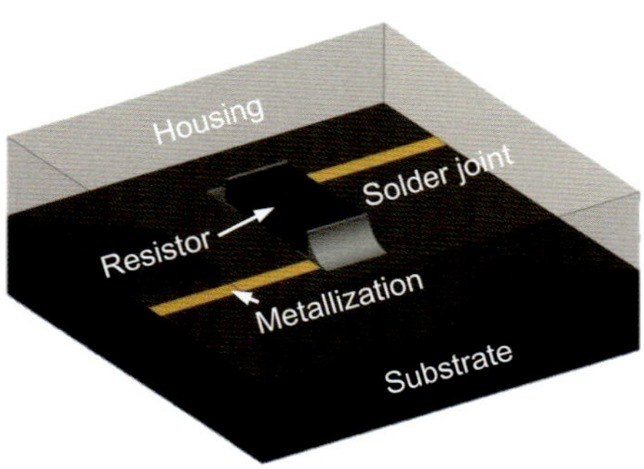

FIGURE 5.25 Sketch illustrating how electronic components are protected by encasing with a thermoplastic in assembly injection molding

from conventional multishot injection molding on account of the metallization and soldering, and the strength of the bond between the first and second shots might not be sufficient for the purpose. Studies have shown that it is advisable to slot plasma treatment into the process chain before the second shot and temper the interconnect devices before soldering. Close to identical coefficients of thermal expansion of the two materials, parallel fiber orientation, and having first and second shots of similar thickness all help toward contribute to good bonding during production and in use. [188]

Other alternatives for protecting the electronic components are casting and painting. Both are standard processes in electronics production that have to be qualified for MID. Once again, good adhesion of the casting resin or paint system to the thermoplastic is important. No studies have been published, but there are some applications with a paint system that have gone into industrial implementation.

Parylene coatings are another possibility. Under vacuum, a pore-free, transparent polymer film is deposited by condensation from the gas phase on to the substrate (e.g., metal, ceramic, plastic). Gas-phase deposition means that the process can coat areas that liquid-based methods cannot reliably protect, including for example sharp edges, narrow, deep gaps, and the underside of large-footprint components. Parylene-coated modules are protected against organic and inorganic mediums, including for example acids, alkalies, moisture, and dust. Another excellent property is electrical insulation with high breakdown resistance. The coatings are free of pores at thicknesses of 0.2 μm and more and are temperature-resistant up to 220 °C. Drawbacks include uncontrolled penetration into connectors and areas that should remain uncoated for subsequent contacting. [72, 157]

6 Quality and Reliability

J. Hörber, P. Daneschwar

6.1 Challenges of Quality Assurance

As explained in Chapter 2, because of ongoing developments in the technology, MID are now used in very large numbers in series-production applications and, to an ever increasing extent, in safety- and security-relevant systems. It is clear from this range of applications that product quality of the three-dimensional interconnect devices is a crucial aspect. This in turn means that the groundwork for end-to-end quality assurance has to be laid in both the development and production of the modules.

The challenge facing those responsible is to align MID and integrative MID design with very widely differing requirements. This applies across mechanical, electrical, and electronic functionality. Very different materials and material combinations have to be used to achieve this, and they in turn call for suitably adapted process technologies. In practice this necessitates close cooperation between individual business units within a company and also across companies working in a variety of technological fields along the process chain. Cooperation cannot work smoothly without clearly defined interfaces and target requirements between supplier and contractor. Furthermore, the stakeholders involved in the production of mechatronic devices must be aware that the individual processes in the production chain interact closely among themselves and have a definitive influence on product quality. Interface definition cannot remain purely hierarchical. Instead, multiple interfaces have to be set up and brought into alignment in accordance with networking complexity (Fig. 6.1).

It is also very important to appreciate the complex interactions of the measures they take and the results they obtain in production with those of all the other subsections involved. For instance, it is not only the direct down-line process of metallization that is affected by a change in the type of plastic. This also has a significant effect on later assembly and connection technology, for example if the new substrate fails to evince the level of temperature resistance required for the soldering process.

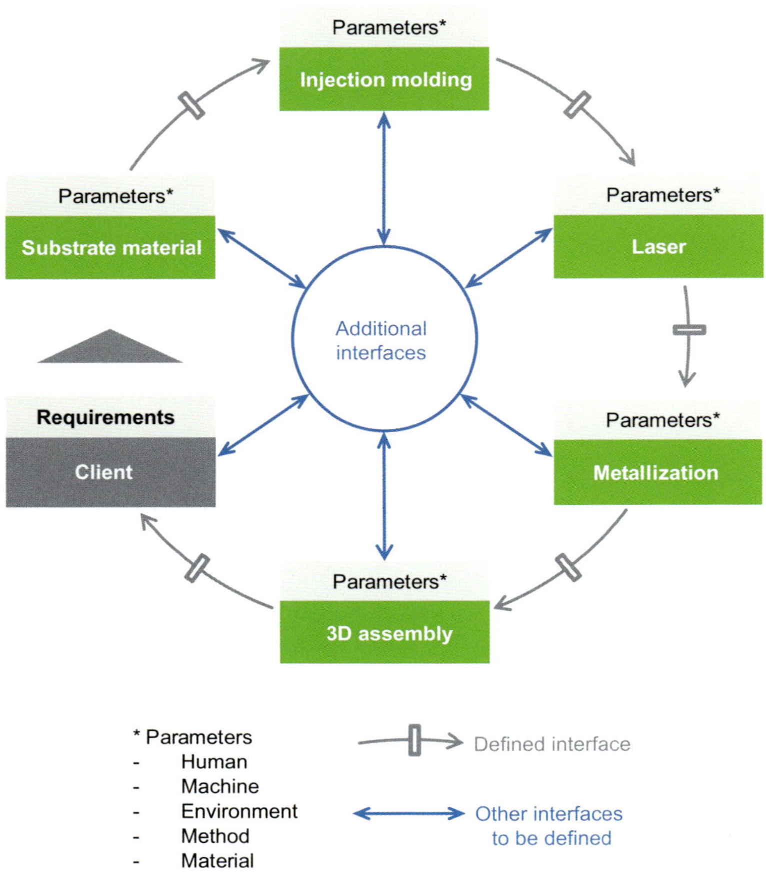

FIGURE 6.1 Networking in quality assurance and derived interfaces; an MID metallized by the LPKF-LDS® as example, as in [50]

Another case in point would be a switch in the injection-molding process to a new release agent that could suddenly prevent reliable structuring and metallization on down the line. Consequently, it is often a good idea for the project team to include a central liaison with an overall grasp of MID technology to notify the various experts in good time of possible interactions. [63]

Over and above complexity as such, precise interface definition is further hampered by the lack of MID-specific standardization and unified procedural policies. In the absence of policies, many manufacturers apply guidelines and design codes they have developed themselves. It is also common practice to adapt examinations and tests from existing standards generally developed for applications in plastics engineering or classic electronics production. MID-specific tests and strategies for quality assurance are then developed by analogy with this existing rule set.

This gives rise to the risk of minimum requirements and key performance indicators being transferred one-to-one, even though they are not suitable for MID and consequently often cannot be met. The way to avoid these difficulties is to define specific quality requirements for each MID product in close cooperation with the client and closely matched to the specific application.

6.2 Simulation-Assisted Quality Validation

Computer-assisted optimization of product quality is necessary because development cycles are becoming faster and on account of the increasingly widespread use of MID in high numbers and in systems with high dependability requirements. As explained in Section 8.10, specially developed software tools are now available for some parts of the MID product life cycle; this holds true for development in particular. End-to-end validation of quality for modules and systems in MID technology, however, will call for further research and development work on computer-assisted modeling of production processes and the behavior of the interconnect devices in the field. At this time, for example, there is a lack of suitable simulation models for adequate mapping of MID soldering processes. Taking existing models for simulation of the soldering process as the point of departure [2], it will be necessary to add the three-dimensional character of the interconnect devices and the extended range of material combinations to the model set. Experiments show the clear differences in heat distribution within a 3D body during reflow soldering in comparison with planar printed-circuit boards. The resulting difference in solder-joint quality can lead to uncontrolled early failures in the field [92]. High thermal input, especially when lead-free solders are worked, leads to the risk of localized thermal overloading causing deformation or excessive aging of the plastic substrate, impairing geometric and mechanical functionality. [130, 131]

In this context, more and more significance devolves on simulations for reliability assessments alongside the simulation of production processes. Considerable progress has been made in recent years with regard to modeling failure under thermo-mechanical loading in service. Recent research studies such as the ZUMIKROSYS project [191] indicate a high level of congruence between forecasts for durability for selected solder-joint shapes with real-life values from parallel experiments using reference modules to explore accelerated aging. One main focus of attention in this project was on detailed modeling of the injection-molded plastic substrate with the SIGMASOFT (from SIGMA Engineering GmbH) and digimat MF (from e-Xstream engineering) software tools. The thermo-mechanical model was generated and ana-

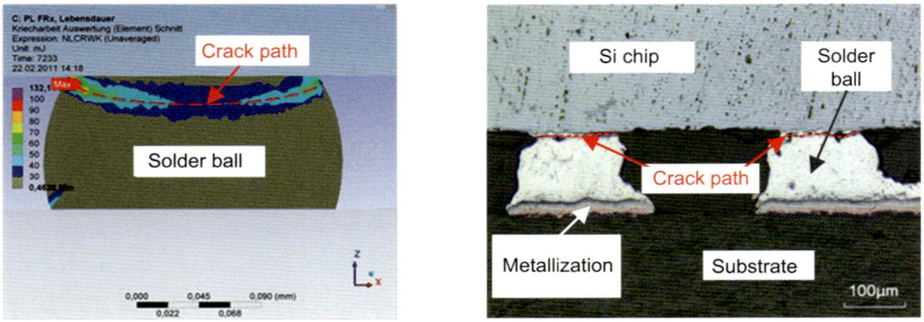

FIGURE 6.2 Left: FEM simulation of crack paths in a solder ball after temperature shock loading; right: comparison with experimental characterization [191]

lyzed in the FEM program ANSYS (from ANSYS Inc.). Figure 6.2 shows a sample result from simulation and the practical experiments.

Upcoming simulation-assisted quality assurance will broaden its focus to include model extensions with other load types, the final aim being to examine nearly real load collectives. The findings obtained in this way will be applied to complex substrate, metallization, and join-point geometries [152]. The ultimate objective will be to combine the various approaches to arrive at viable mapping and boost the user friendliness of the software tools. This will be an essential prerequisite for high market acceptance among MID manufacturers.

6.3 Nondestructive Testing Methods

The variety of MID material combinations and production processes is reflected in the methods of analysis and testing used to monitor defined product features. In addition to the methods of plastics testing explained in Section 2.2, this chapter deals with the specifics of the procedures commonly used to characterize assembly and connection technology and the adaptations necessary to bring these processes fully into line with MID technology. The main focus is on techniques used in mechanical or geometric and in electrical metrology. As far as in-production quality assurance is concerned, nondestructive testing methods are becoming more and more important. They help lay the basis for zero-fault production.

6.3.1 Optical Testing and Inspection Methods

Optical component characterization is extremely versatile and the routines are noncontacting, so the influence of optical inspection is growing. High-end optical systems are becoming more and more widespread in MID production for monitoring geometric variables. Laser triangulation, light sectioning, and white-light or laser-light interferometry are among the more common methods. They can be used to scan a large set of reference points on the object in a very short time so that a whole array of properties can be checked simultaneously [80]. In the MID world these techniques are used primarily to check:

- shrinkage, curvature, and distortion (e.g., of the interconnect device after metallization or soldering),
- planarity and roughness of the function surfaces and metallization (e.g., for bonding),
- conductor track geometry and damage and overmetallization (e.g., in chemical coating),
- shaping of solder/adhesive joints, and
- positioning and orientation of electronic components.

MID designers have to consider the possibility of shadowing effects that could obstruct metrological processes of this nature. Many of these optical methods, moreover, are based on the principle of analyzing the light reflected by the object. The results are heavily dependent on the reflectiveness of the object, so the broad range of materials for MID and the resulting variation in surface properties can be very challenging for optical metrology. Transparent materials of the type sometimes encountered as MID substrates render it impossible to use optical measuring systems for precision measurement. But even if the substrate is opaque, the depth to which the beam of a laser triangulation sensor can penetrate into the material varies as a function of the wavelength used, so the figures obtained for a surface-roughness profile, say, might be of no more than conditional reliability. High-resolution confocal laser microscopy has advantages in this respect, but as a metrological method it is also much slower. As Fig. 6.3 shows, shutters prevent these microscopes from detecting measuring points outside the focal plane, so it becomes possible to take high-precision measurements on transparent surfaces. An alternative approach is to use multisensor coordinate measuring instruments. An instrument like this can measure the object optically, tactilely, and by X-ray examination, which means that the system of sensors can be tailored neatly to the specifics of MID inspection. [168, 169]

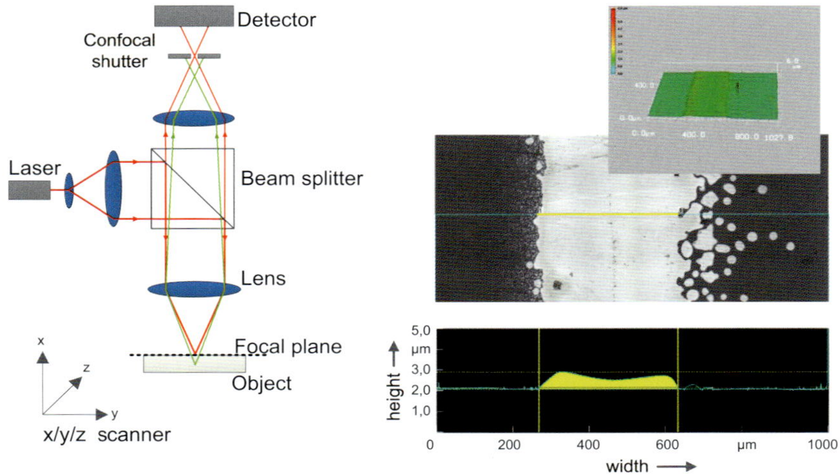

FIGURE 6.3 Left: operating principle of a confocal laser scanning microscope as in [62]; right: measurement of printed silver coatings less than 1 µm thick on glass/polycarbonate surfaces

6.3.2 Automated Optical Inspection

In recent years automated optical inspection (AOI)–integrated into the production chain–has become an established part of printed-circuit board production in the electronics industry. A 100% optical inspection is used to regulate individual production steps and to ensure product quality. In MID production too there are now high-performance AOI solutions available for detecting

- geometric performance indicators for mechanical features on the MID (edges, holes, etc.),
- metallization defects (misplaced deposition, shorts, cracks, etc.),
- electronic components (presence, position, polarity, etc.), and
- shaping of the contact points (geometry, cold solder points, solder outflow along the conductor, etc.).

A camera system produces a multiplicity of single images, generally from several positions and with different light settings. These images are then processed and combined by high-performance image-processing software. In this way the measured values can be compared with specifications and the images from a master module. User-defined limits can be applied to grade each specimen as “OK” or “for reworking” or “scrap”. Individual process parameters such as component offset can be evaluated and used to control the processes [10]. As shown in Fig. 6.4, left, these automated optical inspection systems have to be adapted to the challenges peculiar

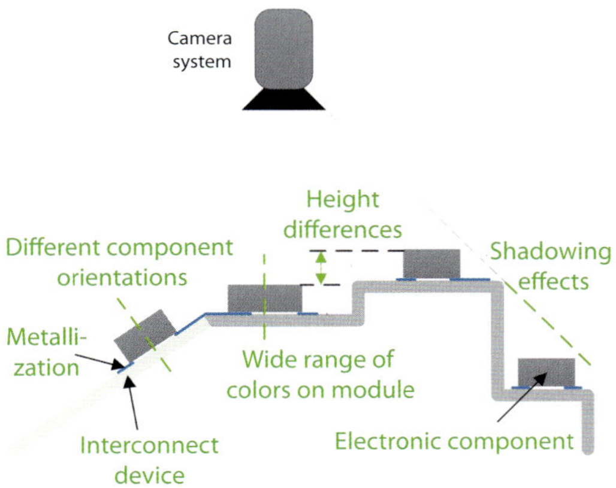

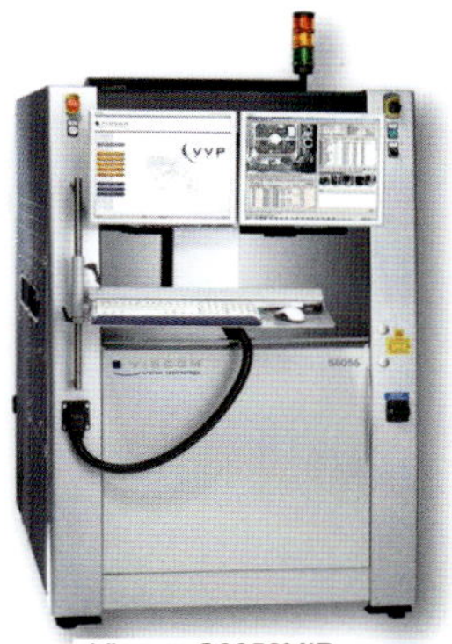

FIGURE 6.4 Left: challenges for the automation of optical inspection for MID; right: AOI system from the Viscom company [10]

to three-dimensional interconnect devices with electronic components set in 3D layouts. There are, broadly speaking, two approaches. Stand-alone AOI machines from Viscom, as shown in Fig. 6.4, right, have a z axis that standard 2D systems lack so that the sensors can be moved to different places for imaging 3D-MID. This approach utilizes orthogonal and tilted cameras to ensure all-around imaging of the MID object [10]. Solutions developed by the Häcker and Xenon companies, on the other hand, are based on integration into the modular placement systems. Much like the pick & place system, the camera and lighting unit can be freely oriented in space relative to the MID. They check the MID sequentially, area by area. [5, 154]

Another MID specific besides geometry that should be mentioned in this context is that there can be a great deal of variation in the range of colors and the optical behavior of the substrate plastics and the conductor layout. Consequently, the software and the camera system for AOI need image correction capabilities, and users have to take these discrepancies into account when configuring the limits for production tolerances.

6.3.3 X-ray Analysis

In this process X-rays are generated, exit the X-ray tube in a directional beam, and incide with a detector. The detector, which can be either an array or a matrix, converts the X-rays into a gray-scale image that can be analyzed by image processing software running on the computer. When an object is inserted between the X-ray tube and the detector, the X-rays reaching the detector are dissipated. By contrast with classic optical techniques, the rays pass through the object. The degree of dissipation depends on the type (atomic number) and the thickness of the mate-

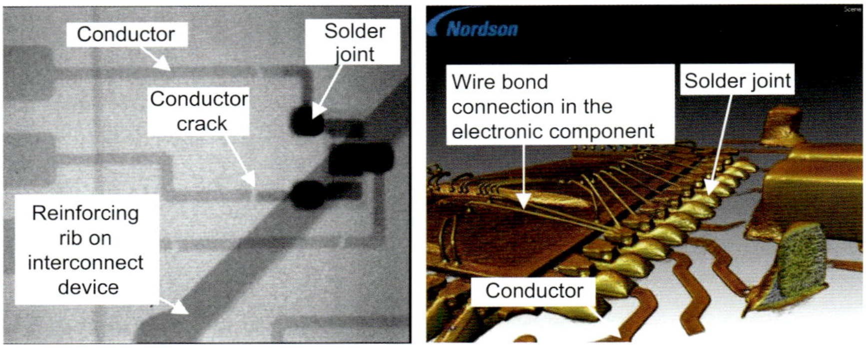

FIGURE 6.5 Left: detection of conductor breaks on a fully encapsulated MID; right: computerized tomographic image of a module detail (graphics courtesy of Nordson Dage company presentation)

rial in the direction of illumination. The areas of use for X-ray analysis this opens up are primarily

- position and orientation of electronic devices,
- geometric analysis of the connection (solder, conductive adhesive, wire bond),
- detection of cracks and voids,
- metallization defects (misplaced deposition, shorts, cracks, etc.).

Figure 6.5, left, shows a typical X-ray image of an MID by way of example. Suitable control and systems technology can extend X-ray processing to automated X-ray inspection (AXI), which can also be combined with automated optical inspection (AOI).

The possibilities for analysis are closely linked to the levels of contrast in the image. For example, it is very difficult to examine thin conductors on the surface of encapsulated metal parts (leadframes) by X-ray. To simplify challenges of this nature some X-ray systems incorporate a swivel unit so that the object can be observed from another defined angle.

6.3.4 Computerized Tomography

Computerized tomography is an evolved method of X-ray analysis with imaging from a number of different viewing angles. Even concealed structures can be visualized in 3D, as the example in Fig. 6.5, right, shows. This frequently simplifies analysis of complicated failures that would otherwise necessitate the complex and time-consuming preparation of micrographic sections. Unlike standard CT practice in medicine, the processes used in electronics production generally image the object as it is turned, with several hundred individual images being generated. This data is then used to create a volumetric reconstruction of the object. The computer accomplishes this by converting the pixels of the individual images into volumetric pixel elements, voxels for short. In practice this entails a certain loss of resolution. Complete tomographic examination, moreover, takes considerably more time than 2D X-ray analysis, so it is only in exceptional cases that this process can be considered suitable for 100% checking. Partial tomography processes are therefore being developed for use in production lines. The trade-off for restricted 3D analysis is a considerable reduction in the number of images and therefore in the time necessary for measurement. [13]

6.3.5 X-ray Fluorescence Analysis

In X-ray fluorescence analysis the object is exposed to high-energy X-ray radiation. This excites the atoms in the object, and each material emits a characteristic secondary X-ray radiation. The wavelength spectrum and the intensity of the emitted radiation are collected by a detector and analyzed by special analyzer software. The process is used in MID technology to measure the thickness of surface plating. One of the benefits of a machine like the X-RAY-Fischerscope (developed by the Helmut Fischer company) for example is the ability to measure the thicknesses of several superimposed metal coatings in accordance with DIN EN ISO 3497 by nondestructive means in a single operation. Uses include testing the Cu-Ni-Au plating laid down

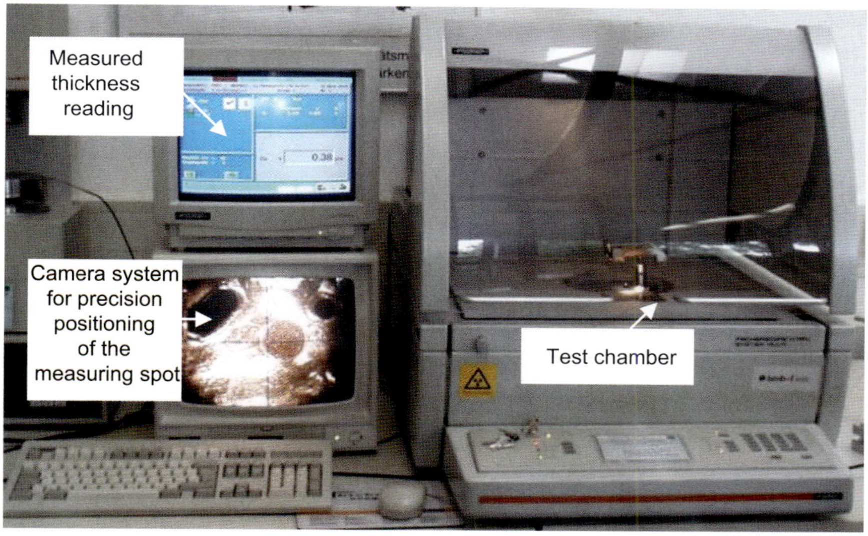

FIGURE 6.6 X-ray fluorescence measuring station with an X-RAY-Fischerscope

when LDS structures are metallized. The smallest possible dot for measurement is only approximately 0.1 mm × 0.1 mm, so this method is also suitable for very fine MID structures. Five measurements are usually taken per measuring point in order to enhance accuracy. Measuring time is about 20 seconds, making this one of the fastest and most accurate methods of measuring the thickness of the metallization.

6.4 Destructive Testing Methods

It can often be practical to supplement the methods described above with destructive testing. This can afford valuable leads under laboratory conditions in particular, so that the causes of failure in the field can be analyzed in detail. In series production too, these processes are used with random sampling as a way of monitoring production parameter settings.

6.4.1 Adhesive Strength

The adhesion of the metallization to the thermoplastic substrate is a key performance indicator in MID technology. The strength of the bond between metallization and body must be sufficient to ensure the mechanical security of the conductors and the electronic components on the carrier, both during production and throughout subsequent use in the field. The metal-and-plastic construct must be able to withstand mechanical and thermo-mechanical loads. Determining the adhesive strength of the bond between the thin layers of metal and the plastic substrates remains a considerable challenge, however. On the one hand there is the practicality of testing, because the three-dimensional layout and the structure sizes involved hamper the reproducibility of adhesion tests. On the other hand there is a paucity of information on analysis of the results obtained by measurement, for example the requisite minimum values for adhesive strength. It is for this reason that, as indicated in Section 6.1, resort is often made to existing standards such as those for printed-circuit board engineering.

6.4.1.1 Peel Test

This is a common test for printed-circuit boards (DIN EN 60249). The strength of the bond is measured by peeling conductors vertically off the substrate material. Figure 6.7 shows this method of testing in schematic form. A means of attachment (eye, clip) has to be affixed for the pull-off hook by soldering, gluing, or clamping, and a steadily increasing tensile force is applied orthogonally to the conductor until

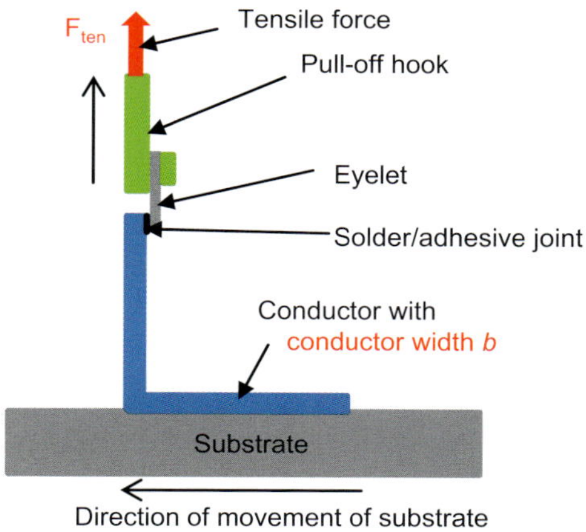

FIGURE 6.7 Schematic representation of the peel test

the conductor peels off at a constant 50 mm/min. The peel-off force is defined as the lowest force measured per conductor. Analysis is on the basis of peel resistance, deriving from the quotient of measured force and conductor width.

When this method is applied to MID it is important to bear in mind that a comparatively large test metallization is needed. Empirical experience indicates that connectors for peel-off testing should be more than 800 µm in width. The metallization should be at least 17 µm thick in order to withstand the necessary tensile forces. The common metallizations such as LPKF-LDS® plating therefore often require galvanic overplating to bring them up to the necessary thickness. The thickness of the metallization influences the result, and this has to be taken into account. Deformation of the metal layer factors into the measurement result. Consequently, the result cannot be applied one-to-one to combinations of materials other than that tested.

6.4.1.2 Pull-Off Test

The principle of the EN ISO 4624:2003 pull-off test requires a dolly to be glued or soldered to the composite panel. An increasing pull-off force is then applied to the dolly until the coating detaches from the substrate. The adhesive strength of the bond is calculated from the maximum force applied and the size of the detached area. If failure occurs elsewhere than between substrate and metallization, it is essential to add a reference to the fracture pattern as a supplement to the measured value.

Porous coatings such as those that prevail in MID technology when conductors are applied by thermal spraying, for example, can allow the adhesive applied to fix the dolly to the sample to penetrate through the coating to the substrate and falsify the result of measurement. It is also necessary to ensure that the substrate is adequately secured. A twist test is an alternative if compliance with this require-

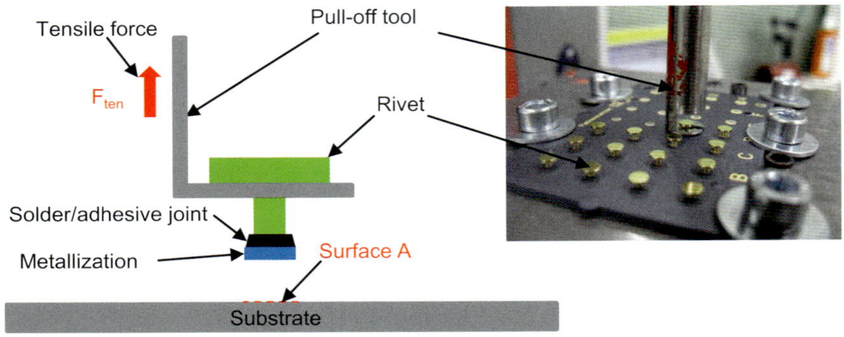

FIGURE 6.8 Adapted pull-off test as in [175]

ment is difficult or impossible. Instead of the dolly being pulled off normal to the surface, the measured variable is the torsional moment applied about the axis of the dolly glued to the panel and required to break the bond [160]. In yet another version of the pull-off test as described in [175], a rivet-like dolly is secured to the metallized face and then pulled off perpendicularly to the substrate by a special puller tool (Fig. 6.8). The rivets can also be used to perform shear tests.

6.4.1.3 Shear Pull Test

Pull-off testing and peel testing both measure forces acting at right angles to the surface. Permissible forces parallel to the surface also have to be measured for comprehensive analysis of adhesive strength. As Fig. 6.9 shows, in a test of this nature adhesive strength is measured by gluing a body to the metallization and applying a force parallel to the surface until failure occurs. Shear strength is calculated from tensile force applied and size of shear failure surface. Restricted accessibility, however, means that this method is little used for near-production MID.

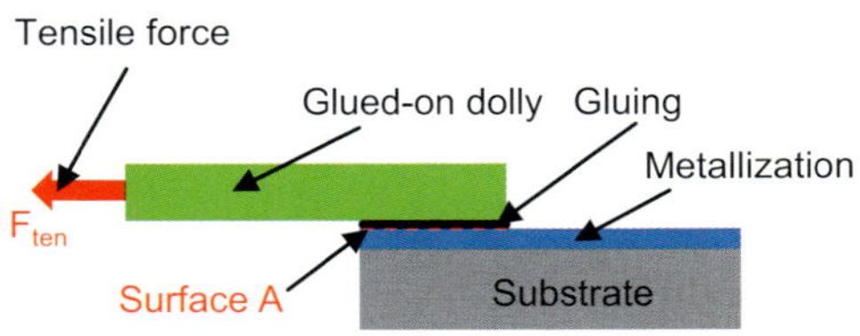

FIGURE 6.9 Schematic view of a shear pull test configuration

6.4.1.4 Chisel Test

This innovative process looks promising as a means of measuring adhesive strength on MID because it is eminently suitable for very thin coatings. A microchisel is used to scan the topography of the conductor. The peeling chisel is then set to the

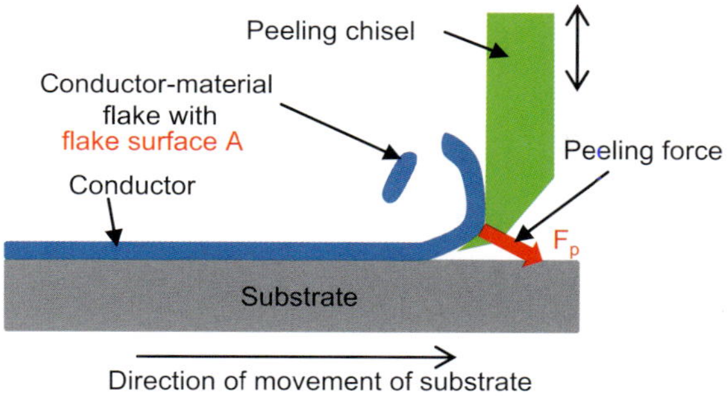

FIGURE 6.10 Sketch illustrating the principle of the chisel test as in [193]

appropriate height above the substrate. The chisel then traces the topography from the scan and the cutting edge scrapes off the conductor. The conductor breaks off in small flakes, as shown in Fig. 6.10. The forces at the chisel are measured and adhesive strength is calculated with reference to flake surface area, approximated from flake length and conductor width. [193]

6.4.1.5 Cross-Cut Test (Tape Test)

This is a qualitative test for quickly estimating adhesive strength, mostly of thin coatings on the substrate material. Six parallel cuts are made through the coating down to the surface of the substrate as shown in Fig. 6.11 and are then intersected at right angles by another six cuts.

The pitch between the cuts is defined by thickness of coating and type of substrate, as set out in Table 6.1.

Once the cuts have been made the grid is brushed lightly several times. The results are compared with the reference values in Table 6.2. The flaked area is analyzed and the pattern assigned to a performance indicator. The cross-cut test is purely qualitative, so no precise conclusions can be drawn about the adhesive strength of the coating.

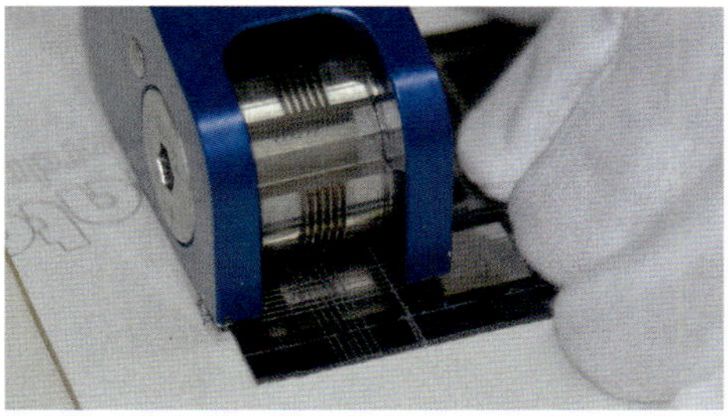

FIGURE 6.11 Cutting tool for producing the grid of cuts for the cross-cut test

TABLE 6.1 Standardized Number of Cuts as a Function of Coating Thickness

Standard	Coating thickness [µm]	No. of cuts × pitch [mm]
DIN EN ISO 2409	0 to 60	6 × 1
	60 to 120	6 × 2
	120 to 250	6 × 3
ASTM D 3359-02	0 to 50	11 × 1
	50 to 125	6 × 2

TABLE 6.2 Performance Indicators for the DIN EN ISO 2409 Cross-Cut Test

Description	Surface	ISO performance indicator	ASTM performance indicator
Cut edges are perfectly smooth. No flaking whatsoever.	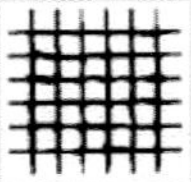	GT 0	5B
Small splinters of coating have flaked off at the cut intersections; flaked area about 5% of subsections.	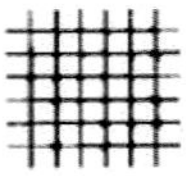	GT 1	4B
The coating has flaked along the cut edges and/or at the intersections of the cuts; flaked area about 15% of subsections.	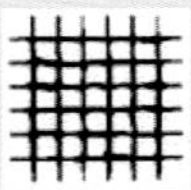	GT 2	3B
The coating has flaked off along the cut edges partly or completely in wide strips and/or the coating has flaked off completely or partly from individual subsections; flaked area about 35% of subsections.	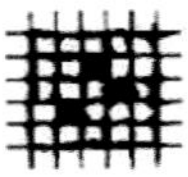	GT 3	2B
The coating has flaked along the cut edges in wide strips and/or completely or partly from individual subsections; flaked area about 65% of subsections.	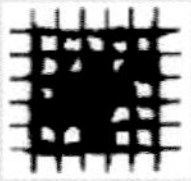	GT 4	1B
Any flaking not classifiable as GT 4.		GT 5	0B

An enhancement to the cross-cut test is to apply adhesive tape to the grid. The tape is then pulled off by hand in a smooth action (preferably at an angle of 90°) and the grid is then assessed as per Table 6.2. In order to maximize the comparability of measurement, the results should always be stated together with the adhesive strength of the adhesive tape used and if possible the angle at which the tape is pulled off.

6.4.2 Measurement of Shear Force and Pull Test

The mechanical strength of soldered or glued bonds is determined indirectly by measuring the strength of components on the metallization. A chisel is used to shear off components parallel to the surface of the conductor. The necessary force is measured and logged. As regards applicability to three-dimensional interconnect devices, adequate and reproducible location of the substrate in the measuring instrument is very important. Otherwise, depending on the geometry of the interconnect device and the clamping points, there is every likelihood that the force applied by the shearing chisel will twist the substrate out of position. If this happens the force measured is not shear force alone but shear force with a proportional peeling force. There is also a possibility of the chisel contacting the surface of the substrate as it carries out its shearing motion. Once the component has been sheared off, the fracture face is analyzed (fracture in the soldered joint, in the metallization, between metallization and substrate, or directly in the component). The result is very helpful for uncovering hidden shortcomings in production. Microscopic inspection focuses primarily on intercrystalline structures such as grain size distributions, phase growth, corrosion, and changes in the metallic structure. This data too can frequently be used to draw conclusions about causes and mechanisms of measured strength. [121]

As described in Section 6.4.1.2, it is also possible to use shear testing to determine the adhesive strength of the metallization on MID [175]. An alternative to shear testing is to use a hooked chisel for pull-testing the legs of components and wire bonds. This can often be accomplished using the same system technology.

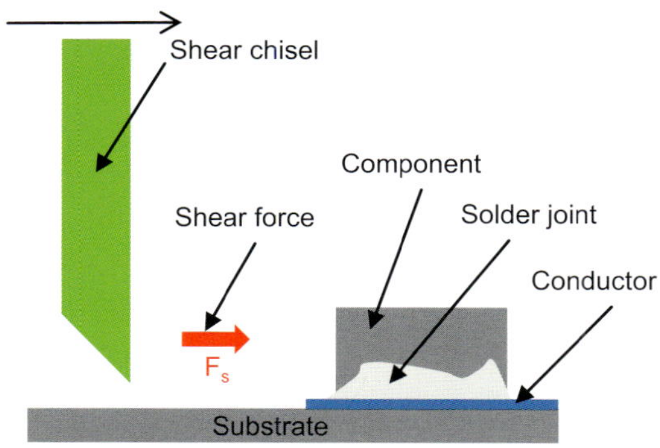

FIGURE 6.12 Sketch illustrating the principle of shear force measurement

6.4.3 Analysis of Prepared Sections

Sections like those often used in metallography can be prepared for in-depth analysis of the assembly and connection technology. This method of destructive testing entails the removal of coupons, by means such as sawing, from selected parts of the MID. The cut faces are then cleaned and, if necessary, deburred. The coupon is then set in liquid resin. When the resin has cured a section is prepared by grinding and polishing and can then be examined under the microscope. The techniques of preparation are described in detail in [119], which also includes in-depth information on the analysis of metal specimens in particular. Oblique sections can be prepared instead of cross-sections for the detailed analysis of very thin layers. The plane sectioning of the coupon at a defined angle broadens the layers for viewing under the microscope [56].

Microscopic inspection of prepared sections can cover a wider analytical spectrum:

- formation of crystalline and amorphous areas in the plastic
- filler distribution and orientation in the substrate
- faults at the boundary between substrate and metallization (blowholes)
- layering of the metallization and thicknesses of the individual layers
- metallization in vias
- height and shape of embossing beads after hot embossing
- faults such as blowholes, cracks, etc., in the connection medium
- filler distribution in electrically conductive adhesives
- intermetallic phases in solder joints
- influence of embossing beads on the position of electronic components
- stand-off between component and substrate at concealed connection structures

By way of example, Fig. 6.13 shows prepared MID microsections. A great deal of work is involved in this method of testing, because embedding and grinding are generally manual or only partly automated at best and only a very small part of the test specimen can be examined, but on the upside even ultra-fine structures can be studied under suitable high-resolution microscopes. Prepared sections can also be subjected to processes from materials analysis, such as energy-dispersive X-ray spectroscopy (EDX), in order to determine the distribution of individual elements in the solder joint.

Some requirements applicable to preparation are general, including for example the need to minimize mechanical and thermal loading when removing the coupons, but there are also certain specifics that have to be borne in mind with regard to MID microsections. The substrates are thermoplastics, so when specimens are prepared the curing temperature of the embedding material has to be taken into

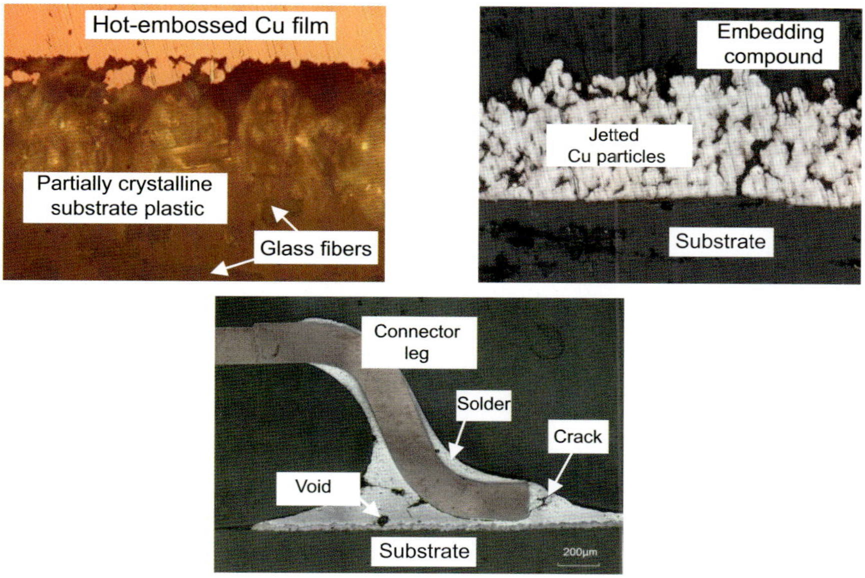

FIGURE 6.13 Examples of prepared MID sections:
top left: hot-embossed conductor on a partly crystalline thermoplastic
top right: porosity and roughness of a jetted metallization
bottom: cracks and voids in a solder joint after accelerated thermal aging

account in order to avoid damage to the coupons and shrinkage cracking between coupon and embedding resin. The plastic of the injection-molded interconnect device is not as hard as other materials, so it has a tendency to grind away more quickly and irregularly. This can result in localized rippling. Fillers such as glass fibers or ceramic additives in the substrate can detach from the plastic matrix and cause grinding tracks and scores. When studying samples under the microscope, therefore, it is a good idea to make sure that the lens has sufficient depth of focus. Another possibility is to image the topography in order to rule out the possibility of misinterpretation.

6.5 Electrical Characterization

To a very large extent, procedures from printed-circuit board technology can be applied for characterizing the electrical properties of assembled MID. An overview of the mainstream analog and digital methods of measurement is given in [25], for example. Once again, however, provision has to be made for some specifics of practical application of these methods to MID.

6.5.1 Resistance

Measurement of ohmic resistance is a basic test method in many sectors of electronics production as a means of checking:

- transition resistance at points of contact to the periphery,
- the quality of metallization on the basis of conductor resistances,
- the electrical quality of connections between metallization and mounted electronic components, and
- the operation of electronic components after working by MID-adapted assembly and connection technology.

The most frequently used methods are two-point and four-point measurement. These two methods are compared in Fig. 6.14.

Four-point measurement makes for higher precision. By contrast with two-point measurement, this method separates power supply and resistance measurement by the voltage drop. With an imposed constant current flow, the four-conductor arrangement allows voltage drop to be measured virtually without current and so to a very large extent independently of contact resistance and line resistance.

In both methods the sample is contacted manually or automatically by fine metallized probes. The probes can be firmly set at the appropriate positions in an adapter, or else guided and positioned individually in a flying probe test. An important point to be kept in mind when testing metallizations on MID is that care is necessary to prevent sharp tips of probes from damaging the surface of the thin conductors on the soft plastic substrates. Using spring-mounted probes is one way of avoiding this risk.

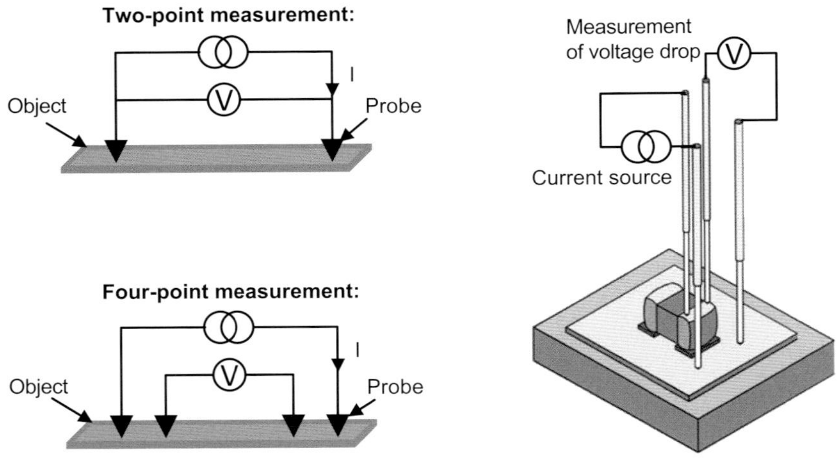

FIGURE 6.14 Left: schematic view of two-point and four-point measurement; right: four-point measurement using spring-loaded probes on an electric component

If possible, the design of the MID should make provision for extra connecting pads for electrical measurements.

6.5.2 Ohmic Heating

Current flowing through electrical conductors causes them to heat up on account of their ohmic resistance. This heat rise can damage the plastic substrate of an MID and cause detachment or destruction of the conductor. The current-carrying capability of printed-circuit boards is set out in the IPC-2221A standard. The values set out in the standard are based on the material characteristics of copper-plated standard wiring boards, so application of the standard to MID technology is viable only subject to severe restrictions, as explained in Section 3.2.4. In the production of conductor tracks, specific conductivity varies depending on the process of metallization used. Furthermore, the calculations in the standard are based on a rectangular conductor cross-section, a shape that is not necessarily ensured by some MID processes. Consequently, it is always best to use suitable test specimens and ascertain ohmic heating separately for each combination of materials and process. The conductors on the MID are electrically contacted and connected to a current source. The contact point must evince a considerably lower electrical resistance than the conductor itself in order to minimize the effects of heating at the point of contact. Using the configuration shown in Fig. 3.32, heat rise in the conductor is measured as a function of applied current and recorded. Figure 6.15 is an example of how ohmic heating graphs on a curve. Using thermographic measuring systems to register temperature makes it possible to monitor the entire conductor and detect hotspots occurring as a result of production defects, for example.

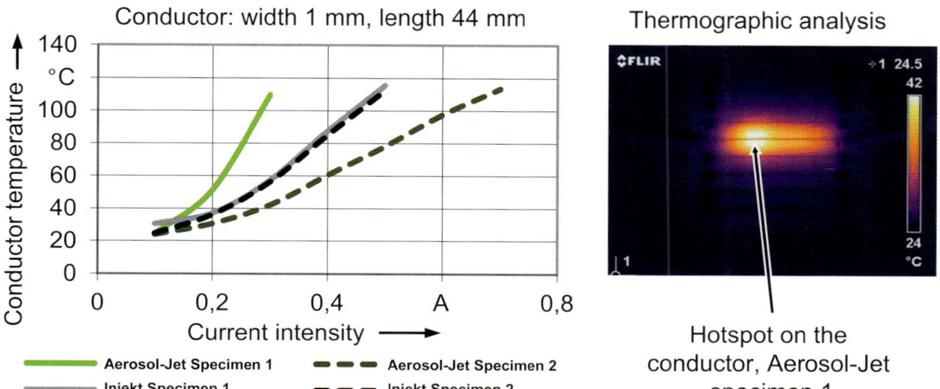

FIGURE 6.15 Thermographic determination of ohmic heating in printed conductors on injection moldings [195]

Depending on the application (e.g., plug connector), derating curves can also be useful in MID evaluation. These curves graph maximum current-carrying capability up to a defined upper temperature limit as a function of ambient temperature.

6.5.3 Insulating Properties

Standardized processes in materials testing are in place for determining the interconnect device material's electrical characteristics such as volume resistance or surface resistance. These tests are set out in DIN EN 61340, for example. Further tests, also used in printed board technology, are viable for ensuring MID compliance with electrical specifications. One example is measurement of insulation characteristics of applied comb structures by analogy with IPC-B-25A. The influences of numerous factors on surface resistance can be analyzed in this way. In addition to purely geometric characteristics, the moisture content in the substrate has a crucial part to play in this respect. Depending on the type of plastic, surface resistance can vary by two powers of ten or more. Measuring comb structures during climate tests can uncover possible weaknesses in the applied electronic circuitry. Contaminants left behind by metallization or soldering processes can also affect the insulating properties of MID. They too can be detected in this way. [78, 160]

6.6 Reliability Analysis

In order to ensure lifetime product quality, MID are reliability-tested by subjection to stress factors replicating the conditions prevailing in subsequent use. Stress factors applicable for a wide range of MID applications include

- thermal loading (high/low temperatures, temperature gradients),
- climatic influences (moisture, condensation),
- mechanical loads (vibration, shock),
- chemical/biological factors (harmful gases, liquids).

The effects of these influences can be checked in adapted environmental tests and the results used to select the suitable MID manufacturing process and the materials, and for optimizing and monitoring production parameters.

6.6.1 MID-Specific Challenges

Special attention has to be paid to the properties specific to thermoplastic substrates in environmental testing.

Coefficient of Thermal Expansion (CTE)

Thermoplastics in general have high CTEs, and MID reliability can suffer as a result. Temperature fluctuations in use and also in production induce mechanical stresses and strains between plastic substrate and metallization plating. The high CTEs of the plastic substrates have a particularly critical effect on the dependability of the solder/adhesive joints of the assembled MID. Electronic components in general and ceramic components in particular have low coefficients of thermal expansion, so there is a significant thermal mismatch between the joining partners. Under thermal cycling this can soon lead to fatigue breakages in the joints. If the thermoplastics are partially crystalline the glass transition temperature has to be taken into account along with the CTE. Above this temperature the coefficient of thermal expansion is generally much higher than the value at room temperature as stated on the data sheet. [34]

Moisture Absorption

The proportion of water in the polymer matrix can be as high as several percent, depending on ambient temperature and type of plastic. This has an effect on the electrical properties of the plastic and also on mechanical properties such as the modulus of elasticity E. Consequently, the interconnect device has to be dried to prevent evaporation of this entrained moisture in the reflow soldering process. This precaution helps ensure uniform quality of the modules. Moisture absorption in use, moreover, causes volume expansion of the substrate. This in turn can create mechanical stresses in the metallization and at contact points to electronic components mounted on the device.

Fibrous Fillers and Directionality of Material Properties

Fillers can improve the mechanical properties and also the thermal expansion of thermoplastics. The use of glass fibers has proved helpful in many MID applications, not least on account of economic considerations. However, it is important to bear in mind that after being admixed, the fibrous fillers produce anisotropic material characteristics in the substrate in injection molding. These characteristics subsequently have to be taken into consideration in mechanical testing and also for testing under conditions of rapid temperature change. Anisotropic filler orientation is all the more significant if the interconnect device is severely three-dimensional on account of the more complex filling of the injection-molding tool cavity.

Weld Lines in the Injection-Molded Substrate

Depending on geometry, it might well be necessary to divide the melt flow in the injection molding tool in order to produce the three-dimensional interconnect device. Weld lines form if the divided melt flows meet up. Material properties in this zone of the interconnect device are often considerably altered. This is due to a number of reasons, including marked orientation of the polymer chains and the fillers at the points where the melt flows come together, or to inadequate welding [33]. A variation in the way the substrate expands when subjected to heat can be expected in this zone. Empirical evidence shows that breakages are often inevitable if conductors track cross weld lines.

Three-Dimensional Arrangement of the Electronics

Vis-à-vis flat printed boards, the three-dimensional arrangement of the applied conductors and the electronic components grants increased significance to other failure mechanisms that have to be taken into account for analyzing dependability on the basis of reference coupons. Depending on the three-dimensional orientation of the electronic devices, it is easy to see that under mechanical loading forces can act on the connection points from different directions, producing varying load cases. Less obviously, the shear strains induced between plastic and metallization by temperature fluctuations, for example, are transformed into compressive stresses at edges and radii. These forces are no longer acting tangentially to the surface and this contributes to detachment of the applied conductors, as Fig. 6.16 shows.

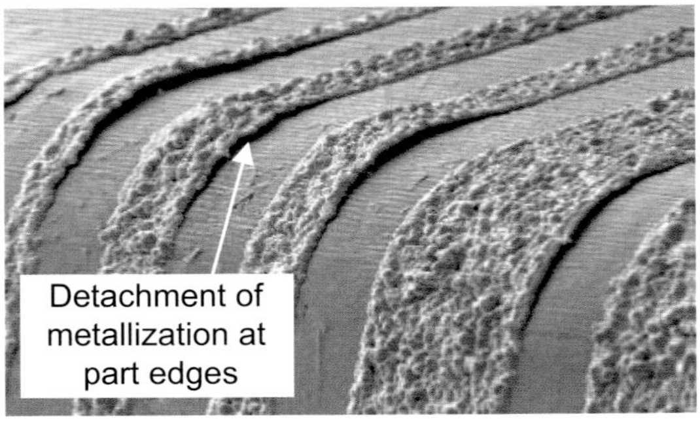

FIGURE 6.16 Detachment of LDS metallization at the component edge after accelerated aging tests (graphics courtesy of HSG-IMAT)

6.6.2 Accelerated Aging

Environment tests and accelerated aging tests on representative test modules are used in practice to estimate long-term reliability. The reliability studies undertaken for MID are often modeled on procedures from electronics production. The test methods described below are in widespread use for basic qualification.

(Rapid) Temperature Cycling Test

This test focuses primarily on influences resulting from the differences in thermal expansion of the joining partners, so given the properties described above it is highly relevant to MID. In accordance with the applicable standard (e.g., DIN EN 60068), high and low conditioning temperatures are defined along with dwell times and transition times. Temperature change is a slow process, whereas a rapid temperature cycling test or temperature shock test compliant with DIN EN 60068-2-14, for example, stipulates short transition times. These times can be less than 10 seconds, depending on the degree of severity. To accomplish these quick changes in the test configuration, the specimens are transitioned between two test chambers, one cold and one warm. After each transition they remain in the appropriate chamber for the selected dwell time, which depends primarily on the thermal capacity of the specimens. This cycle is repeated a number of times. The test focuses primarily on the adhesion properties of the metallization, the change in conductivity of the substrate metallization (due to microcracking, for example), and the electrical and mechanical properties of the connections between conductors and electronic components.

Humidity Heat Test

A test specimen is subjected to elevated temperature and relative humidity over a certain period of time. The parameters frequently used in practice for MID have temperature set to +85 °C and 85% relative humidity for this test. The thermoplastic substrates absorb water under these conditions, so the test enables insulating properties to be checked along with corrosion of the plating. Humidity heat testing is frequently followed up by an extra test of the metallization's adhesion.

Vibration/Shock Test

MID are subjected to different mechanical loads in the course of their product life cycle. Automotive applications for MID are a case in point. Vibrations and mechanical shock loading are common, for example when a vehicle drives over potholes. These load profiles are mapped in vibration tests such as are specified in the ISO 16750 standard. The module is clamped on to a linear motor, known as the shaker, capable of accelerations several times that of the Earth's gravity. The scope of testing includes the adhesive strength of the metallization at the points of connection to (heavy) electronic components. When tests of this nature are conducted, it is important for the load profile and also the clamping of the MID on the shaker to match the specifics of mounting and installed position in the parent system. This is to ensure compliance with the real-life package requirements for the MID.

6.6.3 Application Example I: High-Temperature MID

This brief description outlines the interplay of the possibilities described above to determine MID reliability based on studies carried out in the framework of the HT-MID research project [190]. In the project, the production chain was first extended to include MID manufacture on the basis of hot-emboss high-temperature thermoplastics. PPS with a filler content (glass fiber and mineral) of 50% and PA 6T/6I with a filler content (glass fiber and mineral) of 65% were used as substrate materials for injection-molding flat-board and ribbed samples and a plug cage geometry, which were then metallized with a circuit layout by hot embossing. A test circuit was then assembled by reflow soldering. The test components were used to analyze long-term behavior of the selected combinations of materials and the corresponding production parameters. The objective in this project was to evaluate the possibilities for using hot-embossed MID under elevated long-term usage temperatures occurring in automotive applications. The test schedule with accelerated aging tests shown in Fig. 6.17 was drawn up by analogy with the requirements of the automotive sector.

In addition to the temperature-shock and humidity heat tests common in MID engineering, but with degrees of severity adapted to the automotive application, the project also included steady-state high-temperature tests at 125 °C and 150 °C. The samples were also subjected to mechanical loading in vibration tests, partly because the high proportions of fillers in the plastics were expected to cause reduced adhesion of the metallization. Supplementing the numerous tests for plastics char-

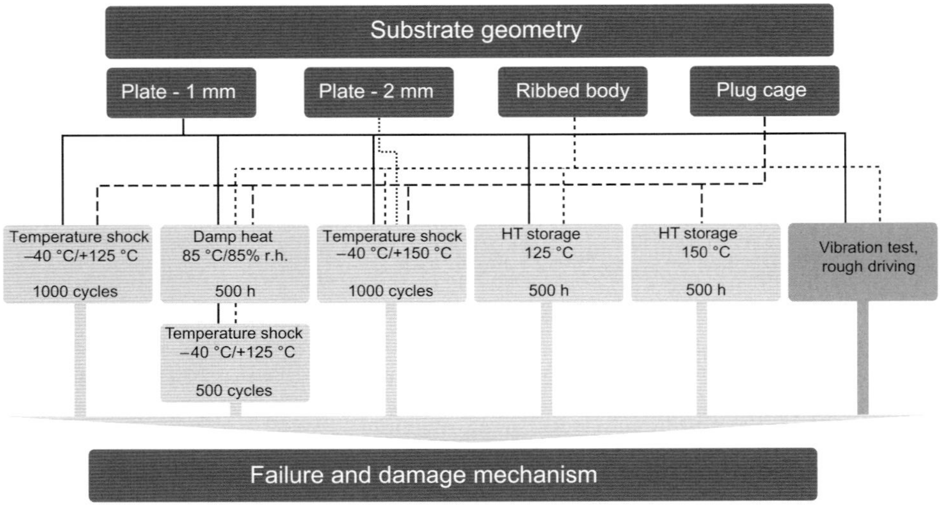

FIGURE 6.17 Test program in the framework of the HT-MID research project for qualification of high-temperature MID [190]

acterization, in order to check the chosen production parameters for assembly and connection technology and to provide a basis for estimating long-term durability of the circuits tested, the following test methods were used:

- preparation of sections and accompanying light-microscopic analyses to determine filler distribution and orientation in the injection molding
- creation of a model for simulating thermally induced stresses with the SIGMASOFT and ANSYS software tools
- measurement of the adhesive strength of hot-embossed conductors on the various substrate plastics by peel testing
- supplementary light-microscope observations of the fractures for analysis of the failure mechanisms
- measurement of shear force at CR 1206, CR 0805, and CR 0603 two-terminal chip resistors before and after the accelerated aging tests
- determination of resistance of daisy-chained chips by four-point measurement during the reliability test comparison of the resistance values obtained by measurement with the results from the shear test
- metallographic analysis of the aged solder joints with regard to delamination, cracking, voids, and growth of intermetallic phases
- computerized tomographic examination of packaged electronic components to determine exact geometric dimensions for mapping in the computer model
- use of optical metrology (laser triangulation, white-light interferometer, autofocus sensor) to check geometric variables of the test interconnect devices and to measure height of hot-embossing beads

These test methods showed that HT thermoplastics with high glass transition temperature such as PA 6T/6I, for example, are eminently suitable as substrate materials for automotive applications. Beneficial from the thermo-mechanical point of view, a high proportion of fillers reduces the adhesive strength of hot-embossed copper films. However, the studies revealed that with the embossing conditions suitably adjusted, materials with a high filler content return an adhesion figure of approximately 1 N/mm in the peel test. This figure is considered adequate for use in electronics. No detachment of the metallization was detected in vibration testing. The use of glass fibers and mineral fillers introduces directionality of the coefficient of thermal expansion. The thermal expansion of the components used is on the low side of 10 ppm/K, a figure achievable with these thermoplastics only if the components are oriented parallel to the flow direction of the plastic with the filler particles oriented in the flow direction across the entire thickness of the substrate. This directionality of module dependability was verified by temperature shock testing of the different sample geometries and on the basis of the accompanying simulation. Consequently, MID substrate geometries should have section thicknesses as thin as possible so that this orientation can be achieved. The tests with the chip resistors also showed that the ratio of cross-section to length of the component has a crucial effect on the module's long-term behavior. In the tests the large component (CR 1206) was invariably the first to fail, whereas the CR 0805 was the geometry exhibiting the best reliability in accelerated aging. This is on account of the component's better cross-section/length ratio. [190]

6.6.4 Application Example II: Press-In Connections

Another research project examined the reliability of press-in connections as a viable alternative connection technique for injection-molded interconnect devices. Unlike the project outlined above, in which the test program and selected methods were tailored to the object of examination, a test program set out in DIN EN 60352 Part 5 was definitive for qualification of the reliability of press-in pins. See Fig. 6.18. The test program is subdivided here into an assessment program for basic qualification of individual connections and an application program for checking connections that are part of a component/parent system.

As well as detailing the number of specimens for testing, the standard sets out permissible geometric, electrical, and mechanical limits for the specimens. Once again, however, from the point of view of MID technology these specifications proved impractical. The diameter specified by the standard for the holes for the pins in the interconnect device is a case in point. Adjustments therefore had to be made, allowing for the complex interactions in the MID production chain set out in Section 6.1. Changes to existing specifications frequently have to be made in order to qualify MID for a multiplicity of practical applications, so one objective of future research into MID technology will be to elaborate on recommendations and guidelines for a systematic procedure of change and to expand existing standards accordingly.

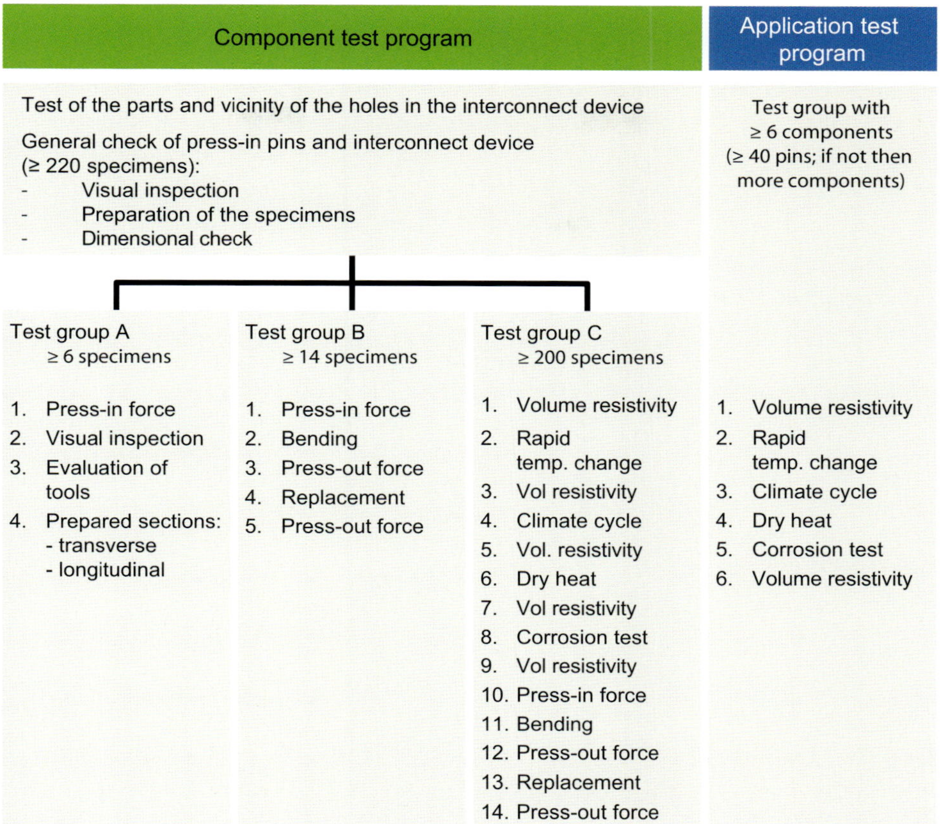

FIGURE 6.18 Test program for ascertaining the reliability of press-in pins to DIN EN 60352 – Part 5 (2001)

7 MID Prototyping

W. Eberhardt, P. Buckmüller, H. Kück

7.1 Classification of Samples and Prototypes

It is always a good idea to have samples or prototypes of a new product available as early as possible in the product development process. They are used for feasibility studies and assembly studies, to validate the intended choice of production technology, and for selection or decision-making processes, for example. The different stages of the product development process are frequently identified by category A, B, C, or D samples. Category A corresponds roughly to the concept models in this chapter, in other words the simplest models. Category B samples are frequently known as demonstrators or lab samples and in the current context are more or less the same as the fully functional samples. Category C samples are known here as prototypes and are generally used in the process of approvals testing for pre-series production. Category D samples correspond to pre-series models that are used for initial sampling by customers and customer validation as essential steps toward the issue of series-production approval.

As far as the production of these samples is concerned, the outlay involved in the lasering, metallization, and component placement processes for MID components is comparable to that for the manufacture of other electronic parts such as printed-circuit boards. The costs incurred for preparing the plastic blanks as the MID bodies in real-part quality, by contrast, are considerably higher because an injection-molding tool is necessary.

MID samples as plastic bodies produced by the familiar rapid prototyping techniques such as stereolithography, laser sintering, or fused deposition modeling can differ radically from their series-production counterparts. Sampling on the basis of rapid prototyping, therefore, generally ideal for assembly package studies, provides no basis whatsoever for qualifying the technology, in other words for judging reliability and durability. So with regard to costs and time outlay, it is important to specify precisely the intended purpose or, more accurately, the findings that testing of the sample or samples should afford. The various ways in which rapid prototyping samples can be used for practical purposes in the development of MID components are discussed and distinguished below. To this end, the rapid prototyping samples are distinguished by purpose:

- visualization samples
- partially functional concept models (A samples)
- fully functional samples (B samples)
- prototypes for pre-series production approval (C samples)

7.1.1 Visualization Samples

The concept model without any working functions is the simplest type of sample. It faithfully reproduces the geometry of the series-production part, but needs no electrical functionality whatsoever. These samples can be made using any of the well-known techniques of modeling, so costs and time to delivery can be minimized. Rapid prototyping of plastic parts made from any base material can be used. The conductor layout can be traced on by hand or printed. The components can also be placed by hand and simply glued into position, for example. A concept model in this form furnishes no direct conclusions about the functionality of the component or glitches in the development process. Mechanical and electrical properties differ widely from those of the later series-production item. Even so, this model is useful in many ways in the conceptualization phase of an MID component's development, including, for example:

- visualization of miniaturization and system integration,
- verification of the conductor pattern,
- verification of assembly,
- verification of feasibility for 3D lasering,
- verification of feasibility for 3D placement with the intended electronic components,
- visualization for those involved in decision-making processes.

7.1.2 Concept Model

The partially functional concept model is useful for verifying important subfunctions and properties of the MID component and does not necessarily have to show the entire geometry of the series-production MID. In medical technology, for example, materials compatibility can be an issue. A part that is identical in terms of material and production process or one that is adapted in simplified form can be used to answer these questions without the expense of manufacturing an injection-molding tool. In other applications, lead-free solderability or moisture/heat exposure can be questions of considerable moment. Here again an existing series-production plastic part made from the actual material and modified for the application can be produced using series-technology resources and evaluated under these specific terms of reference. In lighting applications, say, the issue of heat input can be clarified using a partially functional concept model made of the actual material intended for series production. Samples of this kind can also be used to explore issues concerning reliability, assuming that the critical areas of the series-production part can be reproduced with sufficient precision in the sample. MID manufacturers and the research institutes active in the field frequently have collections of injection-molded parts made of the mainstream materials that lend themselves to use as technology samples of this kind. Laser subtractive structuring techniques can be used to produce samples for clarifying questions relating to two-shot MID components. Sensors are typical of another class of applications in which the particular geometric arrangement of the sensor element with defined position and alignment is crucial in terms of functionality. In these circumstances it can be useful to have a sample based on a rapid-prototyping plastic part without all the electronics of the finished component that can be worked with the ProtoPaint LDS process patented by the LPKF company (Section 7.3.1). A special LDS lacquer is painted onto the part, which is then finished much like an LDS MID. The components can then be fixed in place and contacted in a manual placement process with low-temperature solder such as SnBi or with a conductive adhesive.

Samples from existing moldings made with the actual materials or from rapid prototyping parts can be produced quickly and at comparatively low cost. As these examples show, partially functional concept models can be used to answer a very wide range of questions in the early phases of MID development:

- verification of a subfunction (sensor function, current-carrying capability, heat rise, ...)
- verification of materials compatibility (biocompatibility, resistance, ...)
- assessment of the production technology (soldering, gluing, wire bonding, ...)
- estimation of reliability and durability (temperature/humidity storage, temperature cycles, ...)

7.1.3 Fully Functional Sample

Fully functional samples have to demonstrate the full functionality, geometry, and complexity of the series-production component. They are produced for the purposes of obtaining final approval for development of the MID.

They can be subject to certain restrictions compared to the series-production parts, however, for example with regard to the production process, operating conditions, mechanical strength, or production yield. So less importance attaches to their dimensional accuracy and thermal expansion properties, and any of several processes can be employed to produce these fully functional samples. Broadly speaking it is not necessary to use the actual material of the finished product, assuming that there are no elevated thermal requirements applicable to the fully functional samples. So rapid prototyping plastic parts can be used and structured with the ProtoPaint LDS process and the components placed with low-temperature solder. Clearly, though, fully functional samples made from the same materials as the finished parts will return more reliable information. One elegant approach is to use blanks molded from the actual material, mill them to the correct geometry, and then apply LDS (Section 7.3.4). The electronic components can then be placed in just the same way as in series production. One-shot plastic moldings can be produced using rapid tooling methods for the mold cavity, for example with laser-sintered steel tools and masters. Laser direct structured (LDS) or laser subtractive structured (LSS) samples of this kind can be used to produce fully functional samples of components destined to be manufactured as two-shot MID because producing one-shot moldings involves considerably less complexity than manufacturing two-shot parts.

Compared with visualization samples and concept models, fully functional models are much more useful because they provide information about the component's full functionality and the intended production processes for the actual part.

7.1.4 Prototype

Prototypes are needed in the quality-oriented development of MID components as a means of qualifying the production method and the new product and for achieving approval status for pre-series production. Product validation by the user in field tests, for example, can also call for prototypes. Consequently, there is no leeway for differences between the prototype and the series-production part as regards essential features and properties. Regarding the injection-molding process in particular, the prototype has to be made of the same material as the finished product. The molding tool used at the prototype stage is frequently of series-production quality but with only one cavity. Other cavities are added in the course of ramping up to the start of series production. The dictates of quality management require structuring, plating, and component placement to be identical with the series process. Prototypes are the samples that involve the highest costs and the biggest time outlay. On the other hand, prototypes can furnish dependable answers to all questions in the course of development of an MID part.

7.2 Processes for Producing Plastic Blanks

There are several processes for the production of plastic blanks for samples and prototypes. In addition to the outlay involved in the manufacture of the blanks, the processes described below differ primarily in the materials used and their properties.

7.2.1 Stereolithography

Stereolithography (SLA) is based on the principle of photopolymerization. The part is built up layer by layer in a light-curable plastic, known as a photopolymer, exposed to laser light of a defined wavelength. The 3D CAD data for the part are usually converted into the SLA format, and 2D slices are created. Supporting structures are added as necessary. The materials used are machine- and manufacturer-specific. The resins generally have properties very similar to those of PP, ABS, or PC. The part is produced in a bath filled with the basic monomers of the photosensitive plastic on an epoxy-resin basis. A computer-controlled UV laser builds up the layers by mapping the 2D slice data, causing localized hardening of the plastic material in the bath. The structure produced by this hardening process is then lowered into the bath by the distance equal to the selected layer thickness. Liquid plastic is uniformly spread over the hardened structure by a wiper and exposed in accordance with the

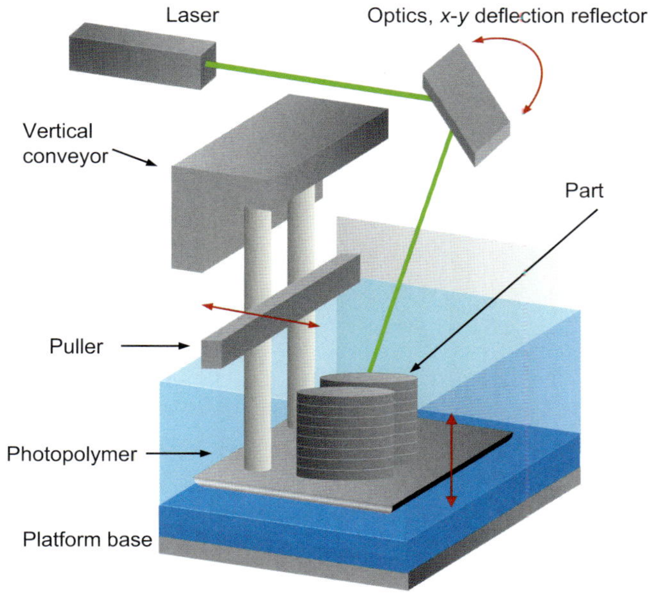

FIGURE 7.1 Schematic view of the stereolithography process [26a]

structure of the next 2D slice, forming the next layer. The process continues layer by layer until the structure is completed (Fig. 7.1). The parts are not self-supporting during the manufacturing process, so cantilevering and ultimately self-supporting elements have to be carried by supporting structures. The part is then removed from the bath and drip-dried before the remains of liquid plastic are removed by application of a solvent. The parts produced by laser-assisted stereolithography generally do not fully polymerize, so special UV ovens are needed for final curing.

The supporting structures are removed either before or after curing in the UV oven, depending on precision requirements and on the material. The supporting structures and the residual materials are removed manually by mechanical drilling, milling, sanding, or abrasive blasting. A good surface quality is then created by finishing with paints, fillers, and grouting compounds [61].

7.2.2 Selective Laser Sintering

Selective laser sintering (SLS) is a process developed by 3D Systems Inc. for localized additive layering with initially powder material melted by a laser beam. Comparable processes developed by other companies include laser sintering (LS) from EOS, selective laser melting (SLM) from Realizer GmbH, and laser cladding. Powdered plastic, metal, or ceramic materials can be used for laser sintering processes such as SLS if they can be softened or melted by heat and then return to the solid state when cool. The materials used are machine- and manufacturer-specific. Polyamides or materials with properties similar to PP or ABS are typically used to produce thermoplastic parts by SLS. The thermoplastic particles are usually 20 to 50 µm in size and roughly spherical in shape. The 3D CAD data for the part are usually converted into the SLA format, and 2D slices are created. The part is produced in a powder bed. A computer-controlled UV laser builds up the layers by mapping the 2D structures, causing sintering of the exposed layer of powder. After exposure and local sintering of a layer by suitable laser scanners, the powder bed complete with the growing part is lowered a distance equal to the thickness of one layer, typically 100 to 200 µm, and a new layer of powder is raked in from a container by a squeegee blade (Fig. 7.2).

Local consolidation of the material takes place, so in this way three-dimensional parts of any shape can be produced. No supporting structures are needed to produce the parts, because the unsintered powder remains in the bed and supports the part. At the end of the production process, the loose powder has to be removed from the parts produced in this way. The sintering process is pressureless, so density is less than that attainable in the processes in which the material is melted completely. Once the parts have been cleaned, the surface can be treated by sanding or by shot-blasting.

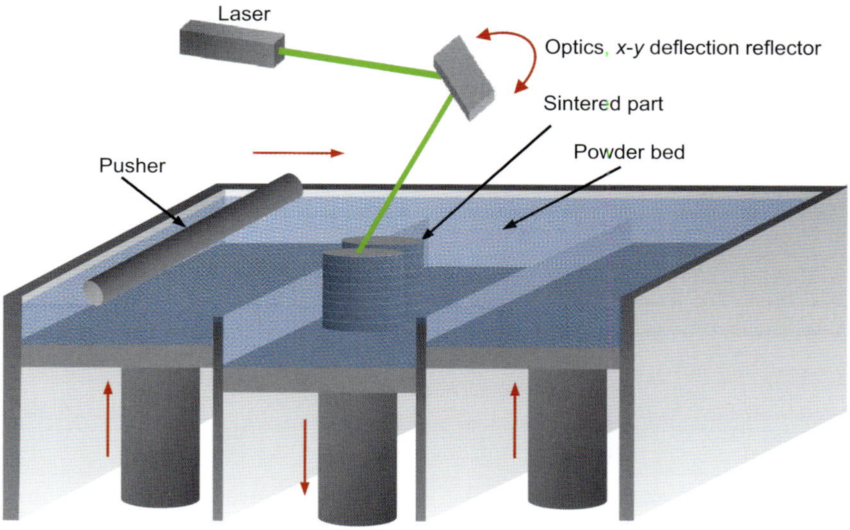

FIGURE 7.2 Schematic view of the selective laser sintering process [26b]

Sintered models are porous, so infiltrating surface-sealing processes can be used. Sintered parts are not as detailed and are rougher than stereolithographic parts, but mechanically they are stronger and can sustain higher loads [61, 186]. Plastics compatible with the LPKF-LDS® process for surface structuring (see Section 7.3) are not available for either of these processes.

7.2.3 Fused Deposition Modeling

Fused deposition modeling (FDM) is an extrusion process developed by Stratasys. Parts are built up in layers from extruded plastic strands. In principle, the FDM process is suitable for all materials that can be melted and extruded. Plastics such as ABS, ABSi, PC-ABS, PC, PC-ISO, and PPSF are typical of the materials used. The material for extrusion must be in filament form, with a circular cross-section and a diameter between 1 and 2 mm. The initial data in 3D CAD form are generally converted to SLA format, and this data is used to create 2D cross-sections as slices through the part to be produced. Although the material solidifies very quickly, structures made of support material are needed for cantilever structures. These supporting structures are subsequently easy to remove. Filaments of different materials are needed for the part and the supporting structure. Drive rollers carry the two filaments to an electrically heated plastifier, where the filaments are heated to just below melting temperature (Fig. 7.3).

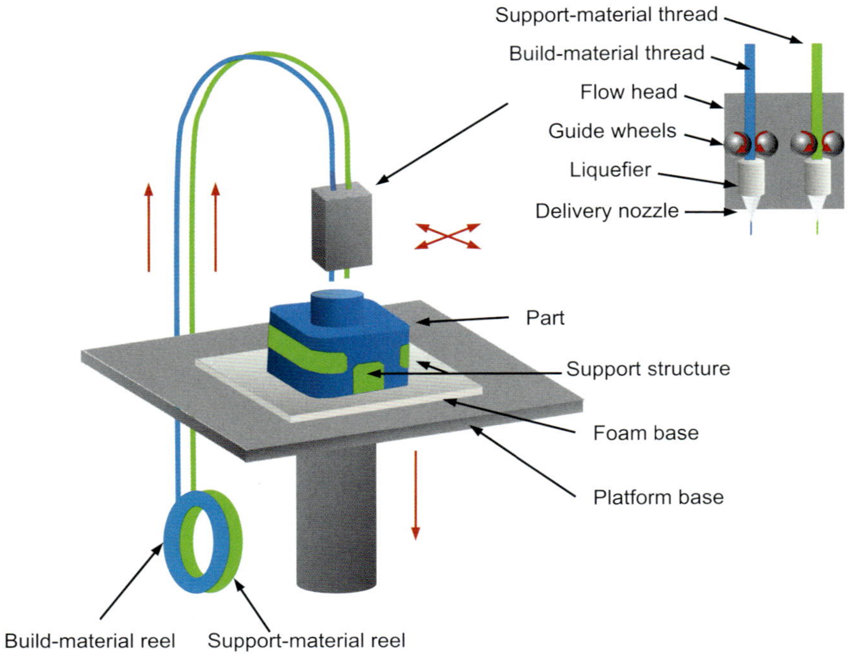

FIGURE 7.3 Schematic view of fused deposition modeling [26c]

The melted strands of material are extruded through nozzles and deposited by the extruder under numerical control to replicate the layer information. The strands and layers deposited beforehand are joined by the local heat input from the newly melted length of extruded material. The newly applied material solidifies, transferring its heat to the layer below. The heat contained in the molten material applied to it is enough to melt the substrate to the extent that a permanent connection is made when the material cools. Steps are produced at the surfaces and edges on account of the geometry and properties of the molten strands. This step effect is a restriction on this process with regard to precision and surface structure. Surface quality can be improved by reworking, for example by sanding with sandpaper, priming, and painting. [61, 186]

7.2.4 Vacuum Casting in Silicone Molds

Vacuum casting is a shaping technique for producing plastic parts using for example stereolithography models as the masters. The master is used as a pattern to cast a mold in silicone, and this mold is used for replication. The material used for replication is a polyurethane casting resin. Resins with a number of different property profiles are available.

7.2.5 Milling Thermoplastic Semifinished Products

Plastic bodies for MID can be produced from thermoplastic semifinished products by conventional milling. The semifinished products are injection-molded sheet materials that can often be obtained from the manufacturers of the plastic granulate. This method is useful only for small MID parts, simply because the supply of large injection-molded semifinished products with thick sections is limited. The thicker the sections of injection-molded semifinished products, moreover, the greater is the tendency for blowholes to form. Blowholes in particular can have a negative effect on the follow-up LPKF-LDS® process, particularly if the milled semifinished products are made of LDS materials such as LCP, PA 6/6T, PBT, PET+PBT, or PPA. Weld lines can cause problems, particularly with semifinished products made of LCP. Machining can relax mechanical stresses and strains frozen into the plastic part, causing distortion. It is always a good idea to temper semifinished parts prior to machining, in order to reduce the distortion that machining can cause and that can also be caused by thermal loading. Parts, moreover, should also be machined from both sides, because this too can help reduce distortion. Milling parameters have to be set such that the surface quality will be as high as possible with low surface roughness and without burrs.

7.2.6 Injection Molding

Plastic blanks for MID are produced in quantity by injection molding, using complex injection-molding tools with hardened multiple cavities. Master tools can be used to achieve part quality on a par with prototypes. These master tools are injection-molding tools set up using mold inserts. The mold inserts can be produced by rapid tooling methods, such as the milling of inserts from aluminum or by laser sintering or by laser beam melting.

Injection-Molding Tools with Mold Inserts Made of Aluminum

Aluminum is not as strong as steel, so it has certain advantages to offer for chip-removing machining. It is comparatively easy to make mold inserts for injection-molding tools from aluminum. Assuming that suitable master tools are available, considerable savings in both cost and time are possible. The disadvantage of aluminum mold inserts is that they are nowhere near as durable as their steel counterparts. If the plastic contains a high proportion of abrasive fillers the molding tool will wear more rapidly, with this accelerated wear affecting the fine elements of the structure in particular.

Injection-Molding Tool Inserts by Laser Sintering or Laser Beam Melting

Mold inserts can be produced using manufacturer-specific processes such as selective laser sintering, direct metal laser sintering (DMLS), selective laser melting (SLM), or laser cladding [61]. Minimum layer thickness achievable at this time is 20 μm. Mold inserts produced in this way can be integrated into the master tool. The advantage of these mold inserts is that they are ideal for producing small, complex cavities. The manufacturing process is additive, so prices rise in proportion to cavity volume, but not to the complexity of the part. Consequently, laser sintering or laser beam melting processes are primarily suitable for manufacturing small, complex geometries. All commercially available thermoplastics can be worked with mold inserts manufactured in this way. However, the working life of a cavity produced by laser sintering or laser beam melting is limited, so the possible number of shots is well below that achievable with hardened steel molds. FIT Fruth Innovative Technologien, for example, rates its DMLS mold inserts for 500 shots. Tool life can be prolonged considerably by using special materials for the tool inserts [47].

Injection-Molding Tools with Mold Inserts Made of Nonhardened Steel

Nonhardened inserts for injection-molding tools can be machined using conventional techniques such as milling. However, the difficulties involved in machining inserts made from nonhardened steel increase tremendously with complexity and diminishing structure size. Mold inserts produced in this way are characterized by very high precision, surface quality, and mechanical strength. The quality of the molded parts corresponds to that of production parts. Mold inserts made of nonhardened steel are good for a service life of several thousand units, depending on the thermoplastic used. The mold inserts made in this way can also be case-hardened using processes that lead to very low levels of distortion, nitrite hardening being one such, and this significantly increases the useful life of the tool.

7.3 Samples and Prototypes Produced by the LPKF-LDS® Process

7.3.1 ProtoPaint LDS Process

LPKF Laser & Electronics AG has developed ProtoPaint LDS, a process by which virtually any plastic body can be structured by the method known as LPKF-LDS® [85, 86, 105]. A paint system based on polyurethanes with an isocyanate hardener and laser-activated additives is applied for the LDS-compatible surface modification of a plastic body produced by rapid prototyping. Known as "ProtoPaint LDS," this paint can be applied as a laser-activated coating to various plastic surfaces. Many plastics can be coated directly with this paint. The range of compatible plastics can be extended even further by applying a primer before the paint. The paint system is available in an aerosol can and is very easy to apply to the plastic parts. There is no need for a separate spray gun. When used in combination with modern rapid prototyping methods, this can significantly accelerate the prototyping of mechatronic components. The CAD data file is used to produce a plastic body, for example by stereolithography, laser sintering, or fused deposition modeling [162] (Sections 7.2.1 to 7.2.3). The plastic body is then prepared for laser structuring by application of a coat of paint approximately 30 μm to 40 μm in thickness. The application of two coats one after the other has proved ideal as a means of building up a layer that is sufficiently thick and homogeneous. If a part is large, only the areas that will actually carry conductors have to be painted. The paint cures at temperatures around 80 °C.

The next step is laser structuring, just as with conventional LDS molded parts. The LPKF company has developed a simple system for laser exposure (Fusion 3D 1100) that can structure 3D parts in a space measuring up to 120 mm × 120 mm × 50 mm (X × Y × Z).

The laser-structured prototypes can then be metallized in a conventional machine for MID. Alternatively, LPKF offers a metallization set consisting of the laboratory containers and equipment, chemicals, and analyzing accessories designed to be installed in a laboratory extractor on the user's premises as a cascade of process containers. Users who do not have laboratory facilities can opt for the ProtoPlate box for chemical copper plating of LDS prototypes. This is a kit in which the electrolyte is ready for use immediately after being mixed and requires no analytical monitoring (Fig. 7.4) [132].

FIGURE 7.4 Metallization set (left) and ProtoPlate box (right) (graphics courtesy of LPKF)

These three components, LDS paint, 3D laser system, and the metallization set, suffice to enable users to make their own MID prototypes in-house. Both the laser settings and the metallization parameters are largely the same as those for components made entirely by injection-molding LDS plastics. The adhesion of the plating to the LDS paint is very good. The design rules of the LDS process apply for painted LDS parts with no restriction whatsoever. However, the paint system imposes a certain limit on the temperature resistance of prototypes built in this way. Conductive adhesive gluing or soldering with low-temperature solder can be used as the assembly and connecting technology.

7.3.2 LDS Process with FDM Plastic Parts

Plastic bodies made by FDM using a suitable LDS material can be laser direct structured. Successful tests have been carried out with PC and PC-ABS [120]. The support materials necessary for the FDM process do not have any negative effect on LDS. The geometric surface structure of the FDM parts, however, can restrict the possible areas of application. Tests of the FDM process using a PBT modified with an LDS additive have shown that, because of the wavy surface structure, an extra smoothing step has to be inserted prior to the LDS operation [134, 186].

7.3.3 LDS Process with Vacuum Castings

The plastic body can also be manufactured by vacuum casting with an LDS-additive-enhanced polyurethane resin. A pattern produced using stereolithography, for example, is used to make a silicone mold in which the LDS-compatible prototypes can then be cast in polyurethane resin [74, 87, 162]. Speedmold GmbH is a company that has specialized in the manufacture of PUR prototypes like this [158].

7.3.4 LDS Process with Milled Semifinished Products

If the MID parts are small, one option is to use mechanical processing such as milling, for example, to machine the bodies from LDS-compatible semifinished products and use these blanks for laser structuring followed by metallization [141].

Before it can be structured in the LDS process, each plastic body has to be thoroughly cleaned in a wet-chemical process and then dipped, in order to remove metallic residues of the machining tools. Dipping is a process in which the parts are immersed first in a strong alkaline and then a strong acidic solution at elevated temperature.

Dipping can slightly roughen the surface of the substrate, so this process has its limitations with regard to very fine structures.

7.3.5 LDS Process with Moldings from Rapid Tooling Injection-Molding Tools

As described in Section 7.2.6, aluminum mold inserts are more easily and speedily manufactured by milling than the corresponding steel inserts, because aluminum is not as strong and is easier to machine. LDS materials, however, contain a high proportion of fillers, and their action against the aluminum molds is abrasive. This can leave traces of metal particles containing aluminum on the surface of the injection-molded LDS parts. In the subsequent metallization process, these metal particles can cause unwanted overmetallization. The way to prevent this is to dip the plastic parts prior to laser structuring, in much the same way as the milled plastic parts are cleaned by wet-chemical dipping as described in Section 7.3.4. In this case, too, dipping can cause slight roughening of the substrate surface, so once again, the process has its limitations as regards very fine structures.

The alternative is to use the laser-sintering process to manufacture form inserts from steel (Section 7.2.6). LDS parts injection-molded with these form inserts can be readied for the LDS process without wet-chemical dipping. On account of the process-related surface roughness of form inserts manufactured by DMLS, there is a slight tendency toward more overmetallization, but this is not generally considered a critical issue.

7.3.6 LDS Process with Moldings from Steel Tools with Nonhardened Inserts

By comparison with the processes described in Section 7.3.5, the manufacture of LDS parts with steel tools with nonhardened inserts is the most complicated. The upside is that MID quality is comparable to that of subsequent series production. The surface quality of the moldings is such that there is no overmetallization. Even ultrahigh requirements with regard to superfine conductor widths and subsequent processes of assembly and connection technology such as chip-assembly processes are fully satisfied.

7.4 Samples and Prototypes Produced by Hot Embossing

Rapid prototyping methods are not currently used to produce embossing dies. Nevertheless, when samples and prototypes are produced using hot-embossing techniques, moldings from rapid tooling methods or from nonhardened steel tools can be used (Section 7.2.6).

7.5 Samples and Prototypes Produced by Two-Shot Molding

In the development of two-shot MID, visualization samples (Section 7.1.1) can be produced using all the methods outlined above. Combining two one-shot injection-molding tools and using a switchover technique can be more cost-effective than a two-shot tool such as a turntable tool for manufacturing prototypes. Geometrical tolerances of two-shot MID are subject to high requirements, so it is important to check very carefully whether one-shot tools with mold inserts manufactured using laser-sintering methods (Section 7.2.6) can be used.

If samples are needed for technological testing in the development of two-shot MID, it might be possible to use a semiadditive laser process in which the part made of the original material (e.g., LCP Vectra E820i Pd) is first fully plated with copper and then structured with a UV laser. Once the cleaning step has been completed, the final layer is deposited from electroless nickel and immersion gold. Alternatively, it is possible to resort to fully subtractive laser structuring of parts metallized all over or to use the LDS technique.

7.6 Aerosol-Jet Printing on SLA Parts

Printing techniques can be an interesting alternative to the selective application of metal layers on plastic bodies. With regard to prototyping, very promising results have been obtained using Aerosol-Jet ® printing on stereolithography parts. [64]

7.7 Overview of the Various Combinations for MID Prototyping

Table 7.1 shows the combinations for MID prototyping that are important with regard to shaping and subsequent structuring of the plastic body.

TABLE 7.1 Overview of the Methods of Manufacturing MID Samples and Prototypes

Shaping	SLA	SLS	FDM	Vacuum casting	Milling semi-finished products	Injection molding with mold inserts of		
						Al	DMLS	steel
Structuring process	Manual	Manual	Manual	Manual	Manual			
	Proto Paint	Proto Paint	Proto Paint	Proto Paint	Proto Paint			
			LDS	LDS	LDS	LDS	LDS	LDS
					HE	HE	HE	HE
								Two-shot

- Visualization sample
- Partially functional concept model
- Fully functional sample
- Prototypes for pre-series production approval

8 Integrative Development of MID

C. Jürgenhake, T. Schierbaum, C. Fischer, R. Dumitrescu

The development of innovative MID products calls for interaction between mechanical and electronics development as equals. The integral design that is typical of MID strengthens interdisciplinary dependencies, because changes in the individual specialist areas influence each other. The integration of mechanical and electronic function carriers in a single component, moreover, leads to numerous interactions within the MID. There are also strong interdependencies between MID product and MID production process. The product concept is determined at an early juncture by numerous restrictions of potential MID production methods. This is why the product and the production system have to be developed in parallel and integratively from the beginning. These are the challenges, but MID technology also affords tremendous advantages over conventional approaches (Chapter 2).

In many instances companies lack the MID-specific systematic approach for developing products. MID experience is garnered over time in an iterative learning process. The expertise of individuals becomes a decisive factor. But the fact is that procedures adapted to MID already exist.

The *Systematik zur Entwicklung mechatronischer Systeme in der Technologie MID* (systematic approach for the development of mechatronic systems in MID technology) proposed by Ingo Kaiser structures the development process and integrates methods and guidelines with the objective of reducing complexity in development. The core is a generic procedural model that defines the phases and milestones of the development process. This is backed by specific procedural models elaborated for particular MID processes. These models cover specific activities and the pertinent interactions, and they cross-reference relevant guidelines. Kaiser's systematic approach is described in detail in Section 8.1.3.

In 2008 the paper entitled *Methodik zur Produktoptimierung mechanisch elektronischer Baugruppen durch die Technologie MID* (Methodology for Product Optimization of Mechanical Electronic Modules by MID Technology) by Thomas Peitz was awarded the year's MID prize for development. It proposes approaches for using MID technology to optimize existing, conventional-design applications. Section 8.1.2 provides a detailed explanation of this methodology.

The development of technical systems is assisted by several procedural systems with different focuses (VDI/VDE 2221, Pahl/Beitz, VDI 2422, and others). The specifics of the development of mechanical electronic modules are largely taken into account in VDI 2206, *Design Methodology for Mechatronic Systems*. This VDI guideline 2206 is discussed in Section 8.1.1.

Section 8.2 to 8.8 contain in-depth explanations of Kaiser's procedure within his systematic approach, with the addition of important aspects from the Peitz methodology.

Section 8.9 follows up with introductions to other MID-specific development methods available on the market. Section 8.10 presents existing software solutions that support developers along the MID process chain.

8.1 Systematic Approach for the Development of MID

8.1.1 VDI Guideline 2206: Design Methodology for Mechatronic Systems

VDI guideline 2206 *Design: Methodology for Mechatronic Systems* is an end-to-end cross-domain guideline for developers. Experience from industrial practice and the results of empirical design research show that the development process for mechatronic systems cannot be mapped as an unbending sequence of individual activities [VDI 2206]. VDI 2206 therefore proposes a flexible procedural model consisting of three elements: **V model as macrocycle**, **problem-solving cycle as microcycle,** and **process modules for recurrent working steps**.

V model as macrocycle: The V model on the macro level was taken from software development and adapted to the requirements of mechatronics (Fig. 8.1). It describes the basic procedure for the development of mechatronic products and defines the major substep requirements, system design, domain-specific design, and system integration accompanied by validation of properties. Modeling and model analysis are both in-process.

- **Requirements:** The point of departure is a specific development assignment. The terms of reference are detailed and described in the form of requirements that also set the standard for subsequent product evaluation.
- **System design:** The objective is to arrive at a cross-domain solution concept that covers the future product's essential physical and logical functions. The overall function of a system is broken down into subfunctions for which the operating principles and the solution elements are determined. The operating principles and solution elements combined add up to the basic solution.

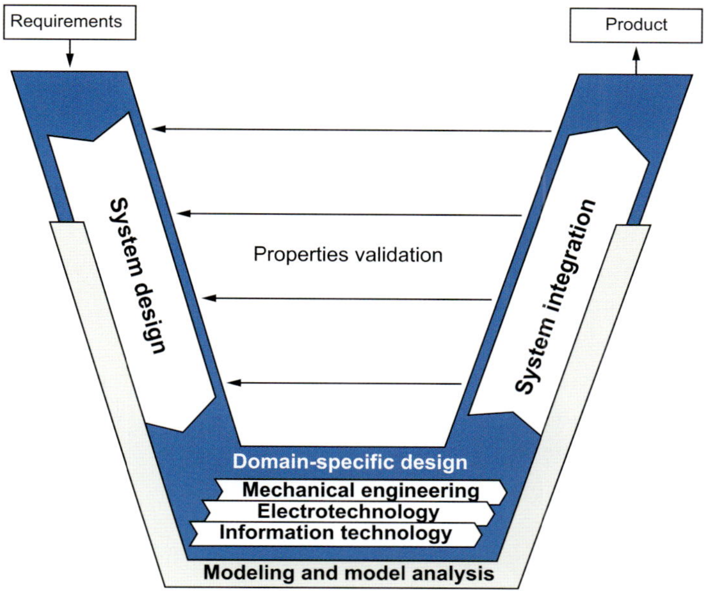

FIGURE 8.1 The VDI 2206 V model

- **Domain-specific design:** The solution concept developed up to this point is further detailed, in most instances separately in the various domains. Detailed design work and calculations are performed to ensure that the functions are as desired, particularly those that are critical.
- **System integration:** The results of development from the individual domains are integrated into an overall system and their interaction is studied.
- **Properties validation:** The results of development have to be checked regularly against the defined requirements. The actual system properties must match the desired system properties.
- **Modeling and model analysis:** The phases described above are accompanied by the mapping and analysis of the system properties with models and computer-aided tools for simulation.
- **Product:** The result of a completed cycle is the product. In most cases, however, a complex mechatronic product cannot be developed within a single cycle. Product, in fact, does not really mean the finished, real item. It refers instead to the increasing level of detail of the future product, such as the laboratory sample, functional sample, or pre-series product.

Problem-solving cycle as microcycle: Problem-solving on the micro level assists the developer in dealing with plannable and also unforeseeable difficulties. It is based on the general problem-solving cycle familiar from systems engineering. Process planning can be adapted flexibly to the specifics of each development undertaking by sequencing and nesting procedural cycles.

Process modules for recurrent working steps: Process modules with typical tasks and activities are defined for frequently recurring steps in the development process. The VDI guidelines describe process modules for the tasks of system design, modeling and model analysis, domain-specific design, system integration, and properties validation.

This macro cycle results in a product that can be a final product or a product of some other degree of maturity. A complex mechatronic system generally requires several iterations of the cycle.

VDI 2206 was the first guideline authored specifically for mechatronic systems. It brings together existing guidelines in a cross-domain systematic approach. It is deliberately general and not restricted to individual technologies. Detail is therefore lacking, particularly in its treatment of domain-specific design. The general approach of the methodology structures the development process and enables results to be applied to other projects. The general nature of this structure means that the principles are also applicable to MID development. However, the strong interactions between product and production technology in the design of MID are not adequately addressed, and there is no explicit treatment of the challenges specific to MID development. The procedural approaches developed by Peitz and Kaiser, which deal specifically with the development of molded interconnect devices, are discussed below.

8.1.2 Thomas Peitz' Methodology for Product Optimization of Mechanical Electronic Modules

The methodology developed by Thomas Peitz aims at employing MID technology to optimize conventional mechanical electronic modules. The strengths and weaknesses of the conventional products are analyzed to identify optimization potential. Once this potential has been pinpointed, the methodology can be applied to develop alternative MID solution concepts. These concepts can be evaluated in terms of benefit and manufacturing costs and are the basis for the follow-on phases of development. The methodology comes into play at the early stage of development and pursues the aim of utilizing the chief potential of MID technology by in-depth conceptualization. The methodology is sketched in Fig. 8.2. It ensures systematic and objectives-oriented product optimization and structures necessary tasks and processes.

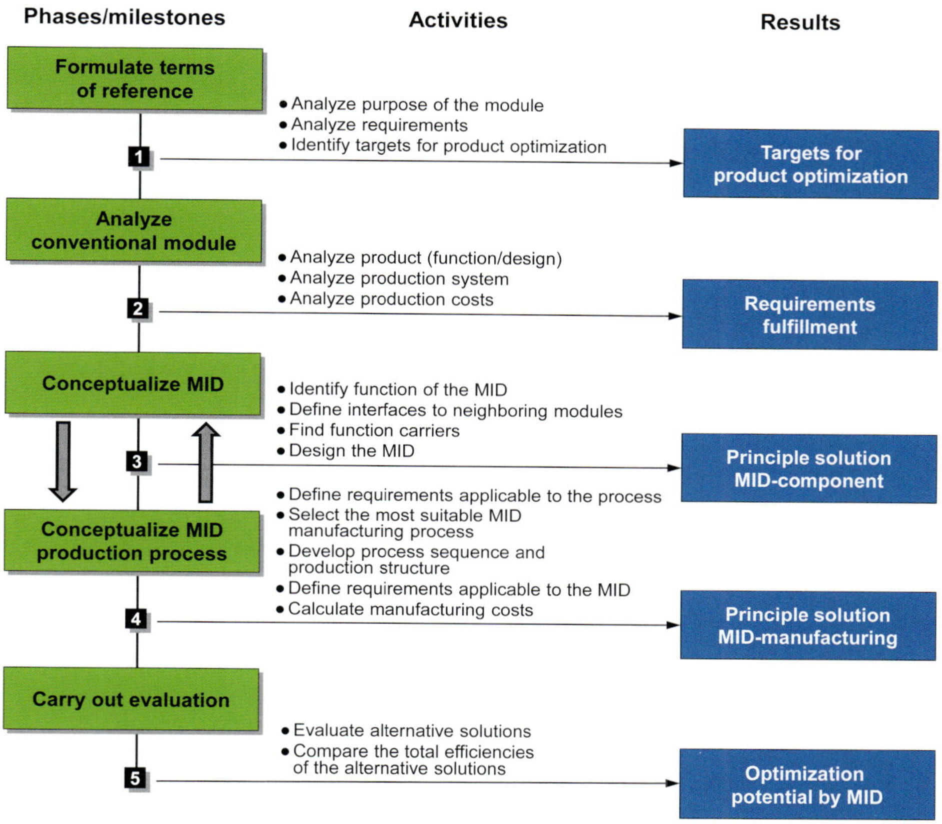

FIGURE 8.2 Model for the product optimization of mechanical electronic modules by MID technology

Product optimization starts on the basis of a conventional mechanical electronic module. The first step is to formulate the terms of reference. This entails determining the use of the module and the requirements it will be called on to satisfy. The developer uses this as the basis for defining the objectives of product optimization. The second phase starts with an analysis of the conventional module. This is characterized by the identification of weaknesses and refers both to the product itself (shape/functions) and the associated production system. Manufacturing costs are also included in the scope of analysis. The developer evaluates the subject module with regard to the objectives defined beforehand and gauges the extent to which the requirements are satisfied. The concepts of the MID and the MID production process are fleshed out in an iterative cycle in the methodology, because the interactions between them are manifold [126]. The objective of component conceptualization is to arrive at a basic solution. This entails defining the functions of the MID.

The developer uses a design catalog to identify a suitable function carrier for the implementation of these functions. The basic solution is then elaborated on the basis of a four-step procedural model. Step one is to position the function carrier with its geometric restrictions. It is important at this stage to take the mechanical or electronic connections duly into account. Step two is designing the layout of the interconnect device. Other function carriers can then be positioned. The concept of the conductor layout then has to be finalized to conclude this step.

The next phase is conceptualization of the manufacturing process. The basis for this step is provided by the properties elaborated beforehand for the basic solution. The most suitable MID manufacturing process is selected, and in the next phases it is defined with increasing detail in terms of precise process sequencing and production structure. The manufacturing costs for the solution are also calculated. The two conceptualization phases are looped until an optimized overall concept is achieved. The solution is then evaluated in a strengths/weaknesses analysis. This affords the developer an overview of the module's optimization potential that can be tapped by MID technology.

The methodology developed by Thomas Peitz enables MID technology to be employed to optimize mechanical electronic modules. The focus of attention is on matured conceptualization of the product and the associated manufacturing processes.

8.1.3 Ingo Kaiser's Systematic Approach for the Development of Mechatronic Systems

Ingo Kaiser has elaborated a systematic approach to development of mechatronic systems using MID technology. His system structures the development process and integrates methods and guidelines with the aim of reducing complexity in development. The interdependencies between geometry and the production system are taken into consideration, as are the processes of assembly and connection technology (ACT). The result is detailed specification of the MID product and the associated production system.

This approach is built around a generic procedural model that the developer can use as a guideline. It defines phases, milestones, and corresponding objectives to be worked through and achieved in the course of the development process (Fig. 8.3). The methods that can be used in the individual phases are also discussed. Ingo Kaiser also goes on to describe a specific procedural model geared for example toward laser direct structuring (LDS). This model details the development tasks of the general procedure.

The phases and milestones of the generic model are characterized below.

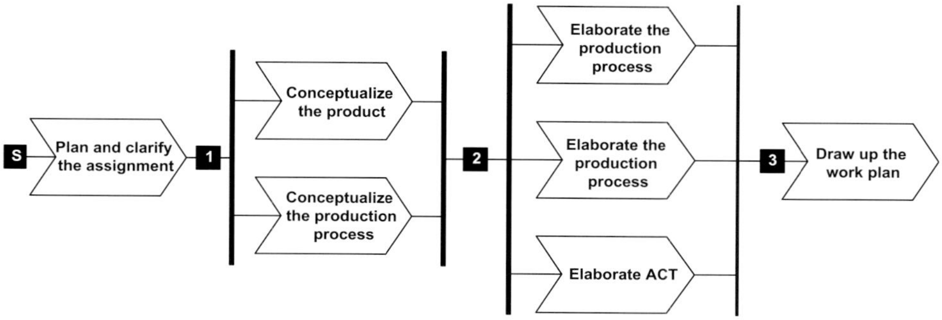

FIGURE 8.3 Generic procedural model as proposed by Kaiser [88]

Plan and clarify the assignment: The product requirements are identified in this phase. They can be classified as technical or economic requirements. A provisional list of requirements is drawn up in this way on the basis of the development assignment and is elaborated in further detail in the subsequent development process.

- Milestone 1: Specified development assignment and provisional requirements list.

Conceptualize the product: The product functions are defined on the basis of the requirements list and then linked in a function structure. Solution elements to satisfy the functions are then identified and combined. This enables the developer to draw up a broad overall solution. This basic solution is then evaluated on the basis of technical and economic aspects, for example in a cost-utility analysis. With production-related restrictions taken duly into consideration, the effects structure and the basic shape of the part are elaborated.

Conceptualize the production process: Conceptualization of the production process parallels product conceptualization. On the one hand, the basic product solution entails restrictions in terms of the possible production processes. On the other hand, there are production-related restrictions deriving from the processes used in production. The developer has to select the most suitable MID production process on the basis of these findings.

- Milestone 2: Product and production concepts as detailed requirements lists with the basic shape and the defined work processes and production resources.

Design the 3D electronics: The three-dimensional MID circuit layout is designed in this phase. Design principles are applied to arrive at an initial draft design, which is then elaborated in detail with regard to geometry, material, and tolerances. The designer selects the electronic components that, when interconnected by the layout, will carry out the electronic functions. Once the circuit layout has been checked against process-specific design rules (e.g., conductor pitch, conductor widths), with adjustments as necessary, a fully developed three-dimensional circuit layout is the outcome of this phase and marks its conclusion.

Elaborate the production process: The production process chain is elaborated in detail on the basis of the product and production concepts. The developer reconsiders the compatibility between product and MID production process. Machinery and handling systems are defined, including those needed for structuring and for component placement. The assembly and connection technology (ACT) is elaborated in parallel and then integrated.

Elaborate the assembly and connection technology: The processes for assembly and connection are selected in this phase of the process. The choice is very heavily dependent on the product properties defined in the conceptualization phase. For example, the temperature resistance of the substrate material has a definitive effect on the choice of connection medium. Once the process has been chosen, the developer moves on to finalize the equipment and the materials and define the requisite process parameters. The outcome is the fully elaborated ACT production process chain.

- Milestone 3: Overall design of the MID product and the developed production process chain (mechanics, circuit layout, ACT).

Draw up the work plan: The final step in this development methodology is to define the work plan, a process that breaks down into work-sequence and work-resources planning. The former entails drawing up work and assembly plans and machine control programs. The machines and jigs are designed in work-resources planning, with number of units to be produced, tolerances, and capacities all taken duly into account.

Ingo Kaiser's systematic approach affords a guideline for the entire MID development process. Integrative development from product and production system is taken into consideration equally with the interdisciplinary reach of the development project. As regards the elaboration of specifics, however, his approach is limited to laser direct structuring.

The Kaiser procedure with the addition of important aspects from the Peitz methodology is described in detail below.

8.2 Requirements

Development invariably starts with the development assignment. The task, the problem, or the product to be developed should be formulated in a way that is as neutral as possible regarding the solution, in order not to restrict the latitude for solutions development unnecessarily at too early a juncture. Outline requirements specifications are drafted on the basis of the development assignment. Kaiser's procedural model places this in the *"Plan and clarify the assignment"* phase. It is part of preparation for development, and it breaks down into the process steps *clarify assignment*, *analyze environment*, and *analyze requirements* (Fig. 8.4).

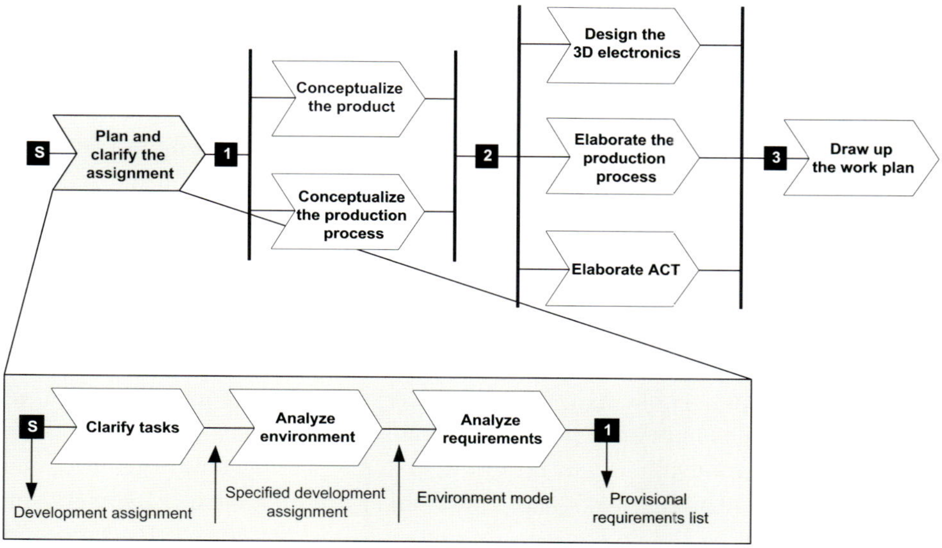

FIGURE 8.4 Process steps in the *plan and clarify the assignment* phase

Modeling languages have become commonplace in product conceptualization over recent years. Model-based systems engineering is a phrase often used in this context. Product and production system are described by different partial models, which come together to yield a consistent entity. There are different modeling languages with different origins. Two languages for early-stage modeling are outlined here by way of example.

- **SysML:** The Systems Modeling Language is a metalanguage developed by the OMG (Object Management Group) and INCOSE (International Council on Systems Engineering) on the basis of UML (unified modeling language). It can be used to describe and interrelate the structure, behavior, and requirements of a system. However, it is not suitable as a language for describing products in MID technology.

- **CONSENS:** CONSENS is an acronym derived from *CONceptual design Specification technique for the ENgineering of complex Systems.* It is a language developed by the Heinz Nixdorf Institute. It can be used to describe advanced mechatronic systems and their production systems from an early stage on in the form of a consistent system of partial models. The other partial models for product description besides the environment model include the application scenarios, the requirements list, the function hierarchy, the effects structure, the shape model, and the behavior models "behavior states" and "behavior activities."

In the conceptualization framework for MID products, the CONSENS language has proved more powerful and is used below for modeling the individual aspects. The process steps are described in more detail below.

As stated above, the development assignment is the starting point for development. In the "clarify tasks" process step the initial examination establishes whether predecessor products exist. If so, a weaknesses analysis can be carried out to identify optimization potential. This can extend from a simple checklist through to in-depth analysis of strengths and weaknesses on the basis of best practices for the sector. In an in-depth strengths/weaknesses study the product can be analyzed and assessed against other products in the framework of an expert survey on the basis of competitive factors such as product quality, functions, or manufacturing costs. The results of the survey are then processed and yield a strengths and weaknesses profile as the basis for deriving specific recommendations for further development.

The next process step considers the environment in which the product will subsequently be used in the field. All relevant influences on the product to be developed are modeled in the environment model. The influences can be modeled with three different flow relationships: material flow (e.g., spray), energy flow (e.g., heat), and information flow (e.g., control signal). Flows can be unidirectional or bidirectional. A distinction can be drawn between wanted and unwanted flows. The latter are termed parasitic flows. Some of these parasitic flows are defined in VDI 4005. They include vibration, temperature, and relative humidity. Along with the influences on the product as such, the geometrical boundary conditions and interfaces to other module parts are important constituents of the environment model, particularly so in the case of product development in MID technology. Initial requirements can be identified on the basis of the environment model.

In *analyze requirements,* the third step in the process, the major requirements applicable to the product are identified and documented with the aid of the requirements list. These include technical requirements originating for example in legal boundary conditions and the strengths and weaknesses analysis. It is also important to factor in the needs and preferences of customers. These are obtained by means of surveys. Pahl/Beitz draw distinctions between needs (Forderungen, "F") and preferences (Wünsche, "W"). Needs must be satisfied at all costs. They include performance data and quality requirements, for example. Preferences should be taken into account whenever possible. Pahl/Beitz suggest structuring needs by main characteristics (e.g., geometry, production, or costs). As with the development assignment, the needs should be formulated in a way that is as neutral as possible regarding the solution, in order not to restrict the latitude for solutions development unnecessarily at too early a juncture.

The first phase closes with a specified development assignment and a provisional requirements list. The requirements list is the basis for the next phase, product conceptualization.

8.3 Product Conceptualization

Product conceptualization parallels conceptualization of the production process, and the approach must be integrative. Invariably, both these phases are iterative cycles. Even at this early stage there is a multiplicity of interacting relationships and restrictions between product concept and production technology. They apply, for example, to the geometry of the part (e.g., maximum permissible inclinations) or take the form of general, technological restrictions. The *conceptualize product* phase breaks down into the process steps *identify functions*, *generate solution ideas*, *create effects structure and draft shape*, and *document restrictions due to production* (Fig. 8.5). The point of departure is the provisional requirements list from the *plan and clarify the task* phase. The process steps are next described in more detail.

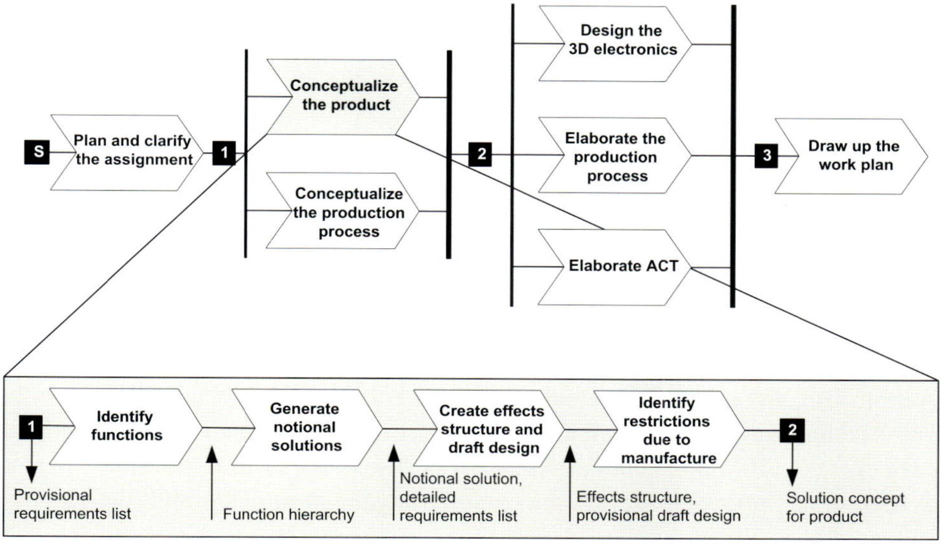

FIGURE 8.5 Process steps in the *conceptualize product* phase

A function hierarchy is mapped out so that the functions can be defined. A function describes the desired relationship between input variable and output variable. A function is always described by a combination of a verb and a noun (e.g., “transfer material”). The partial model of the same name from the CONSENS specification language can be used to compile the function hierarchy. Functions can be distinguished as conventional functions (e.g., “create torque”) and MID-specific functions (e.g., “channel energy,” “protect part”). The process starts with derivation of the product’s overall function from the development assignment. This is generally too complex to be implemented with one technical solution, so it is broken down into subfunctions depicted as a tree structure. A function hierarchy results. The higher-order functions are repeatedly subdivided to the point at which technical solutions can be found for the subfunctions. Creative methods such as the morphological box can be used to find the solutions. The box is a table; the functions of the lowest level of the hierarchy are placed in the rows, and consecutive numbers are written into the columns. The cells are then filled with possible solution elements. The MID design catalog proposed by Peitz is useful for finding solutions for electrical and mechanical MID functions. Solution paths are then traced through the filled morphological grid. Each path represents a possible variant of the product. The individual solution variants are then evaluated on the basis of technical and economic criteria. Possible evaluation criteria are the degree of miniaturization, function integration, manufacturing costs, and degree of innovation. The most promising product idea is selected, and the provisional requirements list is drawn up in detail. Product properties are defined on the basis of performance indicators so that compliance can be deliberately targeted as work progresses.

The next process step is *create effects structure and draft shape,* and it focuses on detailing the selected solution. The effects structure, also a partial model of the CONSENS specification technique, describes the effective interrelationships between the system elements (solution elements) with the aid of flows. As in the environment model, flows are distinguished as material, energy, and information flows. The effects structure maps the basic structure of the system. Alongside the description of the effect relationships, the shape is an important parameter that has to be taken into consideration at an early juncture. The shape of the product has a definitive effect on the choice of MID production method. The first drafts include basic information about geometry and size of the product. The interfaces to other parts or modules defined in the environment model have to be taken into consideration. The effects structure yields information about the electronic components. These components are positioned more or less accurately in the draft shape, and the basic tracks of the conductors are roughed in. The result is an initial shape model of the interconnect device with the basic arrangement of electronic components and neighboring modules.

TABLE 8.1 Examples of Restrictions that can be Imposed on the Product by Production Methods

Production technique	Resulting restrictions
Injection molding	**Ensure demoldability**
	Avoid different material thicknesses
Laser structuring	**Ensure accessibility for laser beam**
	Note maximum laserable component sizes
	Avoid allover metallizations
	Reduce clamping positions
Metallization	**Note acid resistance of the material**
	Provide connections for electrical metallization
	Note increase in metal deposition at edges/corners
Assembly/placement	**Ensure accessibility for placement robot**
	Ensure component placement on flat surfaces
	Facilitate placement of chips in recesses
Connection technique	**Note temperature resistance of the material**
	Provide for homogeneous temperature distribution

The restrictions deriving from the production process then have to be documented. Developed in parallel with the product concept, the production-process concept is the basis for identifying the restrictions for product design, which can then be matched to the product concept. For laser direct structuring, for example, there are restrictions deriving from the process steps of injection molding, structuring, (chemical) metallization, component placement, and connection technology. By way of example, Table 8.1 shows some restrictions for the laser direct structuring process. The restrictions shown bolded in the table could be criteria for rejection in the event of noncompliance.

8.4 Production-Process Conceptualization

Conceptualization of the production process parallels product conceptualization. On the one hand, the basic product solution entails restrictions in terms of the possible production processes. On the other hand, there are production-related restrictions deriving from the processes involved in production. The interrelationships dealt with superficially in the remarks on the preceding process step are analyzed in detail here. Invariably, the two phases *product conceptualization* and *MID production-process conceptualization* are iterative.

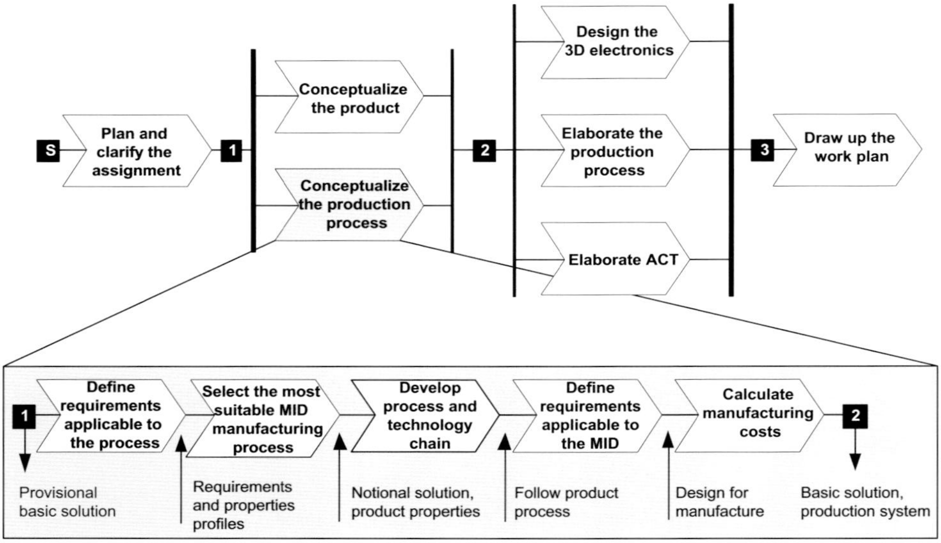

FIGURE 8.6 Process steps in *MID production-process conceptualization*

The first step is to define the **requirements for the MID production process**. They can be derived from the properties of the basic solution. Certain features of the basic solution are particularly important in this respect. *Conductor tracking* is a case in point. It refers to the geometric arrangement of the individual conductor segments on the interconnect device. If conductors are tracked entirely on a single plane, the requirements for the MID production process are low. However, if the conductor segments are distributed across two or more sides of the interconnect device, higher requirements apply to the production process or, more precisely, to the technologies for structuring and metallization. Some MID production techniques permit a high level of 3D design freedom.

Properties cards (Section 8.2.2) characterize the processes on the basis of eight features. Each manufacturing process is evaluated on a scale (1 = weak; 5 = strong) for each of these features. The results are plotted in a diagram. This analysis yields a specific properties profile for each manufacturing process. The properties cards also map the process chain and provide an outline description of the features. Properties cards help the developer to select the most suitable process for the MID product by elaborating a requirements profile for the product based on the eight features. This requirements profile is then compared to the profiles of the candidate technologies.

To a very large extent, the choice of MID manufacturing process also defines the associated **process or technology chain**. The necessary production resources then follow on from this definition. Production resources are machines, tools, and jigs.

For laser structuring as an MID production process, for example, the production resources are the injection-molding machine including the molding tool, the laser unit including the parts handling system, and the galvanizing unit. The final choice of production resources depends on the **product properties**. Crucial in this respect are the product properties geometry, circuit layout, and material. The physical size of the MID is important as regards the choice of injection-molding tool, for example. The number of mold cavities that can be integrated into a single tool depends on this. It is also important to know whether the part will have back-cuts or threads. The molding tool will be of ordinary design, or of the split-follower or ring-follower type, depending on part geometry.

This awareness of the interrelationship between product properties and production resources is important because the product concept has a direct influence on the production system. Manufacturing costs, consequently, are decided on at an early juncture in the conceptualization phase.

There are various machines and tool types for all the technologies that make up the MID reference process. The interdependencies between product properties and production resources are many and diverse. Table 8.2 provides an overview of the most important interrelationships between product properties and production resources.

These relationships have to be taken into consideration during conceptualization of the MID. The design should be such that the MID can be manufactured economically and with cost-effective production resources.

In addition to arriving at conclusions about the nature of the production resources to be deployed, the **costs** of these resources also have to be gauged. Manufacturer and procurement costs have to be stated for new resources to be acquired. Machine-hour rates can be given for resources that are already in place. These figures are needed in a subsequent step for calculating manufacturing costs.

The production resources have to be placed in a logical relationship before the basic solution of the production system can be described in its entirety. The "process sequence" and "production structure" partial models are used for this purpose. The **process sequence** partial model describes the logical sequence of work processes that stem from the production technologies used to manufacture the product. They can be subdivided into work processes for manufacture of single parts and work processes for assembly. The "production structure" partial model describes the production resources needed for the production of a product, complete with their properties and relationships. Relationships are logical relationships and material flows. The **production structure** is influenced primarily by the batch size of the MID product. The batch size derives from the requirements list. Calculating the numbers required for each production resource is then a matter of applying the information about the cycle times for the various machines. Processes for handling and transport do not belong here.

TABLE 8.2 Relationships Between Product Properties and Production Resources

	Product property	Production resource	
Geometry	Part without back-cuts	Ordinary mold	Injection molding tool
	Cup-shaped part	Stripper-plate mold	
	Flat parts with back-cuts	Bar mold	
	Long, wide parts with back-cuts	Split-follower mold	
	Part with threads	Ring-follower mold	
	Flat part	Rolling process	Hot embossing
	3D part	Stroke process	
	Flat parts without high walls	Film transport roll to roll	
	High walls	Manual insertion of film	
Circuit layout	Three-dimensionality and extra-narrow conductors	Beam guidance systems and handling unit (e.g., multi-axis system or 3D focal plane control)	Laser
	Thickness of coating and surface properties	Chemical or electrolytic (pretreatment, galvanizing, or post-treatment)	Metallization
	El. components on one plane	Conventional pick & place robots	ACT
	El. components on two or more planes	MID assembly robot with integrated handling system	
	El. components on freeform surfaces	Industrial robot	
	Number and position of el. components	Dispenser or stencil printing	
Material	Plastic	Bath sequence	Metallization

Design for manufacture (DFM) is the generic term applied when all of the many interactions are taken duly into account in conceptualization of product and production system.

The **requirements for the part** deriving from the technologies in the MID process chain now have to be formulated. Each individual production technology has to be considered separately. The technologies involved in the laser-structuring link in the MID production chain are *one-shot molding, lasering,* and *galvanization*. The drawbacks of the integral nature of MID soon emerge here. Requirements deriving from several different technologies have to be taken into account in a single part. This severely restricts freedom for designing the MID.

There are DFM guidelines for many technologies in the MID reference process. Plastic injection molding is a case in point. The makers of machines and tools publish design notes and guidelines of this nature (Section 8.2.3).

The objective in design for manufacture is to take all the requirements deriving from the production technologies duly into consideration. The design of the MID is refined until all the requirements are met. This phase closes when the basic solutions for the product and for the production system are ready.

The choice of a given basic solution has a crucial role to play in terms of **costs**. Ehrenstein has calculated that about 70% of the costs are finalized when the basic solution is developed [33]. There is very little leeway for reducing costs in the subsequent phases of the product development process. Manufacturing costs are a major factor in product development. The choice of production technology and therefore the question whether MID technology will in fact be employed is generally decided on the basis of cost. This underscores the importance of systematic solution development and evaluation in the conceptualization phase. Fixed costs and variable costs are incurred in the manufacture of any product. Fixed costs are the costs incurred for the maintenance of production capability. They include, for example, depreciation for machinery and plant, rental, labor costs, and the like. Variable costs are traceable back to an item to be produced or a service provided, for example manufacturing wages and materials. Unit costs are always variable costs, but overheads can include both fixed-cost and variable-cost components. The break-even point is the number of units at which it is worthwhile to deploy a given technology, and it depends on how fixed costs and variable costs add up [129].

It is extremely unlikely that any one company can accommodate the entire process chain for an MID on its own (see the reference process) and handle all the accounting, so it is worthwhile calculating the costs for the individual process steps. **Process cost calculation** is the practical choice of partial model for characterizing costs. The process as such is the central element in this model. The basic concept of process cost calculation is the breakdown of overheads across the individual processes, so overheads as such no longer appear. The process cost model is suitable for economic assessment and advance calculation for the introduction of MID technology, because indirect services are very much a necessity in the development phase in particular. More importantly, however, the domain-spanning breakdown of overheads to individual processes as implemented in process cost calculation is very much in tune with the integrative approach underlying MID technology. It forestalls the delimiting practice of accounting to individual cost centers that is common in many companies.

Figure 8.7 shows the basic procedure for process cost calculation. The costs of the individual parts are added together and give the total costs of the module.

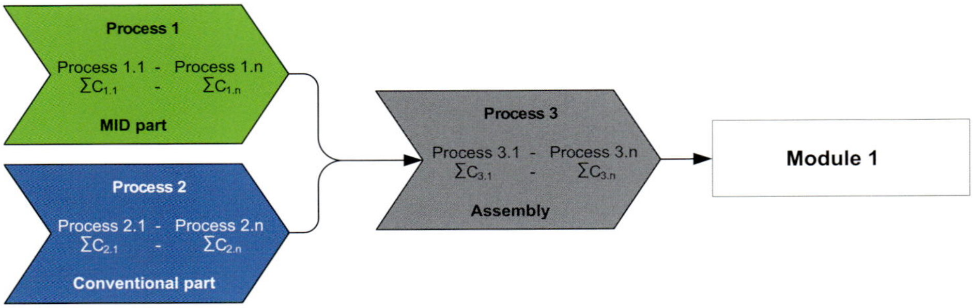

FIGURE 8.7 Breakdown of process unit costs by analogy with [38]

Each part, in turn, is broken down by manufacturing processes. It is important to include the assembly costs incurred for assembling the individual parts into a module. The process unit costs (C_{m}) for a process (lm) are the sum of the resources (R_{lm}) multiplied by the cost factor (c_{lm}).

$$C_{\mathrm{m}} = \sum R_{\mathrm{lm}} \cdot c_{\mathrm{lm}} \tag{8.1}$$

All individual process costs added together over the number of processes give the process costs C_{p} for the MID. A process hierarchy diagram or a process plan is useful for obtaining an overview of the individual processes. The next step is to set up a total cost model (C) for an MID over the phases of the product life cycle.

$$C = C_{\mathrm{de}} + C_{\mathrm{sh}} + C_{\mathrm{me}} + C_{\mathrm{st}} + C_{\mathrm{as}} + C_{\mathrm{us}} + C_{\mathrm{log}} \tag{8.2}$$

Where

C = overall costs

C_{de} = development costs (proportional)

C_{sh} = shaping costs

C_{me} = metallization costs

C_{st} = structuring costs

C_{as} = assembly costs

C_{us} = costs during use

C_{log} = logistics costs (proportional)

Broadly speaking, an early cost estimate, even if only rough, is useful for assessing alternative solutions. Absolute accuracy in cost calculation is not essential at this stage. Details can be filled in later in a subsequent costing phase.

8.5 Electronics Design

More detail is added to the chosen solution concept for the three-dimensional MID in this phase. This detailing extends to the layout of the interconnect device and to the design of the electronic circuitry it will carry. These are activities to be undertaken by specialists from different domains so they are carried out as parallel processes, one primarily mechanical for the interconnect device and one primarily electrical for the circuitry. The two processes have to proceed in close concurrence with each other because the result is a part common to both. The work on the interconnect device proceeds in accordance with a process described by Pahl/Beitz, whereas Gajski and Rose developed the process for circuitry design. The interconnect device is drafted as an initial design, then precision-designed and then elaborated. The steps in circuitry design are *draft circuitry, create circuit diagram,* and *create layout.* Interactions with the manufacturing processes have to be taken into consideration throughout [57, 122, 145].

In the first step in designing the interconnect device, the broad, shape-defining requirements are taken from the requirements list. They include size-defining and arrangement-defining requirements, e.g., size and position of other parts, and also the maximum working reaches of the production systems. Reflow ovens, for example,

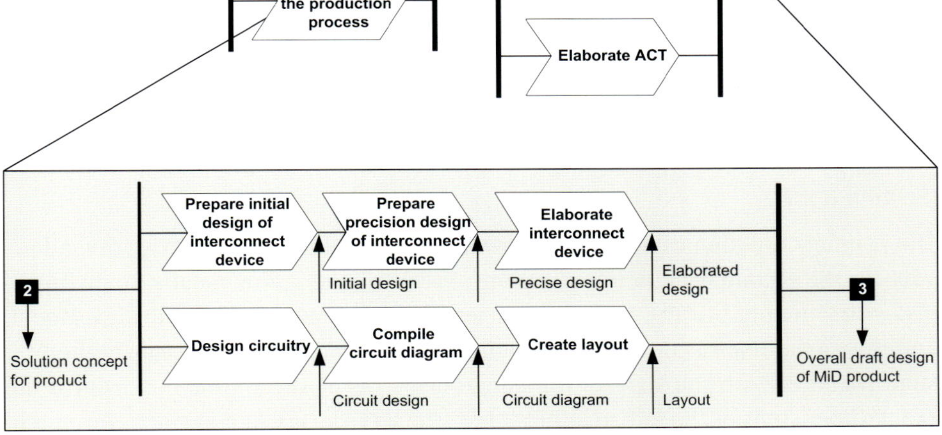

FIGURE 8.8 Process steps in the *3D electronics design* phase

can solder only modules that are not more than 3 cm in height. Ergonomics and safety aspects also have to be factored into the considerations. The design freedom afforded by MID technology is extremely useful with regard to ergonomic requirements. Once the shape-defining requirements have been identified, the material has to be selected. The properties of the material will influence the product (wall thicknesses, bevels for easy demolding, shrinkage, price of the material) and the production process (working temperature and pressure, resistance to temperature and acids) and other processes such as predrying for LCP (liquid crystal polymer). The primary and secondary function carriers have to be identified before the work of designing the interconnect device can proceed. Primary function carriers carry out the principal functions of the module, for example carrying components, connecting components, or protecting components. Examples of secondary functions include sealing, supporting, and cooling functions. The task of designing the interconnect device can begin.

The primary function carriers are roughly shaped on the basis of the draft design from the basic solution for the product. Design solutions have to be found for each function carrier. These solutions can then be combined to produce the overall design of the interconnect device. Areas where conductors are tracked and electronic components are placed merit special attention. They depend on the design of the electronics. Even though the basic layout of the conductors was determined in the basic solution, it is the three-dimensional solution that is definitive. Obviously, therefore, requirements such as interfaces, number of wires, and so on that apply to these areas have to be in agreement with the parallel process of circuitry development. VDI/VDE guideline 2424 *Industrial Design* is useful in this context. Circumstances might well require a number of drafts of the **initial design**. With the concurrence of the production specialists, one of these drafts is then selected for further design.

The next step is to find solutions for the remaining secondary functions. Standard parts or catalog parts can be used as secondary function carriers. Some secondary function carriers can be integrated directly into the package, for example metallized surfaces for heat dissipation. Design catalogs for plastic parts and MID design catalogs are good sources of ideas for the design of secondary function carriers (Section 8.2.1).

The primary function carriers then have to be precision-designed, with due allowance for the secondary function solutions. Precision-designing means finalizing all necessary details. All the function carriers are integrated into the design model, and the model itself is designed in detail. Defining section thicknesses is the first step. Together with the sizes of the neighboring parts, these thicknesses give the exact dimensions of the entire interconnect device. The contact surfaces for neighboring parts are then designed, along with the shape and size of the surfaces for component placement. Information from the circuit diagram, which is a parallel development, has to be integrated at this stage. All the electronic components are defined in the circuit diagram (except bridge resistors without electrical function). Size and number of electronic components have an effect on the surfaces for placement; rough positioning of these components on the interconnect device also affords the basis for the subsequent layout. The secondary function carriers then have to be precision-designed. The *variation of design* method helps in implementing best-fit partial designs. The interconnect device is precision-designed until a consistent design model emerges that satisfies all defined requirements. This is the **precision design** of the interconnect device.

The final step is the **elaboration of the precision design** of the interconnect device. This starts with adaptation of the interconnect device to the layout, which is developed in parallel: size and positioning of the components are checked, as are the exact tracks of the conductors. Surfaces that have turned out too small are adapted, and obstacles for conductors and components are removed. Holes for THT components are added, as are connecting elements such as snap hooks, flanges, and centering guides. The interconnect device is then optimized in terms of shape, surface, and tolerances. Applicable tolerances are defined in DIN 16742 *Plastics Moldings - Tolerances and Acceptance Conditions*. The objective is to achieve design for manufacture: thickness transitions must be smooth, edges have to be radiused, and so on. Demolding lines have to be worked in and the type and position of the sprue have to be discussed with the manufacturer of the mold. It would be unacceptable for the impressions left on the surface by the ejector pins to coincide with the layout area. Injection-molding simulations are of assistance in achieving design for manufacture. The part is then rechecked for faults, which have to be eliminated, if found.

Circuitry design (also known as functional design) starts with structuring the circuit for development. It is advisable to break down large circuits into function blocks. Control electronics for data processing and communication, signal transformation for sensors, indicators and controls, power electronics for actuator control, and energy supply are all typical function blocks. Structuring is based on the effects structure. The electronic system elements are analyzed and assigned to the function blocks. Within the function blocks the system elements are iteratively refined by function modeling until the individual elements can be replaced with known basic functions (e.g., send clock), stock circuits (e.g., adder), and components (e.g., transistor). The usual notation in circuit engineering is the block circuit diagram, as defined in general, not MID-specific, terms in DIN EN 61082 *Preparation of Documents Used in Electrotechnology*. Effects structure and block circuit diagram are equivalents, however. Circuitry design has to allow for subsequent testing of the product in series production. Thus each test point has to be defined at this stage.

Test points are contacts for testing equipment, frequently contact pads or pins. They are produced later on in the layout. All that is required at this stage is to define which signals will be tested. For MID circuitry the system elements from the effects structure are replaced by electronic functional elements, for example infrared LEDs.

Once completed, the circuitry design is transferred to an ECAD program. The result is the **circuit diagram**. Notation is in line with the generally applicable DIN 60617 *Graphical Symbols for Diagrams*. The first step in creating the circuit diagram is to find components that will implement the function elements. Empirical knowledge, solution element libraries, and manufacturers' catalogs are all of assistance for selecting the components. The data sheets issued by the manufacturers frequently include sample circuits. They indicate which additional parts are necessary for operation of a component and to complete the circuit. When this step concludes, all the components in the circuit have been defined, complete with component values whenever possible. The following information is required as a minimum:

- for passive components (resistors, coils, etc.), shape
- for active components (diodes, transistors, etc.), shape and type of assembly (flat, upright, with or without heat sink)
- for integrated circuits, the exact package shape
- for electro-mechanical components, the exact type designation

Directive 2002/95/EC *Restriction of the Use of Certain Hazardous Substances in Electrical and Electronic Equipment (RoHS)* requires the components to be free of toxic substances, with only a few exceptions. Only lead-free solder can be used. Lead-free solders generally have higher melting points, so soldering operations have to be hotter and the components correspondingly temperature-resistant.

The components are then positioned in the circuit diagram, whenever possible complete with the associated parameters such as resistance, capacitance, etc. The inputs and outputs of the components are then linked and unequivocal signal names are assigned to the connections. The positioning and linking of the components are purely logical in nature; there is no direct relation to the layout on the interconnect device. Nevertheless, the precision design of the interconnect device has to be taken into account in preparation of the circuit diagram: as regards their size and type, the components have to fit in with the design of the interconnect device. Information on component type, size, and shape can be derived from the available space. The defined test points are then added. A design rule check (DRC) is run to ensure the correctness of the circuit diagram: the computer checks that all links have been drawn in correctly (output to input, digital to digital, etc.) and flags mistakes with error messages as appropriate. These mistakes are corrected and the DRC repeated until no more error messages are output.

The behavior of the circuit can be checked by simulation. What are known as SPICE tools are used for this purpose. If behavior (e.g., run time, line termination, etc.) does not tally with the specifications from the requirements list, the circuit has to go back for redesign. The final step is generation of the network and components lists. The network list contains the designations of all components with component type, package shape and value, and the connections between the components. It is the connecting link between the circuit-diagram program and the layout editor. The components list contains all the components used in the circuit.

The shape of the interconnect device is defined in the first step in **layouting**. The layout support available in mainstream software tools is generally only for planar wiring boards. A three-dimensional MID has to be "planarized" for layouting. Spherical interconnect devices distort and split edges, and walls interrupt the conductors in the planarized layout. The result of this step is a two-dimensional map of the MID in which all the surfaces on which components cannot be placed (walls, holes, reinforcements, ejector-pin marks, and sprues) are flagged as no-go areas. All the remaining surfaces can be used for components and conductors. The electronic components are then positioned on the interconnect device. It is generally advisable to work from the outside in: the fixed elements (controls, connects, switches, indicators, etc.) are positioned first, then the function blocks (see the section on circuitry design), and the final stage is optimization within each block to make the connections as short as possible and distribute the components uniformly. The conductors are then routed. The following are important points for MID:

- Only the surface carries conductors. Conductor intersections have to be jumped by bridge resistors (0 ohm "resistances").
- The metallization is generally thinner than on planar wiring boards. Line resistance is higher, and conductor widths have to be adjusted accordingly.
- Galvanic metallization makes for thicker plating; the requisite connection pads have to be allowed for in the conductor layout.

The test points are also positioned in this step. VDI/VDE 2428 *Equipment Engineering; Design for Testability* is useful for this aspect of layouting. The test method used and the grid specifics have to be agreed upon with the specialists for ACT. The layout is then subjected to another design rule check. The circuit is checked for missed connections, open terminals, and below-minimum spacing between conductors. The electromagnetic compatibility of the layout can be simulated, but in the absence of three-dimensionality the results have to be critically examined.

There is as yet no established **rapid prototyping** process for MID technology. However, near-MID, three-dimensional prototypes can be used for assembly checks and function tests (Chapter 7). Typical faults identifiable with the aid of prototypes include

- layout position of a THT part does not match the holes in the interconnect device,
- conductors tracked over split edges (see above, planarization of the interconnect device) do not line up.

Once the design has passed as satisfactory, production documents are compiled for all the parts (technical drawings, circuit diagram, layout, components lists, etc.).

8.6 Elaboration of the Production Process

The preceding phase of *production process conceptualization* (Section 8.1.2) produced the concept for the production process. Together with the data from product design (precision design of the interconnect device, circuit diagram) it provides the input variable for subsequent elaboration. The objective is to produce fully elaborated production processes and test procedures. The compatibility between product and production process is verified, and then the production concept of the interconnect device is elaborated in more detail. This phase is closed by selection of the test procedures for quality assurance. These procedures are then integrated into the elaborated production process.

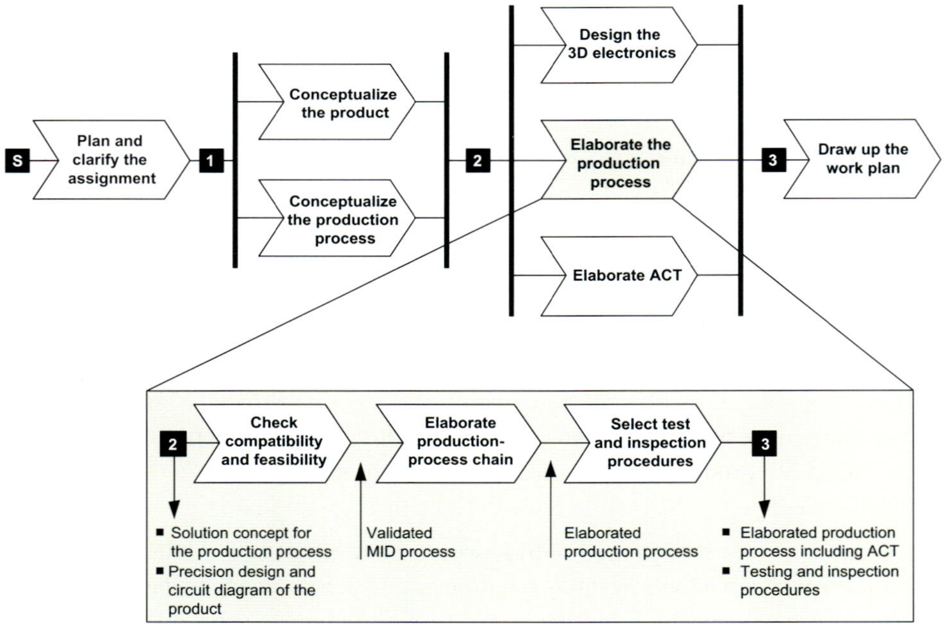

FIGURE 8.9 Process steps in the *elaboration of production process* phase

The product's properties emerge in increasing detail in the course of product elaboration (see design the 3D electronics). So the **compatibility** between product and the MID processes selected in the production concept and their **feasibility** can be rechecked with more precision before the production process chain has been elaborated. Typical questions include:

- Is the selected MID process (e.g., laser direct structuring) still suitable for manufacture of this interconnect device (e.g., regarding three-dimensionality and conductor width)?
- Is tool demoldability ensured? Does the design comply with specified minimum structure sizes?
- Is accessibility ensured for lasering?
- Is the selected material compliant with temperature-resistance requirements (use and (soldering) process)?
- Are the specified metallization thicknesses viable?
- Is the selected material acid-resistant, for example?

If there are no obstacles to manufacture of the MID, the MID production process is acceptable as selected.

On the basis of the solution concept for the production process and the validated MID process, the **process chain** of MID production is elaborated in this step. In the case of laser direct structuring, for example, this includes injection molding of the body, laser structuring, and metallization as the main steps. There are also intermediate steps such as cleaning (e.g., to remove ablation residues after laser structuring) and drying (before injection molding and after metallization). A suitable plant has to be selected for the individual process steps. Selection entails taking into consideration all the applicable requirements (e.g., injection-molding pressure required, working space and resolution of the laser, bath size and type of metal for galvanization). The parameters of the selected process steps can then be finalized, including for example pressure and temperature for injection molding. Some of the guidelines are the same as those for process conceptualization (DIN ISO 12165, DIN 60825, VDI/VDE 2421). Handling and materials-flow systems between the individual process steps also have to be defined. The use of a handling system depends primarily on the numbers of MID units, but it might also be necessary to provide for removal of the molded blanks from the tool and multi-axis handling for laser structuring. Detailing of the production process continues until each production step has been fully specified.

In order to ensure the quality required by the customer, the production expert works together with the mechanics and electronics developers to define the tests and test procedures. The following are the primary test criteria for MID (see Chapter 6):

- dimensional precision of the parts before and after working
- optical cleanliness of the metal structures (sharply defined structures)
- insulation resistance of the plastic (because of water absorption during metallization)
- presence of short circuits due to overmetallization
- electrical conductivity of the conductors
- adhesive strength of the conductors
- rapid temperature change

Integration of testing into the production workflow is also considered. Depending on the test equipment selected, it may also be necessary to optimize the product in terms of testability. The final step is integration of the ACT process developed in parallel.

8.7 Elaboration of Assembly and Connection Technology

In much the same way as the production process is elaborated, the assembly and connection technology (ACT) is elaborated in this phase on the basis of the solution concept for the production process. Progress parallels and closely interacts with design of the 3D electronics. Initial clarification focuses on ascertaining whether, as detailing increases, the product is still compatible with the production process. The ACT process chain is then elaborated and test procedures are selected. The result is the elaborated production process of the ACT.

As is the case with elaboration of the production process, prior to elaboration of the ACT process chain the **compatibility** between the detailed product and the assembly and connection technologies selected in the production concept is rechecked. For example, large, solid components act as heat sinks in the vapor phase and impair solderability. The product data are analyzed and compared with the solution concept for the production process. It is important to clarify the extent to which the choice of material and the design of the interconnect device and the components to be used are compatible with the selected assembly and connection technologies. The following are product properties that particularly require consideration in this context:

- **Solder-application and component-placement surfaces:** The way in which medium is applied and components placed depends largely on the accessibility of the surfaces (part design). If the process originally selected is found to be unsuitable, other processes can be used instead, for example dispensing or pin transfer.

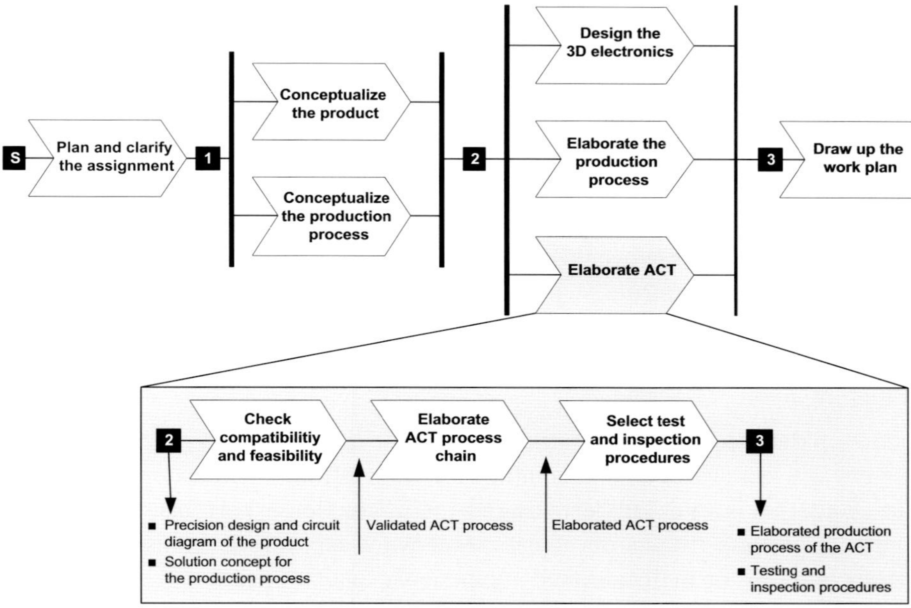

FIGURE 8.10 Process steps in the *elaboration of ACT* phase

- **Degrees of freedom:** Additional degrees of freedom might have to be taken into consideration because of the three-dimensionality of the MID. Vertical walls are an example. A rotatable handling system and repeat process passes with the MID appropriately repositioned for each successive pass can counter this difficulty. It is important to ensure that solder applied beforehand is not remelted in subsequent passes. Solders with different melting points or vapor-phase soldering are possible solutions to this difficulty.
- **Types of electronic components:** Both component size and the type and arrangement of the "legs" (THT, SMT, etc.) place high requirements on the ACT, particularly with regard to tolerancing and process control.
- **Types of solder materials:** The RoHS directive stipulates the use of lead-free solders, so interconnect substrate material, the electronics components, and the process all have to be checked to ensure that they are suitable.

The resulting requirements and restrictions for the ACT have to be clarified jointly with the mechanics and electronics developers and the production specialists. Once compatibility has been established the selected processes can be elaborated.

The work processes from the **process sequence** partial model are detailed, with the work steps being defined on the basis of the production concept and the intermediate steps such as cleaning inserted. In the reflow soldering process the work steps are solder-paste application, component placement, and reflow soldering.

The handling system suitable for the production conditions (e.g., existing product lines and planned number of units) also has to be selected. Guidelines for these process steps are VDI 3712 *Determining the Machine Capability of Dispensing Systems* and VDI/VDE 2251 *Precise Mechanical Elements; Connections; Survey*. The **production structure** partial model is also detailed at this stage. Equipment available in the company is analyzed and selected. The choice is restricted by product requirements such as the size of the module and the temperature resistance of the electronic components. Vapor-phase ovens, for example, make for much better precision and uniformity of temperature distribution. Process parameters such as placement sequence and temperature-transient control are defined for the individual items of equipment. Criteria include the nature and mass of the electronic components and the melting point of the solder.

Much as in the process step from the *elaborate production process* phase, the requisite tests and test procedures are defined for assembly and connection technology. Once again, this is a joint undertaking by the production expert together with the mechanics and electronics developers. Important aspects for testing assembled MID are missing or malpositioned components, poor solder joins, and function testing of the assembled module (Chapter 6). The purpose of these tests is to ascertain whether the electronic components have survived the high soldering temperatures undamaged. Guidelines applicable to the testing of electronic circuits are VDI/VDE 2628 *Automated Visual Inspection; Description of the Testing Task* and VDI/VDE 3715 *SMD - Surface Mount Devices - Process Measurement and Testing for Printed Circuit Boards (PCB)*. The result is an elaborated ACT process complete with test procedures.

8.8 Work Planning

Compilation of the production documents is one result of the concluding phase of *work planning*. The initial step is matching production processes to product requirements. This includes, for example, aligning process parameters (e.g., conductor width) with permissible tolerances from the requirements catalog. The objective is to ensure that the production process satisfies the requirements. Two parallel strands then emerge, namely job-workflow planning and job-resource planning (Fig. 8.11). Job-workflow planning is the process of developing work and assembly plans, control programs, and test plans.

Work planning deals with detailing of the blank (type of blank and its dimensions), definition of process sequencing, choice of production resources (machines and tools), and calculation of specified times for the process steps. Assembly planning

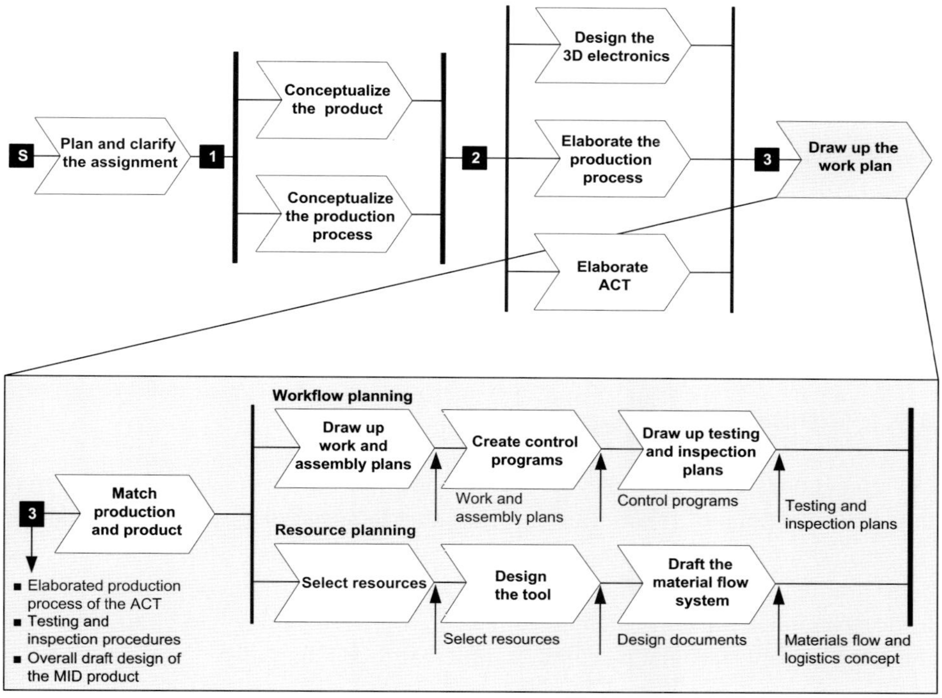

FIGURE 8.11 Process steps in the *work planning* phase

maps the sequence interrelationships between the assembly processes and also the buffer times. Control programs for the machines are created on this basis. All geometric, technological, and flow-oriented information has to be collected for this purpose. The third process step entails defining the test procedures and integrating them into the manufacturing process. The test plans are then generated. The first step in the framework of job-resource planning is selection of the resources. The resources pool includes plant, equipment, and jigs necessary for fulfilling the assignment. Alternative work processes and resources are calculated. Then decisions on in-house production or outsourcing are made on the basis of technological and economic criteria. It is highly unlikely that any one company will be in a position to cover the entire process chain in MID production. Process cost calculation is therefore the partial model frequently applied for reliable calculation (Fig. 8.7).

The necessary jigs and tools are designed and optimized for the product in the second step. The third step is conceptualization of the materials flow system. Deadlines, capacities, and quantities of the production program are figured. The control concept is developed on this basis. Then the materials-management concept is developed. The results are the complete production documents, resources, and tools.

8.9 MID-Specific Instruments of Development

Development methodologies can be applied to the conceptualization and elaboration of MID products. These methodologies structure the process and facilitate systematic and solution-oriented development. The process is generally described in a flexible guideline rather than rigid roadmaps; being flexible the guideline does not restrict the creativity of the developer. The sections below provide an overview of established methodologies and innovative new approaches.

8.9.1 MID Design Catalogs

Design catalogs are collections of tried and tested solutions for given design challenges. Collections of this nature are intended to assist developers in the design process. There are many catalogs, differing in content, mode of presentation, and degree of detail offered. They include for example collections of physical effects, basic principles, machine elements, or standard parts and are to be found in the industry literature, company catalogs, and standards manuals.

Developers searching for solutions place requirements on design catalogs. They should facilitate rapid, task-oriented access to alternative solutions. The range of solutions, moreover, should be mapped in full, and content should be usable within both conventional and computer-aided design. On account of the differing degrees of detail, catalogs can be used in many sections of the development process. As the solution becomes progressively more detailed, the use of catalogs can be more direct or, conversely, more limited. Roth provides guidelines on how to use and compile design catalogs. He also proposes a basic structure, suggesting that a work of this nature should always consist of classification section, main section, and ready reference section, plus an appendix. The classification section contains organizational aspects that consistently structure the content presented in the main part of the catalog. The system of classification has a great deal of influence on practical usability and the ease with which information can be retrieved. The main section is the body of the catalog with its content, presented for example as sketches or drawings. The degree of detail and form of presentation depend on the development phase in which the catalog is intended to be used. The ready reference section contains descriptions of the properties of the individual solution. The appendix is reserved for comments, cross-references to external sources, and other supplementary information.

An MID design catalog originally compiled by Peitz has been enlarged in the interim. It contains various MID solution alternatives for mechanical and electronic functions.

	Function \ Solution	Solution 1	Solution 2	Solution 3	Solution 4	Solution 5	Solution 6	Solution 7
Electrical MID functions	Transport electrical energy	Conductor	Antenna					
	Contact neighboring module	Plug	Contact spring	Direct plug-in principle	Contact pad	Press-in pin	SMD plug	Contact bump
	Contact electronic components (ACT)	Soldered connection	Glued connection	Bond			Flip-chip sold.	Flip-chip glued
	Shield electromag. fields	Shield	Grid					
	Protect against current spikes	Fuse						
	Close/open the circuit	Snip-action switch	Silicon switch					
Mechanical MID function	Carry components	2 ½ D	n x 2D	Shell structure	3D			
	Transmit force	Snap conn.	Sawteeth	Gearwheel	Clamp conn.	Plug conn.	Push conn.	
	Protect against external influences	Package	Casting	Painting	Jacketing			
	Conduct thermal energy	Metal. surface	Cooling fins	Heating structure				
	Stabilize components	Stiffeners	Spacer	Casting				
	Mark components	Barcode	Refference point	Colors	Labeling			
	Contact module	Cylinder	Hole					

FIGURE 8.12 Expanded MID design catalog by analogy with Peitz

The alternatives are presented in the form of illustrative pictograms. The ready reference part of the catalog contains detailed descriptions of the individual solutions.

Alternative solutions for subfunctions of technical products are presented in easy-to-use, structured form in design catalogs. Methodical reference to catalogs enables the developer to go about the job of finding solutions quickly and purposefully. Consequently, they are valuable as an aid in the development process. The MID design catalog developed by Peitz was originally domain-spanning, and it has been steadily expanded over time.

8.9.2 Properties Cards for MID Processes

Different MID manufacturing processes have different properties profiles. The intrinsic strengths and weaknesses of each process are determined by the underlying technology. The properties cards developed by Peitz describe the characteristics of the individual processes.

Peitz suggests a total of eight features that make up the properties profile of an MID basic solution. There are predefined forms of each feature. The requirements are evaluated in the column on the right. Evaluation is on a scale from 1 (low requirement) to 5 (high requirement) (Fig. 8.13). The strengths and weaknesses of the manufacturing processes are determined by the underlying technologies. For example, the *laser direct structuring* process is intrinsically strong as regards the "complexity of change" feature, because lasering and control of the laser make for flexibility in modifying conductor layouts. The *two-shot molding* process used in MID production is weak in terms of this particular feature, because changes necessitate complex alterations to the injection molding tools.

These **properties cards** are helpful for selecting the best-fit process for the MID basic solution. Direct correlations can be set up between the requirements features of the basic solution and the properties features of the production processes. This is the case, for example with "conductor tracking" as a requirement feature and "3D design freedom of the conductors" as a property feature. Requirements profiles for specific MID can be matched directly to property profiles of particular manufacturing processes in this way.

This identifies the MID production process best suited to the requirements profile of the basic solution. The features for which requirements are particularly high are naturally those on which attention should focus. The number of MID production processes is limited, so requirements profiles cannot always be satisfied in their entirety. Compromises are often necessary. In some cases, for example, the design of the part has to be altered or the conductor layout modified to soften individual requirements applicable to the production process.

MID production process: laser direct structuring

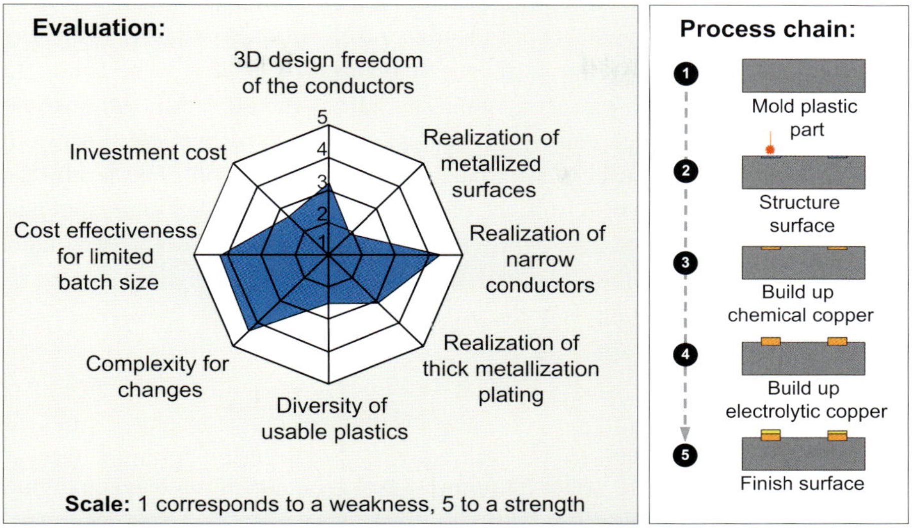

Feature	Property of the manufacturing process
3D design freedom of the conductors	Metallization is possible on all sides, conductor layout is restricted by accessibility for the laser
Realization of metallized surfaces	Considerable complexity is involved in plating large surfaces (many parallel passes)
Realization of narrow conductors	Conductor widths less than 80 µm are possible
Realization of thick metallization plating	Maximum plating thickness is 50 µm
Diversity of usable plastics	Choice of plastics is restricted; this is dependent on the process (LDS, additive, subtractive structuring)
Complexity for changes	Very low; the control program for the laser has to be changed
Cost effectiveness for limited batch size	Laser control is flexible (batch size of 1 unit is not a problem)
Investment cost	High investment costs; laser unit (possibly including handling equipment) and galvanizing unit

FIGURE 8.13 Properties card of the MID production process *laser direct structuring*

Selecting a particular MID process also simultaneously determines the associated process chain. This is outlined below for laser direct structuring. In the first process step the 3D interconnect device is produced in a conventional process by injection molding. In the second step the surface is structured by lasering. The conductor layout is determined in this process step. The part is metallized in the next two process steps. Metallization can be a chemical or an electrochemical process. In step five the conductor surfaces are plated with nickel and gold. The production resources required are derived from the process chain. The number of technologies available is limited, so not infrequently it is not possible to achieve a 100% match between requirements profile and production process. It might well prove necessary to make corresponding changes to the product concept.

8.9.3 MID Guidelines

Companies active in the MID sector frequently develop application-related guidelines and standard operating procedures for their proprietary processes. These documents, although tailored to company specifics, often contain higher-order, generic process knowledge as well. These guidelines, some of which are freely available in the public domain, can be consulted to gain insights into particular terms of reference. Two examples will suffice to explain the layout and structure of these guidelines.

HARTING Design Guidelines [70]

HARTING Mitronics is a service provider working in the field of implementing customer-specific MID products. The range covers many different sectors, including for example sensor technology, telecommunications, and logistics. HARTING offers a complete value-adding chain for MID technology, extending from development through to series-production maturity. The company has developed a guideline for product design that covers the two processes of two-shot molding and laser direct structuring.

Two-Shot Molding

The HARTING design guideline is available in tabular form with illustrations for clarification. It is structured as follows.

General requirements for injection molding: This part sets out requirements that apply to the tool, the molding itself, and process control. For example, no release agent should be used, and when the mold opens there should be no burr at right angles to conductor structures.

Process boundary conditions: This part defines conditions relating to part size and the materials that can be used (for the nonmetallizable and metallizable shots).

Basic design rules: Values for the design of the layout structures are defined.

Geometric specification: The aspects set out in the basic design rules are elaborated and the values for the pads are given as well.

Metallization: Contains information about the buildup of the copper, nickel, and gold plating.

Electrical properties: Automotive engineering is one of many areas of application that place high requirements on the product's electrical properties. HARTING provides guide values for resistivity, dielectric strength, current-carrying capability, and so on and indicates the standards and tests that the two-shot molding process has to satisfy.

Laser direct structuring: The design guideline for LPKF laser direct structuring is also subdivided in the same way. The information is based on the machines and tools used in the company and on knowledge acquired in the implementation of various applications. It does not differ in any significant way from the LDS MID guideline issued by the LPKF company, so it not discussed in more detail here.

LPKF-LDS® Guideline [104]

The LPKF-LDS® guideline compiled by the LPKF company facilitates the design of innovative MID to be manufactured by the LDS process. The guideline published by the company provides designers with reliable figures and recommendations for product design for manufacture. The rules are footed on technical parameters for the plant used in the process and on company experience from applications implemented in the past. The content deals with both the development of the product itself and the production resources involved. LDS MID developers therefore have a guideline covering the process steps of injection molding, structuring, metallization, and ACT.

Materials: The LPKF-LDS® process is compatible only with certain materials. The overview of materials approved by LPKF comes right after the introduction to the guideline.

Design rules: The guideline supplies very specific design rules that assist developers in designing molded interconnect devices. Recommended conductor widths and spacing are explained, along with edge radii and conductor positions.

Injection molding: The MID are one-shot injection moldings with postmolding laser structuring, so the injection-molding process is to a very large extent coresponsible for the quality of the finished MID product. Shrinkage of the plastic has to be taken into account in designing the basic body, but so too does the material of the injection-molding tool. The mold should not be made of aluminum, for example, to avoid tool-related inclusions when the LDS MID are metallized. The surface quality of the mold and the position of the gates are also crucial factors for the quality of the part and are therefore discussed in detail in the guideline.

Laser structuring: The guideline contains detailed information on laser parameters, parts holders, and other process parameters for structuring.

Metallization: The developer needs a good understanding of certain process parameters in order to ensure correct metallization. These too are explained in detail in the guideline.

The LPKF design guideline provides a very detailed and user-friendly overview of all the processes associated with the LPKF-LDS® method. The design rules and parameters for the laser direct structuring process itself are dealt with in detail in Chapter 3.

8.9.4 MID Features

MID features are specimen solutions for MID products developed on the basis of the preceding chapters and a product analysis of existing applications. They assist developers in that they provide integrative descriptions of design possibilities and the applicable rules for manufacture. Comparing the features with product requirements is a way of deriving design concepts and possibilities for the implementation of individual product functions.

The contents of each MID feature are stated clearly and unequivocally in a profile. The elements that make up the profile are description, design, MID functions, and process-specific requirements. Essential aspects of the feature are summarized briefly in the description. Small pictograms illustrate the design of the feature. The MID function element refers to the developed MID design catalog, so it points to electronic and mechanical product functions implementable to a qualified extent with the feature. The process-specific requirements deal with issues that have to be taken into account in engineering implementation of the feature. The development of new features is currently a subject of research; the feature used here by way of example is “alignment.”

Alignment

MID technology enables electronic components to be positioned directly on the surface of the structured interconnect device. Inclining the functional surfaces means that electronic components can be given a defined alignment.

The feature refers firstly to alignment of the component relative to external reference points. For example, SMD connectors for establishing electrical contact with neighboring modules can be given the ideal orientation to bring them into perfect alignment with their off-module counterparts. LEDs or switch elements on the 3D-MID can be aligned in accordance with application-specific requirements. Secondly, the feature permits defined angles to be created between electronic components set on the MID.

MID feature: Alignment

Description:
The feature refers to the alignment of electronic components on the MID. When the components are placed on inclined process surfaces they can be aligned with external reference points. The individual components of the module can also be positioned at preset angles to each other.

Design:

• Inclined functional surface	• Defined angle between components	• Component alignment on level functional surface
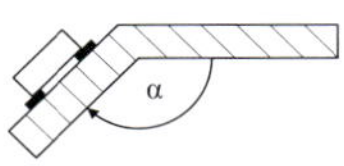	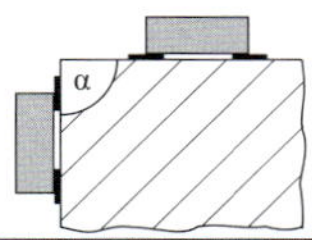	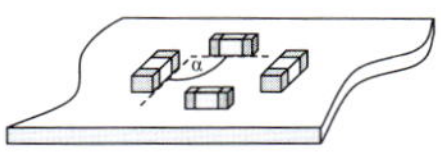

MID functions:

• Carry components	• Achieve optimum alignment for LEDs or sensor chips on inclined functional surfaces	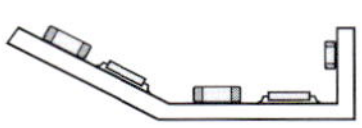
• Close/open the circuit	• Alignment of switch elements	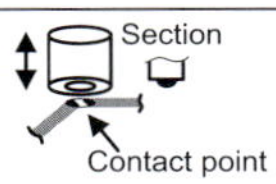
• Contact neighboring module	• Adapt alignment relative to other modules for electrical contacting by SMD plugs, for example	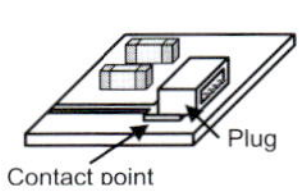
• Transport electrical energy	• Antenna structures acting in defined direction	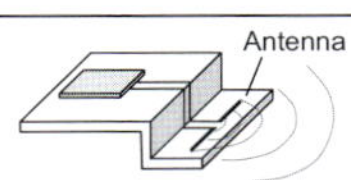

Process-specific requirements:

- Allow for restrictions of the placement system with regard to geometric classification of the functional surfaces.
- Carrier for alignment of the components:
  If a six-axis robot is not used for placement an adapted carrier is necessary in order to rotate the part so that it is normal to the placement head.
- Note maximum size of electronic components:
  SMD plugs and switch elements are generally among the larger SMD parts.
  Additional gluing to secure the parts might be necessary.
- Adapt connection processes to suit the alignment of the functional surfaces:
  Different connection steps and connection temperatures necessary, depending on how the process surfaces are positioned.

FIGURE 8.14 Layout of a feature, "*alignment*" as example

Some restrictions apply with regard to design. They relate primarily to the positioning of electronic components on inclined function surfaces, because special requirements apply to the ACT process. If the pick & place system is of conventional design, special carriers have to be used so that the part can be secured and held in correct alignment for component placement. An alternative is to use a six-axis industrial robot for component placement. However, these robots are generally slower than high-tech fully automated pick & place machines, so usually the concept is economically viable only if there are several inclined function surfaces and high-volume production is planned.

An MID with planes inclined at different angles will generally necessitate a change of clamped position when handled by a conventional pick & place robot. Using a multifunction carrier is an alternative. The carrier automatically changes the position of the MID relative to the placement head once all the components have been positioned on a surface.

8.10 Computer Assistance

Development of MID products is very complex. It often makes sense to use software to structure the data and solve specific problems.

In technical product development, computer assistance has long been established practice in the areas of mechanical design and electronic layout design. There are numerous advantages, including validation of product properties on the basis of digital models analyzed with the aid of simulation programs. Nor is there any denying the benefits of realistic visualization of the product for company-internal communication and discussion with the client. Last but not least, the production department benefits from early availability of product models and can fine-tune processes in advance without the need for physical prototypes. Modern parlance includes terms such as concurrent or simultaneous engineering to designate arrangements in which different departments work on the product or the production processes at the same time, basing their endeavors on unified virtual models [155].

Another advantage of computer assistance in designing mechatronic products like 3D-MID is that the task, although complex, is not overwhelmingly so. The benefit is all the more tangible when interdisciplinary cooperation is absolutely essential. Nowadays, individual departments cooperate more and more through the interplay of what have become very powerful software tools and also to an increasing extent through MID-specific development tools.

8.10.1 MID-Specific Requirements for Development Tools

The requirements applicable to development tools for MID products were identified back in the 1990s, in other words during the early days of this technology. The direct substantial integration of mechanical and electrical or, more accurately, electronic functions on a three-dimensional interconnect device stands out clearly [51]. Existing software for product development can therefore be evaluated in terms of its suitability for MID technology. This section explains the properties required of the programs and the additional functions rendered necessary in this way.

The numerous technological explanations and examples in this book illustrate the many and diverse benefits of MID technology, foremost among which are design freedom and the miniaturization of parts. In order to render the benefits of this technology available in the form of a product, the software tools too must be able to serve the following terms of reference in the individual phases of product development:

- assessment of existing and new products and product technologies by application and cost-effectiveness analysis and comparison of a possible product optimization or new product ideas with the specifics of MID technology taken duly into account
- function-oriented concept design of the product:

  integration of thermal, optical, fluidic, mechanical, and electrical functionalities
- integrated design of the mechanical and electrical structures in a three-dimensional computer-aided design (CAD) system for electronics
- design for manufacture (DFM) with computer-aided simulation systems (especially process, kinematics, discrete system simulation)

The computer-aided design of the product and the related simulation processes are discussed in detail below.

Necessity for a 3D Development Environment

One result of the computer-aided development of electric circuitry for printed circuit boards was that the software products had a two-dimensional development environment. Rectangular-shaped planar boards were the norm, so for decades this mode of representation was perfectly adequate. Expansions were allowed for the number of layers in a multilayer board construct, the placing of electronic components on both sides of the board, and the stacking of multiple flat boards. The differences in the complexity of the software tools for electronic computer-aided design (ECAD) can be explained by the difference in focus. The more layers in a printed circuit board (PCB) or the more specialized the applications (high-frequency circuits are a case in point), the larger are the quantities of data to be handled and the more specialized are the functions needed.

Only the first steps in the procedure typical for the development of electric circuits for MID are comparable with that for PCB. The latter process is dealt with in Section 8.1.3, but the following remarks also apply. It is essential to have the function of the target circuit defined in the system specifications. The next step drafts the architecture and behavior designs, which define the interfaces to the environment, for example, and the timing. The logic design covers aspects such as control signals and arithmetic operations, distinguished by data and control flows. Hardware description languages (HDL) such as Verilog are used to realize computer-assisted automation of the logic design. Layout synthesis then follows, and in this step the circuit design is transposed into electronic components and their electrical connections so that geometric design and arrangement can be detailed. The layout is finalized with placement and wiring of the electronic components on the interconnect device, the latter process commonly referred to as routing. Concurrently, technological and production-related requirements in the form of design rules and manufacturing rules have to be complied with, and this is ensured by automated rule checks (DRC, MRC). As indicated above, these are the steps that apply for arriving at the 2D layout. This is illustrated in Fig. 8.15. The graphic shows the current demonstrator of the Institute for Factory Automation and Production Systems (FAPS) in the Eagle 2D development environment. [51, 99]

As regards designing electronic circuits on MID, existing ECAD tools suffice for the steps in system specification up to logic design. Programs with the ordinary range of functions are generally powerful enough for this, because until now MID circuits in general have been much less complex than the circuitry on printed circuit boards. The electronic design automation (EDA) programs for 2D layouts described in Section 8.10.2 are not suitable for 3D-MID, however, because they cannot handle the three-dimensionally structured surface. From the layout synthesis step onward, special functions are necessary for MID. Workarounds that planarize the three-dimensional surface of an MID work only with simple geometries, which are only very rarely encountered. In other words, stopgap solutions that unwind the 3D body in 2D software tools do not produce the desired results. Consequently, a **three-dimensional development environment** is essential for MID layout. [51, 177]

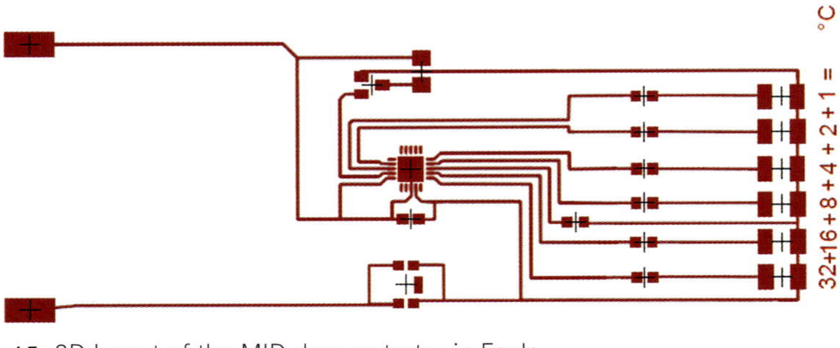

FIGURE 8.15 2D layout of the MID demonstrator in Eagle

Integrated Software for Mechanics and Electronics

Alongside the two-dimensional layout of flat printed circuit boards, parallel studies in three-dimensional mechanical CAD systems (MCAD) have become established practice. The advantage is that the installation specifics of the electronic module can be mapped. The shapes of the printed circuit board and the package can be modeled and defined in MCAD, along with attachment points and screw holes. Dual modeling such as this enables the electronics to be slotted neatly into the later product as an entity. The usual procedure for parallel ECAD-MCAD development is shown in Fig. 8.16. The designer responsible for the mechanical layout specifies the size of the available space and therefore the shape of the printed circuit board. The layout specialist responsible for the electronics positions the circuitry on this surface, routes the conductors, and runs the necessary simulations for thermal loading and electromagnetic compatibility (EMC) to verify the design. The designer who receives these data in the form of position and orientation of the module package can check for potential problems such as collisions at the installed position and the accessibility of plug connectors. At this point the designer also has to run simulations for structural mechanics influences, for example. The designer's changes to the layout, if any, have to be checked by the electronics specialist. This necessitates iterative loops between the two, the channel being file interchange. [51, 177]

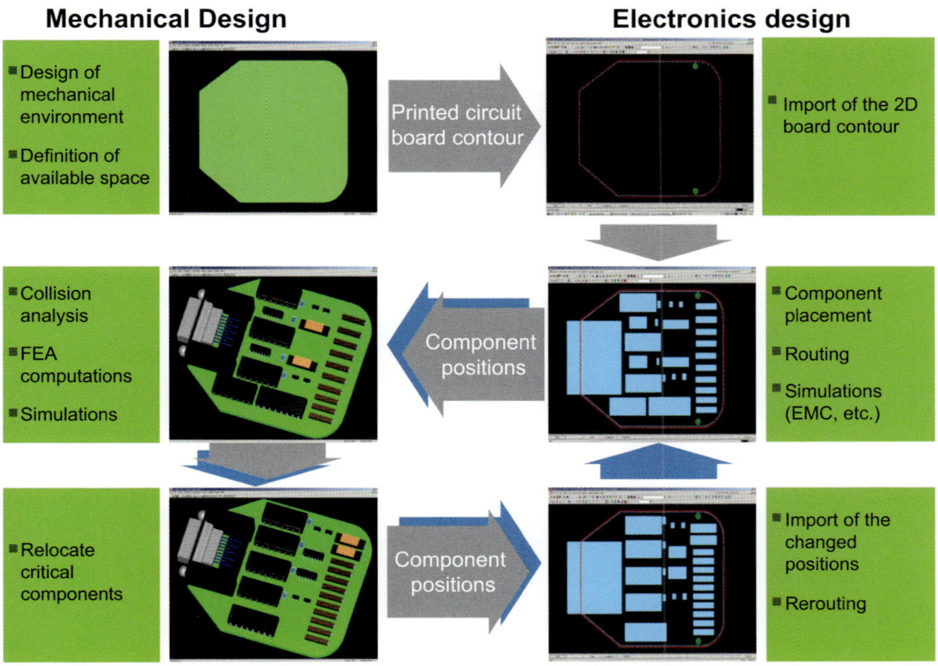

FIGURE 8.16 Mechanics and electronics development are separate, with file-based data interchange (graphics courtesy of Mecadtron)

The drawbacks of separate development in ECAD and MCAD are apparent in Fig. 8.16:

- time-consuming and therefore costly loops between the separate development departments for mechanics and electronics on account of possibly incompatible requirements, knowledge baselines, and file formats
- introduction of unnecessarily wide tolerances to accommodate the other party
- innovative product ideas hampered by the wealth of detailing

Section 8.10.2 deals with the joint endeavors of software manufacturers in the ECAD and MCAD sectors to eradicate these disadvantages. However, the question remains whether this procedure should be adopted for MID. And the answer is no. The unsuitability for MID emerges in the lack of support for three-dimensional interconnect devices. Even incorporating 3D MCAD tools does not make up for the shortcomings of the 2D ECAD tools. In fact, the conclusion can be taken even further: MID design needs a unified and therefore **integrated 3D development environment** that combines mechanical design functions and electronic layouting and maps the substantial integration of mechanical and electrical or electronic functions within a 3D-MID (Fig. 8.17). Direct and indirect interactions between mechanics and electronics can then be taken adequately into consideration. Integrating simulation software for design for product and design for manufacture through MID modeling, which under certain circumstances might in fact prove impossible, also becomes considerably more straightforward. Other disciplines such as optoelectronics can also be incorporated into this model without difficulty. [177]

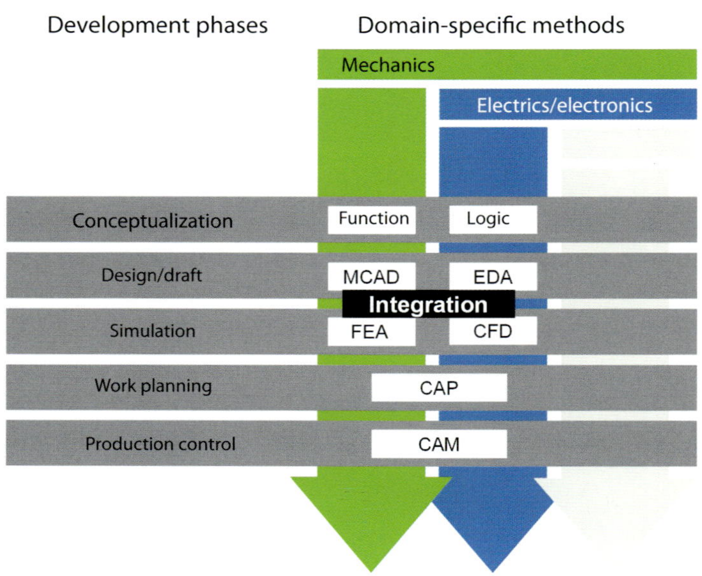

FIGURE 8.17 Identification of the development phases for MID-specific integration

Integrated MID Product Model

The baseline for product development consists of the internal data models of the software tools and higher-order, product-describing data formats that must include all the relevant information for modeling and processing. This includes in particular a given product's design, materials, functions, internal connections, external interfaces, and manufacturing processes. Alongside the proprietary file formats of the software manufacturers there are usually generally applicable data structures for the tools of a class intended for data interchange.

The sets of information specific to the individual disciplines are termed partial models. For MID the partials of relevance are the mechanical and electrical/electronic partial models, with content deriving from the structural design of the MID as an electromechanical product. Of note too is the fact that despite the differences there can still be redundancies.

Mechanical partial model:

- geometry of the 3D-MID and the electronic components
- position and orientation of the components and 3D routing of the conductors
- information about materials
- production-related parameters, for example for injection molding or computerized numerical control (CNC) data for drilling holes

Electrical/electronic partial model:

- circuitry from the logic design
- netlist from logic synthesis: list of electronic components and the electrical signals passing between them (nets) for pins and conductors
- specifications and results from placement and routing algorithms
- rules for design rule checks (DRC) and manufacturing rule checks (MRC)

An integrated 3D development environment is needed for MID. By analogy, there must also be an integrated product model that combines the mechanical and the electrical/electronic partial models (Fig. 8.18). This has long been a focus of research in the Institute for Factory Automation and Production Systems (FAPS) of the Friedrich Alexander University, Erlangen-Nuremberg, based on the ISO 10303 standard for the exchange of product model data model (STEP). STEP application protocol 210 in particular is used to describe electromechanical products. This would serve as the basis for implementing integrated electromechanical modeling and electrical or electronic elements in a mechanical modeling system. This was conceived of and realized as the STEP application system. [94, 177]

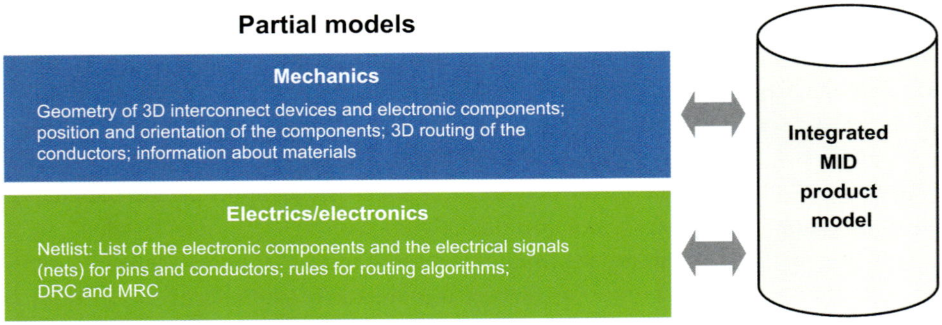

FIGURE 8.18 Integrated product model with mechanical and electrical partial models

Interfaces to Associated Programs

An integrated development tool for 3D-MID cannot and need not cover product development in its entirety. All the phases adequately covered by conventional programs for the development of mechatronic products can be excluded. As indicated in the section dealing with the procedure for the design of the electronic circuitry, MID-specific software functions are not necessary for development of the logical functions and logic synthesis. Accordingly, the software for MID products must have interfaces to the preceding and subsequent programs:

- Import of geometry: This applies for three-dimensional interconnect devices from specialized 3D-MCAD systems and the 3D-CAD models of the electronic components. Along with the proprietary file formats of the CAD manufacturers for which import and export modules have to be purchased, there are also neutral exchange formats like STEP and Initial Graphics Exchange Specification (IGES).
- Import of circuit structures and netlists from 2D ECAD: Here again each software manufacturer has developed proprietary file formats that require import and export filters. In some instances the data exist in a text-only format. The principal difficulty, however, is that cryptic designations for components have to be traced back and matched to the descriptions in the component library in 2D ECAD. This is discussed in more detail in Section 8.10.2.
- Export to simulation programs: Examples for injection molding are explained in Chapter 3 of this book. Data interchange takes place either in the file formats described above for MCAD systems or by a direct link between CAD system and simulation software.
- Export to preproduction and to production control: CNC machining as well as structuring and metallization are examples. Data exchange by CAD formats is one possibility and is used by LPKF. Another possibility is to use the CAD data as the basis for defining further information that can be saved in separate formats. Section 8.10.4 contains an example for a metallization process.

MID-specific software in line with the development steps in Section 8.1 is available for designing the 3D electronics and for elaborating the assembly and connection technology and the corresponding production processes. To date there are no separate programs for planning and clarifying the purpose of an MID product or for conceptualization of the functions. The integration of MID-specific evaluation tools and design catalogs into software should be a prime field for scientific development. Accordingly, the remarks below focus primarily on development tools with regard to mechanical design and electronic layout and on the connectivity between these tools and simulation programs and production.

8.10.2 Software Tools for Design and Layout

Many of the companies that concern themselves with the development and manufacture of MID come from classic sectors of industry such as plastics or electronics production. Naturally, these companies have a marked tendency to apply their departments' proven software products for MID. It is for this reason that the ECAD and MCAD programs currently in use are discussed here. As the production volume of MID products increases and their complexity develops, attempts are being made to use MID-specific software because the return justifies the outlay and cost of procurement, training, support, and licensing. MIDCAD comes from the world of research, while NEXTRA® is a commercially available program; these two are discussed below.

Current Development Tools Used in the Industry Environment

The classic MCAD and ECAD systems are still very much in use in MID product development. It is important to distinguish whether ECAD tools are used at all, or if a simple circuit diagram is to be developed directly in MCAD. The MID manufacturers frequently do not undertake development of the electronics because their clients supply predefined circuit diagrams. But if an **ECAD tool** has to be used, resort is made to relatively simple programs like Eagle from CadSoft Computer GmbH [21], or a spreadsheet editor is used to calculate electrical parameters. To date, the electronic circuits have not been complex, so there has rarely been a need for high-end ECAD systems. The big players in this sector include Cadence Design Systems, Inc. [20], Mentor Graphics Corporation [115], and Zuken GmbH [180]. Efforts are being made to increase the complexity of the electronic circuits on three-dimensional interconnect devices, so these ECAD tools merit mention in the framework of this book.

There are diverse **MCAD systems** available for mechanical design. The design remit includes not only the three-dimensional injection-molded body but also the modeling of electronic components and conductors. Practical experience has shown, however, that layouting the electronic circuit on the three-dimensionally structured substrate in MCAD is both difficult and time-consuming. Special functions for simple component placement are lacking, and the same applies to functions for routing conductors. The effect of these inadequacies is particularly negative whenever changes have to be made to the layout because the existing conductor tracks have to be completely deleted and the conductors rerouted. From the point of view of the user, the need for MID-specific layout functions in MCAD systems is urgent.

MID manufacturers generally use mainstream MCAD systems like Pro/ENGINEER (Pro/E) or Creo from the Parametric Technology Corporation [125], NX from Siemens Industry Software GmbH & Co. KG [156], or, in the automotive sector, CATIA from Dassault Systèmes [27]. For historical reasons the system of choice for the FAPS institute is Pro/E, and this system is discussed in more detail below.

The examples for **ECAD-MCAD collaboration** offer potential for improvement in adapting electronics to suit package space. The ECAD-MCAD Collaboration Extension (ECX) module is embedded in Pro/E for data-transfer connectivity with software from Mentor Graphics and Cadence. This module is based on findings obtained by the ProSTEP iViP Association from investigations into a new, neutral exchange format for ECAD-MCAD [133]. The association postulates a common data model that permits the definition of collaboration objects with descriptions and attributes. The terms and the hierarchical structure were taken from STEP (ISO 10303). Unlike the intermediate data format (IDF), which always transfers the complete assembly and does not make changes transparent, this format, which is based on the extensible markup language (XML), allows each element to be defined as a collaboration exchange object and the changes displayed simultaneously on the mechanical and electronic sides [164]. The layouter now also has a three-dimensional view of the assembly and the mechanics expert has a visualization of the circuits path on the PCB. The component libraries from ECAD and MCAD are standardized, moreover, so the 3D master models can be tagged with both mechanical and electrical information [111]. Collaboration thus becomes structured, transparent, and coherent.

The solutions of the big-player software manufacturers simplify cooperation between designers and layouters, but they do not support three-dimensional interconnect devices. So even though program interfaces have improved, the conclusion is the same as in Section 8.10.1. ECAD-MCAD collaboration works very well for planar boards but not for three-dimensional electronic circuits.

MID-Specific Software in the Research Sector

Research and other considerations undertaken over years by the FAPS institute geared toward the requirements for an MID-specific development tool and have resulted in two programs. The software tool **MIDCAD** for designing 3D-MID is discussed first [177]. On account of the requirements for the design of 3D-MID, it made sense to use 3D-MCAD systems as the basis for an MID tool. MIDCAD was implemented as a plug-in for Pro/E Wildfire through the proprietary Pro/Toolkit programming interface. MIDCAD's internal data structure is arrived at by adding the electrical information for electronic components and their signal connections. This led to an MID product model as described in Section 8.10.1 that has a functional view and also a physical view.

The 3D geometries of the electronic components are stored in a device library. Pin mapping, land patterns, and electronic (e.g., functionality), mechanical (rigidity, elasticity, etc.), and thermal properties (heat conductivity, etc.) are all included in the dataset. The components can be read from the library on the basis of the netlist from the ECAD tool Eagle. The components can be positioned on the three-dimensional interconnect device freely on analytically describable surfaces, and keepout areas (e.g., dummies for holes to be drilled) can be defined in advance.

The placement step is followed by routing, made possible by the underlying netlist and the electrical properties of the component geometries, as mentioned above. Once the netlist has been imported from Eagle, the direct connections of the individual pins are indicated by colored lines representing the shortest routes, providing the designer with visual assistance for tracing the best routes through this complex web. As regards optimum conductor routing, it is important to identify the surface properties on 3D interconnect devices and typical MID features such as vias between the different sides of the interconnect device. Routing can be done manually with commands implemented specifically for the purpose or, for prototype implementation, automatic by an integrated function that adds a third dimension to 2D routing. The evolved operating procedures are applied in this respect, and both grid- and line-based or gridless and topological routing methods for two-point nets can be used. Multipoint nets, by contrast, are simplified by creating minimum spanning trees or minimum Steiner trees and broken down into two-point nets. In addition to routing, there are functions for parametric solids modeling of the conductors with production-typical conductor cross-sections as presets (Fig. 8.19).

MID-specific design rule checks such as design for metallization at edges and shifting of join lines are also implemented, and typical MID features (vias in molding body, pin headers) are available in parameterizable form. CAD/CAM chains for 3D component placement by robot or modified pick & place machine are implemented in exemplary fashion in MIDCAD and illustrate the wide-ranging possibilities for interfacing an MID design tool to production control. [177]

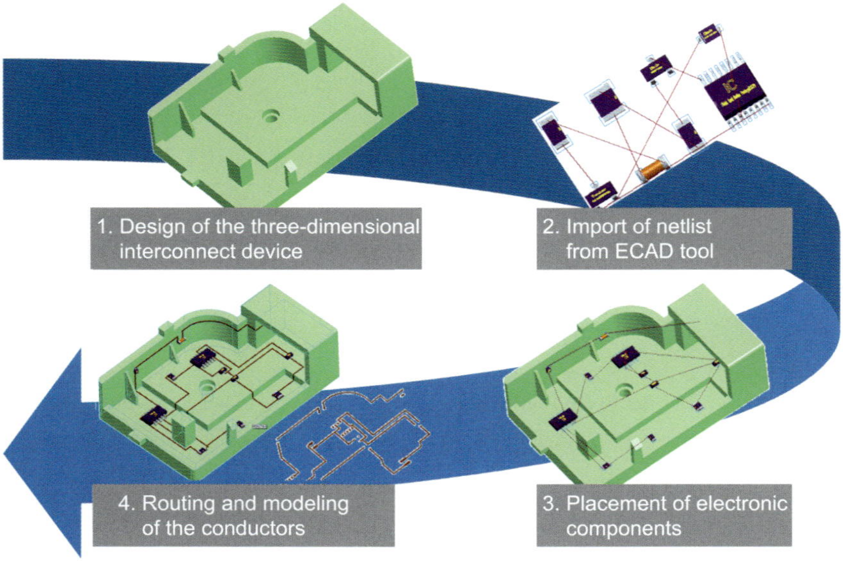

FIGURE 8.19 Typical procedure in MIDCAD

Commercially Available MID-Specific Software

Research endeavors also led to a second tool, the basics of which were elaborated in the framework of a doctoral thesis [94]. This tool is now maintained by and can be purchased from a dedicated company. This solution is implemented in the **NEXTRA**® software marketed by Mecadtron GmbH. Although attributable to the ECAD sector, contrary to the classical approach it offers a three-dimensional development environment on the basis of an MCAD kernel (3D-ECAD) [112]. This environment allows the designer access to functions essential for designing electromechanically integrated products. These functions include import netlists, three-dimensional positioning of electronic components, and manual routing of electrical conductors on the surface of the three-dimensional circuit carrier. As in MIDCAD, both layout methods are intuitive, eliminating the difficulties of complexity in designing 3D-MID that are inherent to MCAD systems. The advantages vis-à-vis the use of 2D ECAD tools and MCAD-only tools have been put to good use in the development of several products and are evidenced for example in the FAPS institute's MID demonstrator depicted in Fig. 8.20.

Unlike MIDCAD, NEXTRA® can directly integrate the EDA tools from Cadence, Mentor Graphics, and Zuken. Imported directly into NEXTRA®, the circuits can be placed on a three-dimensional interconnect device and routed. There are also device libraries containing electronic components as 3D geometries to make the user's job easier. Mapping between NEXTRA® and the EDA program ensures consistency across component libraries for both sides. Import and export filters for native

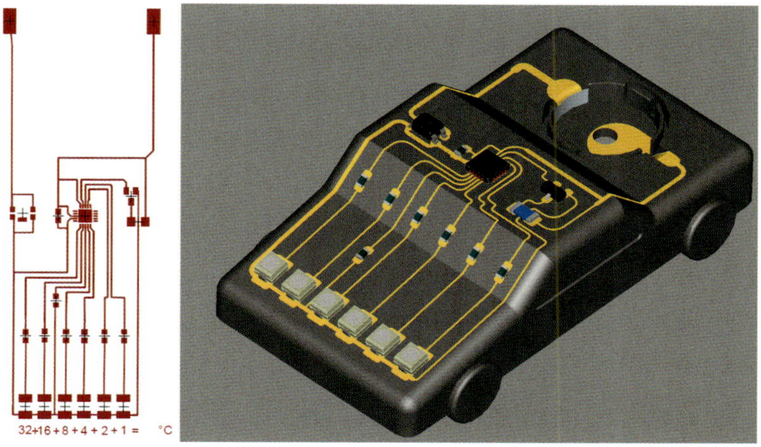

FIGURE 8.20 Comparison of a 2D layout in Eagle (left) and a 3D layout in NEXTRA® (right)

CATIA, NX, and Pro/E formats and exchange formats such as STEP and IGES allow for connectivity to MCAD-only systems.

Additional technological information such as typical sizes for conductors and landing zones is kept in a database and can be accessed as the basis for DRC, for example. The resulting geometries can be derived for use by the production department. There is an export provision for laser direct structuring (LDS) from LPKF, for example (see Section 3.1.1.1.1). [95]

A module for creating electronic assemblies on flex foils and a module for creepage analysis are among those available. NEXTRA® is used in numerous sectors of industry, including transport, medical technology, and IT.

8.10.3 Software Tools for Simulation

As for other mechatronic products, MID design necessitates virtual validation with diverse simulation programs. The same also applies to the manufacturing processes. Specialized simulation tools are available for various discipline-related terms of reference. With regard to injection molding of plastic in particular, simulation helps avoid design flaws and reduce the number of possible corrections to the mold making process. The section below deals with this subject and was authored with the close cooperation of Mr. Mann from the PLEXPERT GmbH company.

Simulation of Injection Molding of Plastic

In accordance with VDI 2234 "Basic Economical Information for Design Engineers" and as explained in Section 8.1.2, the designer of the mold is responsible for a large part of the costs for the entire production process. This is true even if the mold design only accounts for mere 10% of costs [VDI 2234]. The rule of tens, when applied to calculation of failure costs, also places a similar burden on the shoulders of the designer. This rule states that an error in one phase of the production process results in costs that are higher by a factor of ten in the next phase [149]. Due to this responsibility, the use of simulation in parallel to the MID product development is advisable.

As regard to the simulation of the injection-molding process, it is good that many of the MCAD systems available on the market have integrated modules for the **simulation of cavity filling**. These modules automatically generate a mesh necessary for analysis, representing the model subdivided spatially into small parts on the basis of the current CAD geometry. If the mold designer does not have access to an integrated simulation program, the CAD data for the mold can be exported as Parasolid, STEP, or surface tesselation language (STL) file. External calculations run on STL data representing a CAD model as a mesh of triangles and therefore capable of describing the geometry with sufficient precision. Prior to the computation run, this mesh is refined locally and further processed to satisfy the requirements of the simulation. In 90% of all cases the default settings of the mainstream CAD systems are adequate for the generation of STL data. In isolated instances it requires the specialist who is going to run the simulation to generate the STL data from the geometry information of the molding.

Once the mesh has been generated, and this is an automatic background process for integrated modules, it is up to the user to set the boundary conditions. The first step is to examine how the part-cavity fills (Fig. 8.21). This is of particular importance for the designer of the mold. If necessary this specialist can change the geometry to optimize cavity-filling conditions. The choice of a suitable material is important

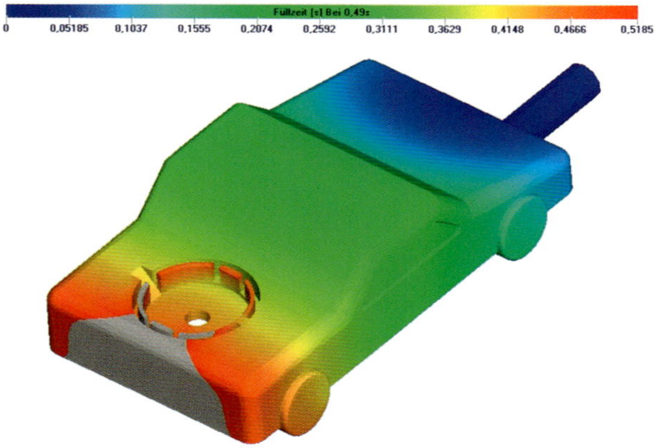

FIGURE 8.21 Progress of the flow front during cavity filling, with the MID demonstrator as example (graphics courtesy of PLEXPERT)

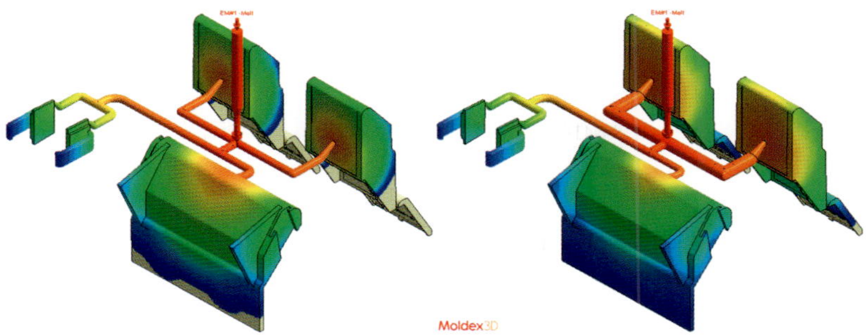

FIGURE 8.22 Nonoptimized (left) vs. optimized (right) cold runner manifold (graphics courtesy of PLEXPERT)

in this respect, as are the injection time and the position of the gate. The simulation program can usually propose a cycle time. Users can experience difficulty in positioning the gate, a task that requires considerable experience. It is advisable to consult the client and the tool designer in these matters. If no information on gating is forthcoming, the software user can position the gate in such a way as to have uniform, balanced cavity filling (Fig. 8.22). All empty spaces are filled together toward the end of cavity filling. Specifications regarding visible surfaces and possibly ejector pins have to be taken into consideration when positioning the cut-off. Some systems on the market have a function offering automatically optimized gate positioning. However, practical considerations often contradict this supposedly optimum siting of the gate.

The analysis enables the user, ideally the molding designer, to draw conclusions about

- the positions of weld lines,
- the locations of airtraps,
- the type of cavity filling (especially balanced filling),
- the filling pressure required, and
- expected mold clamp force.

It is important to make sure that weld lines are not located in a critical area. Critical areas are visible surfaces, subbases for electric conductors, and points that will subsequently be subjected to external forces. If necessary the weld line can be shifted out of these areas by relocating the gating point or changing wall thickness in the flow path.

If the wall thickness of a part duly is changed, the cavity filling pressure has to be taken into account. The maximum filling pressure allowable for technical parts, for example, is 800 bar. Thinning down walls is preferable in terms of weight savings and resource economy.

The positions of airtraps can be changed in the same way. If there are technical reasons why this is not possible, the user should notify the tool designer accordingly. Making sure that the area in question can vent correctly is in the tool designer's remit. If the cavity does not fill uniformly the position of the gate has to be altered accordingly. The objective is to have flow paths of uniform length to the ends of the molding. If this can be accomplished it generally entails the lowest filling pressure. The number of gates can be increased if filling pressure still exceeds the specified limit.

Once molding design has been completed and the mold is optimized in terms of cavity filling and strength properties, the process continues with the next step of **design of the injection-molding tool**. The tool and die maker's first job is to describe the path of the molten shot from the machine nozzle to the gating points defined in the cavity-filling calculations. This designer uses either cold-runner or hot-runner nozzle manifolds (Fig. 8.22). Simulation enables the specialist to calculate the pressure loss in the nozzle manifold system for both variants. The aim is to ensure good cavity filling with the lowest possible pressure loss. Manifold cross-sections can be adapted to ensure uniform filling of all the cavities in a family tool.

The total costs can be reduced due to a change in the cross-section of cold-runner nozzle manifolds. The smaller the diameter, the less material is left over for post-molding disposal or granulation.

Along with pressure loss, it is also important to consider shear and shear stress, which can be critical in the gate region (transition from manifold system to the mold). If the gate diameter is too narrow, friction can occur here and a fibrous material, if used, can be damaged. Narrow cross-sections initiate additional energy input because of the friction in the molten material. This produces a temperature rise local to this area. A hotspot at the gate is indicative of a need for a change in gate geometry. This is also significant in terms of expected packing pressure time. As soon as the material at the gate has solidified, there is no point in maintaining packing pressure because there is no longer any positive effect on shrinkage of the material inside the cavity. The gate has to be checked to ascertain that packing pressure can be applied long enough to compensate for shrinkage in the mold.

Regarding the study of the manifold system as opposed to the other sections of the tool, there is no practical benefit to be gained by importing CAD data. It is easier for the user to model the planned manifold segments in the simulation software and tweak the design as necessary. All simulation programs include the menus for quick changes. Once the manifold has been optimized in the simulation program, its key points and diameters are exported so that the manifold system can be implemented in the CAD tool.

With the manifold system brought to the point of best-possible design, it is time for the designers to turn their attention to cooling system. This is one of the most important points in use of the simulation, because about 70% of overall cycle time is accounted for by heat control (cooling). So in terms of estimating unit costs, cooling is right at the head of the list. Common practice these days, unfortunately, is to focus on the ejector pins first and then squeeze temperature-control channels into the remaining space. The outcome is that many injection-molding tools are not as cost-efficient as they should be. Good temperature control in the tool can reduce cycle time to a minimum so that more parts of better quality can be manufactured in a given time unit. Optimized positioning of the cooling channels is one option, another is using cores with different heat conductivities or providing close-to-contour cooling. Expressed in terms of unit costs, the higher investment outlay for the tool generally turns out to have a positive effect.

As in the case of the manifold system, it is advisable to map the cooling channels in the simulation program. In this way it is much quicker and easier to make and evaluate changes. The most important results of simulating cooling channels are

- pressure loss in each cooling channel,
- heat rise in the cooling medium,
- effectiveness of the cooling circuit, and
- position of hotspots in the mold.

If pressure losses are high in a cooling channel it is a good idea to check whether the results could be improved by splitting the channel into two separate channels, each with a lower pressure differential. Simulation also has to take into account hose bridges planned in production. Bottlenecks increase pressure consumption, and this has a negative effect on the delivery power of the cooling device. Splitting a cooling channel might also be advisable if the cooling medium experiences heat rise from the point of entry into the channel to the point of discharge. Particularly in the case of technical parts the cooling medium should not experience a heat rise exceeding 3 °C.

Visualizing efficiency makes it possible to dispense with cooling channels that contribute little to heat dissipation from the tool. At other points, repositioning the cooling channels can increase the efficiency of the cooling system. The objective is to achieve uniform temperature distribution at the surface of the mold.

Once temperature control has been optimized in this way in the simulation program, the key points of the individual cooling channels are calculated and the holes positioned accordingly in the CAD system. It is important to note that the simulation program maps only the parts of the holes in which the cooling medium flows. The holes from the outside of the mold plate, complete with all the necessary plugs, etc., have to be incorporated in the CAD system.

Following the optimization of cavity filling, the manifold system, and tool temperature control, the final step is the **calculation of warpage**. The parameters for injection time and packing pressure as optimized beforehand are now used to calculate distortion of the part. The most important results of this study are

- total deformation of the part,
- proportionate distortion in the *x*, *y*, and *z* directions,
- proportionate distortion due to cooling, shrinkage, fiber orientation, depending on the program used for simulation.

The total warpage enables the user to decide whether the part can still be used for the intended purpose in the field. In addition to its effect on the part's visuals (weld lines in particular), if it is allowed to exceed critical limits, distortion can hamper the installation of other components or make it difficult to interconnect multiple parts.

The simulation program enables the user to estimate warpage in advance and adopt suitable measures to reduce distortion. The program outputs evaluations indicating the contributors to overall warpage. The effects of measures taken (change in process parameters, introduction of ribs, change of cooling, etc.) are directly visible after the next simulation.

All programs are able to output the deformed geometry at the end of distortion simulation as a points cloud or STL data set. The underlying idea is that negative deformation equaling the warpage values will change the initial geometry to the extent that distortion settles to zero. This can be achieved by changing the mold on the basis of the distortion values. At this time, however, no simulation tool on the market is capable of calculating warpage as an absolute value with sufficient accuracy. The values must be taken as qualitative–an indication of whether the selected measure improves or worsens the situation.

There is also a certain technical difficulty involved in creating a CAD geometry from the STL geometry. This can be done by reverse-engineering a surface or by building a new surface allowing for the boundary conditions given by the STL geometry. It is always advisable to allow a certain latitude in the tool because the absolute values from the results of warpage simulation cannot be transferred in their entirety to the physical model. Despite this residual uncertainty affecting distortion, concurrence is high (approximately 99% for cavity filling, approximately 90% for temperature control) between theory (simulation) and practice (injection molding).

Consistent use of simulation programs helps mold and tool designers discharge their responsibilities for laying the groundwork for the production of cost-effective, high-quality moldings. The literature contains detailed information on the various disciplines [17, 82, 114, 159].

8.10.4 CAD/CAM Chains

Designing MID products and validating their properties with simulation programs involves not inconsiderable outlay. Instead of being used merely for visualizing external appearance and details and ensuring operability and durability, simulation should also lead to templates for controlling production and assembly machinery. The capabilities of CAD/CAM chains for MID technology are illustrated here on the basis of a structuring and metallization process.

Status of Research

The research outlined below is based on observation of the thermokinetic structuring and metallization process Flamecon® (Section 3.1.4.3.1) from LEONI AG [161]. This process is used to apply metallization directly to a substrate. A plug-in integrated into the CATIA V5 MCAD system calculates the sizes of conductors for thermal applications on the basis of physical models. The results are incorporated into an automated design that can be projected on to any structured surfaces of macro-MID. The design rules of the Flamecon® process have to be implemented in the plug-in for a production link to be mapped in the form of a CAD/CAM chain. The aspects this covers include the correct angling of the applicator nozzle, the uniformity of the gap between nozzle and substrate surface, and a balanced point density for defining the trajectory. Control data for the production process based on an industrial robot are derived from the geometric model, with the process-specific properties being parameterized in the program. The LEONI company has conducted a number of successful test runs [46].

This project yielded valuable experience for CAD-based production control that is also applicable to other robot-assisted production and assembly processes. A case in point is Aerosol-Jet printing (Section 3.1.1.2.1). Other research undertakings build on the findings on integration and processing of production data as a supplement to the design of the product. [49]

Developments in Industry

In industrial development the system manufacturer generally provides appropriate CAD/CAM chains. The following is an example from the LPKF Laser & Electronics AG company, whose CircuitPro3D software is used in LDS systems [102] for offline programming and calculation of cycle times. The explanatory input contributed by Mr. Sparrer from LPKF is greatly appreciated by the authors.

CircuitPro3D was intended purely as a CAM software solution. It helps the user prepare data and supports the generation of structures for lasering. In order to accelerate the process and enable lasering from different angles, the laser structuring systems from LPKF work with as many as four lasers firing simultaneously. LPKF

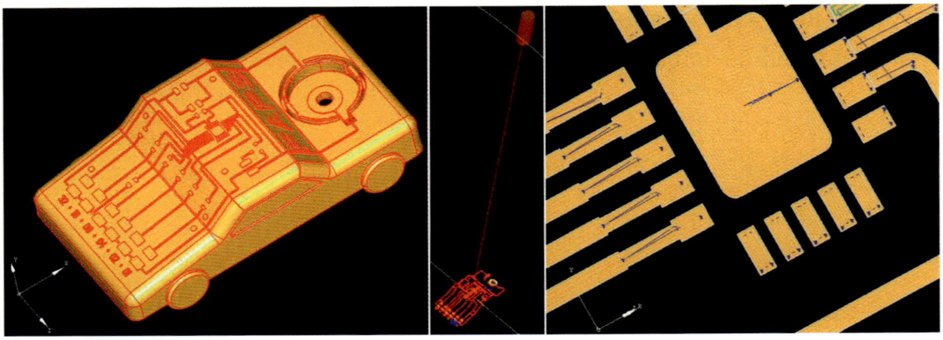

FIGURE 8.23 MID demonstrator project in CircuitPro3D (left) with separate view for the processing unit (center) and fill structures for the conductors (right)

CircuitPro3D is used during preproduction to prepare and optimize the structures and arrange them in lasering positions divided over the individual lasers (Fig. 8.23, center).

The main purpose of CircuitPro3D is to prepare the input data for lasering and assist the user in this task. The creation of hatching structures is an essential function in this respect. CircuitPro3D has a multitude of options and settings for optimizing these hatching structures (Fig. 8.23, right) in terms of hatching strategy, heat transfer to the material, and cycle-time minimization. Experienced designers can identify potential for optimization by studying more closely the fill structures they generate and they can tweak their designs in problem areas. In the early phase of data preparation, moreover, it is easy to focus on areas that are obscured by the part or the part carrier, making them inaccessible to the laser beam.

In preparation for laser direct structuring of MID, it is advisable to ensure that the LDS design recommendations (Section 8.9.3) are applied during product development. It can also be helpful to import CAD prototypes into CircuitPro3D, for example to check the layout and ascertain whether optimization could save a lasering position and shorten the cycle time (see Fig. 8.23, left). It also makes sense to integrate the processing volume for laser structuring and the nominal position of the laser into the CAD project during product development, because all this helps optimize the subsequent layout for lasering.

LPKF CircuitPro3D can import and process data in the IGES, STEP, and boundary representation (BREP) formats. Data processed with the software can also be exported in these formats or backed up as LPKF-internal CBF project files. CBF is the preferred file format because extended information such as the parameters for the laser and the image-processing system, if any, is saved along with the CAD/CAM data. Broadly speaking, there are no special requirements for the data to be imported, but editing can be much simpler and faster if some preparations are made before the data are exported from the CAD system. For example, it is extremely helpful to create the areas for subsequent structuring as separately grouped data. The parlance frequently used in this context refers to faces and face collections. LPKF CircuitPro3D offers the user the alternative of creating these groups or face collections manually, but in most instances this often unnecessarily doubles the work involved because these groups can be created during editing in CAD. It is also advisable to create the conductor tracks in the CAD tool purely as surfaces, in other words without thickness or with height zero. CircuitPro3D uses special algorithms to fill conductors with structures of individual laser tracks.

CircuitPro3D handles both data processing and production control for the parts, so it constitutes an end point in the information-processing chain. Consequently, there is generally no need for further conversion of the data or further-reaching interpretation.

The automated placement of electronic components on MID is described in Chapter 4. As in the case of structuring and metallization processes, a CAD/CAM chain is an objective; the comments below are no more than a brief outlook on the general procedure. In a CAD/CAM chain, the positions and orientations of the components, consisting of the x, y, and z coordinates and rotations about the three axes, should be read out and converted, if possible automatically, into the data required by the control software of system in question. This necessitates coordinate transformations between the coordinate systems of MID product, part carrier, and system. If position and orientation of a component inside the machine are known, the corresponding axis configuration for the placement head can be determined by inverse kinematic calculation of the handler. This has to be converted into system-specific syntax and made available to the system controller. CAD/CAM chains for 3D assembly are being developed by the system manufacturers and are also the subject of university-level research.

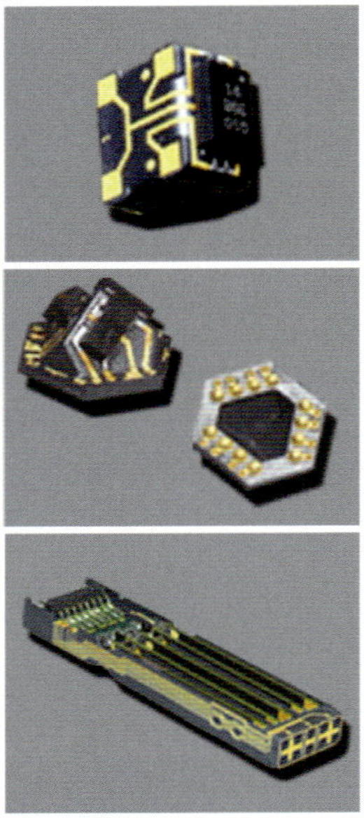

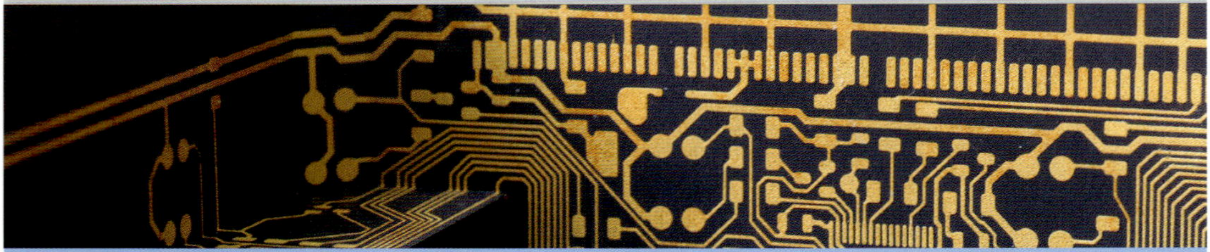

LDS Laser Structuring and Metallization Services
• Job-shop service
• Process evaluation and process enhancement
• Layout development together with the customer
• Prototyping and mass production
LaserMicronics offers the full service spectrum in the field of laser machining of micro materials and process development. Let us find the solution for your product idea together: +49 5131-90811-0.
LaserMicronics
MICROMACHINING SERVICES
LaserMicronics GmbH info@lasermicronics.de www.lasermicronics.com

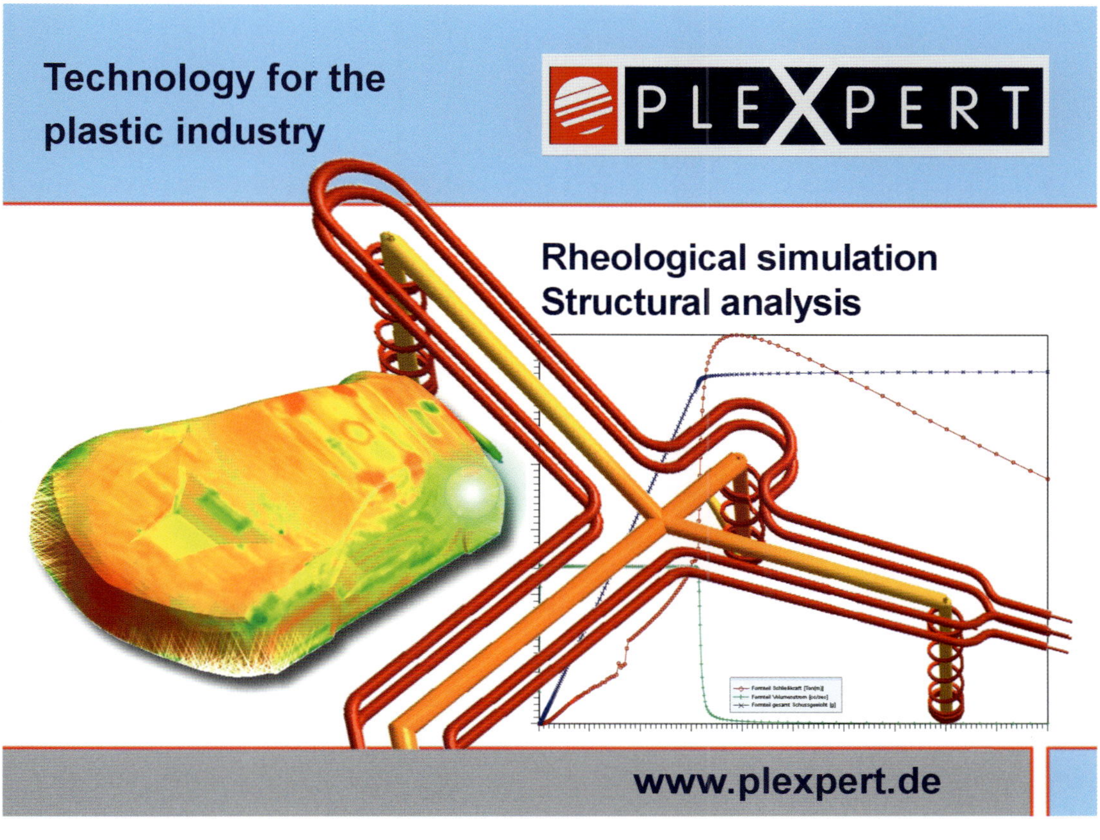
Technology for the plastic industry
PLEXPERT
Rheological simulation
Structural analysis
www.plexpert.de

9 Case Studies

C. Goth

The multiplicity of series-production applications underscores the positive development that MID technology has undergone along the entire process chain: this applies to substrate materials, production processes, systems technology, and 3D assembly. There has been a significant increase in the number of these series-production applications. The MID Survey 2006 identified only eight new series-production applications for the period 2003 to 2006. The MID Survey 2011 counted more than forty MID applications in series production. Nondisclosure agreements, moreover, keep many other applications out of the public domain. Most applications utilize the LPKF-LDS® process, while hot embossing is known to be used in only two cases. The applications described below are examples of series-ready implementation of MID technology and serve to illustrate the diversity of possible areas of use. Each of these studies outlines the purpose of the product and the technological, economic, and organizational aspects of the case. The applications are arranged in sequence by start of series production, with the most recent projects first. [50, 58, 60]

The information for this chapter was obtained through intensive research in the literature. The literature is cited at the appropriate points. The sources for the graphics are also stated. Manufacturers, moreover, supplied a great deal of information without which this presentation would not have been possible. Explicit mention of this is not made at the points in question, but the manufacturers or the lead companies for the individual projects are named for all the applications.

The following individuals, listed here in alphabetical order, provided definitive assistance in the preparation of this chapter:

- Albert Birkicht (HARTING AG Mitronics)
- Stephan Huttenlocher (2E mechatronic GmbH & Co. KG)
- Dr. Wolfgang John (LPKF Laser & Electronics AG)
- Christian Köhler (TE Connectivity)
- Adolf Kraft (Eaton Industries GmbH)
- Jim Liddle (Select Connect Technologies)
- Hartmut Rohde (Robert Bosch GmbH)
- Robert Süß-Wolf/Stefan Süberkrüb (Kromberg & Schubert GmbH & Co. KG)

9.1 OLED

Manufacturing process	Laser direct structuring
Material	LCP
Process chain ▪ Injection molding ▪ Structuring ▪ Metallization ▪ Assembly (without OLED) ▪ OLED placement	 ▪ 2E mechatronic GmbH & Co. KG ▪ 2E mechatronic GmbH & Co. KG ▪ MID-Solutions GmbH/LaserMicronics GmbH ▪ 2E mechatronic GmbH & Co. KG ▪ By customer

FIGURE 9.1 OLED (graphics courtesy of OSRAM Opto Semiconductors)

Purpose of the Product

The MID OLED module was developed to lower the market-entry hurdles for organic LED (OLED) technology. By contrast with individual OLED tiles, the complete module integrates the constant-current source, OLED contacting, heat management, mechanical mounting points, and a standard plug-in connector. This makes the module considerably easier to use. The product also accepts paint finishes for design purposes.

Technological Aspects

Slim and elegant and eminently suitable for OLED technology, this MID solution is only marginally larger than the OLED tile as such (Fig. 9.1). Smart MID design was the key to implementation of all the features listed above. The basic body was designed and molded in accordance with the LDS design rules. Once the body has been molded the conductors are structured by the LPKF-LDS® process and then metallized and the components are placed. All participatory electrical solutions can be implemented using conventional SMT processes.

Economic and Organizational Aspects

An extensive prototyping phase accompanied development at 2E mechatronic. This enabled initial electronics and mechanical concepts to be tested directly on prototypes made of the same material as the finished product.

9.2 Flow Sensor

Manufacturing process	Laser direct structuring
Material	LCP
Process chain ▪ Injection molding ▪ Structuring ▪ Metallization ▪ Assembly	 ▪ 2E mechatronic GmbH & Co. KG ▪ 2E mechatronic GmbH & Co. KG ▪ MID-Solutions GmbH/LaserMicronics GmbH ▪ B&F Bonding GmbH
No. of units	25,000 p. a.
Start of series production	2012
Area of application	Air-conditioning

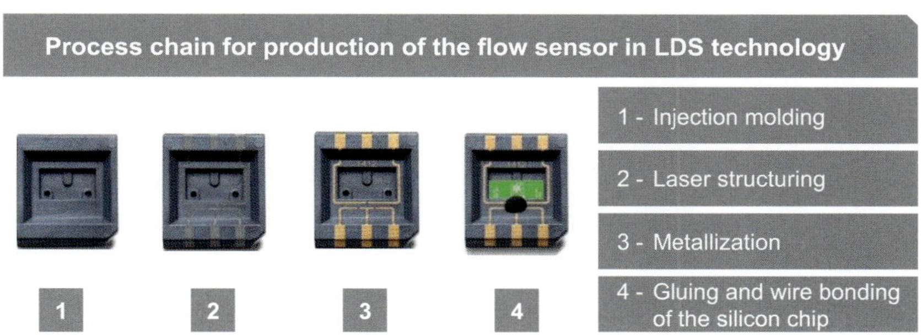

FIGURE 9.2 Flow sensor for volume flow controller (graphics courtesy of 2E mechatronic)

Purpose of the Product

The flow sensor is at the core of volume flow controllers used in air-conditioning. These controllers are important components in modern HVAC applications. They measure, analyze, and control room-air flow in order to supply air in defined quantities to entire buildings or individual rooms. Other target markets for this flow sensor include test facilities, gas handling, and medical technology.

Technological Aspects

The 3D-MID is manufactured by laser direct structuring. The thermal membrane flow sensor (MEMS chip) is ideal for measuring minute pressure differentials

(< 1,000 cP) and flow (< 10 ml/min air). The miniaturized MID package carries the fluidic, mechanical, and optical connections between the silicon chip and higher-order systems. Modular adaptation of the MID package for different applications is possible. The MID package can take variants of the MEMS chip, and as an SMD it can be modified for other processing purposes in electronics production. Other microchips can also be used, significantly extending the measuring range to suit requirements.

In terms of fluidics and mechanics, the chip-to-package connection is made by a standard gluing process for which only the duct opening and the seal have to be taken into consideration. The electrical connections are made by wire bonding. The wire bonds are glob-topped for protection against mechanical influences (Fig. 9.2).

Economic and Organizational Aspects

The MID package is the key to miniaturization of the overall system. Assembly is streamlined by the use of MID technology, because in this case the entire component is an SMD.

9.3 Multiband Antenna for Smartphones

Manufacturing process	Two-shot molding
Material	PC/ABS
Process chain ▪ Injection molding ▪ Metallization ▪ Colorization ▪ RF testing	 ▪ TE Connectivity ▪ TE Connectivity ▪ TE Connectivity ▪ TE Connectivity
No. of units	30 million p. a.
Start of series production	2012
Area of application	Telecommunications

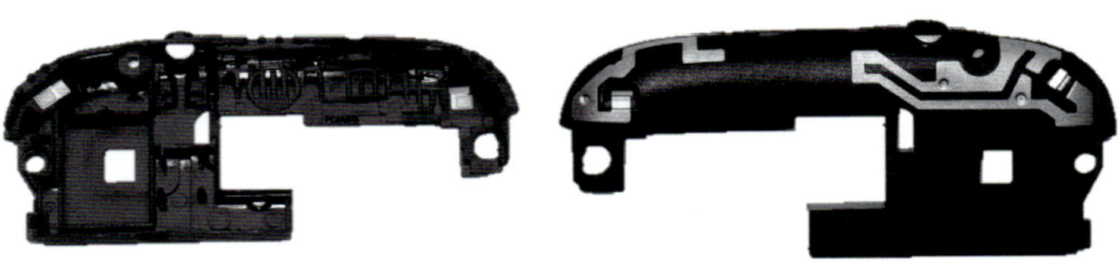

FIGURE 9.3 Multiband antenna for smartphones (graphics courtesy of TE Connectivity)

Purpose of the Product

This antenna is used in smartphones. The basic body furnishes the basis for a high-end, well-tuned RF structure in 3D. There are two different RF functions for cellular (GSM/UMTS) and complementary (WLAN/GPS) purposes.

Technological Aspects

Making smartphones slimmer is important; the *x* and *y* sizes are of less significance. This entails minimization of the entire unit, consisting of casing, display, and printed-circuit board, including the active and passive components, acoustics, and HF components. Another point is that the number of frequencies is increasing, so every available space inside the casing has to be utilized. It is always good to place the antenna as near as possible to the outer skin. Consequently, the antenna structure should match the predefined contours of the 3D shape (Fig. 9.3). MID technology, therefore, is ideal for implementing these specific items and many other integrated antennas as well.

The entire antenna module has to be made with an average thickness of 0.8 mm, which allows about 0.4 mm for the metallizable shot and the same for the nonmetallizable shot. Good adhesion and high reliability in drop testing are important requirements for these thin layers. Good cavity filling in the injection-molding process is also a must. 3D-MID structuring permits an RF path to the underside of the module by two through-holes vias that are an integral part of the two-shot design. The connection to the printed-circuit board is established in this way.

Economic and Organizational Aspects

MID was the technology of choice on account of the challenging three-dimensional design requirements. Even though it had to satisfy a raft of regulatory requirements, the HF antenna structure was developed within a matter of weeks. LDS technology was used in the early phase of development so that layout adaptations could be incorporated quickly. Once RF capability had been verified, the LDS structure was transferred to a two-shot injection-mold design for a fast ramp-up to series production.

Within a limited time window of only seven weeks, 30 multiple-cavity tools for two-shot molding were manufactured at acceptable investment cost. No further jigs or adapters are necessary; the parts go straight from injection molding to metallization. The reproducibility of RF performance is unparalleled. The "first time right" strategy necessitated close cooperation between the mechanical and RF design engineers right from the start.

9.4 ACC Position Sensor

Manufacturing process	Laser direct structuring
Material	LCP
Process chain ▪ Injection molding ▪ Structuring ▪ Metallization ▪ Assembly	 ▪ HARTING AG Mitronics ▪ HARTING AG Mitronics ▪ HARTING AG Mitronics ▪ HARTING AG Mitronics
No. of units	200,000 p. a.
Start of series production	2010
Area of application	Automotive

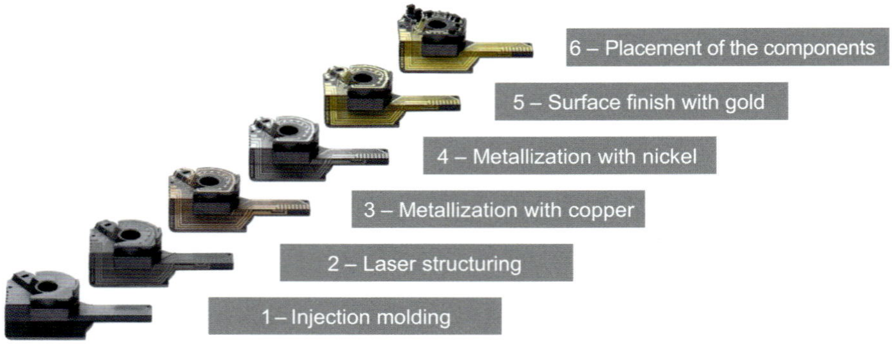

FIGURE 9.4 MID sensor holder for detecting the position of a rotor for propulsion-unit control (graphics courtesy of HARTING Mitronics)

Purpose of the Product

The MID is the sensor in an Adaptive Cruise Control (ACC) system. The ACC position sensor assists in propulsion-unit control. It is definitive in terms of the quality of the radar signal shaped by the rotation of a metallic drum. The drum has a defined surface structure and deflects the radar beam as a function of position.

The driver uses ACC to maintain a preset constant speed without coming too close to the vehicle ahead. Unlike current systems, on account of the wide sensing angle and better resolution, this newly developed radar system can also control the brake system to brake a vehicle from high speed to a complete stop. The system is also able to detect when the vehicle starts moving again. The ACC system warns the driver in good time of possible rear-ending situations. The shorter stopping distances made possible by predictive conditioning of the brakes can save lives.

Technological Aspects

A conventional pick & place process is used on a specially adapted assembly line for component placement. Contacting is by vapor-phase soldering. Solder-stop paint dispensed at critical points prevents solder migration across the part. The MID part carries three Hall-effect sensors and an opto-coupler. It also integrates a plug connector for signal tapping to the periphery (Fig. 9.4). The solder pads are supply lines for the propulsion unit. At the end of the assembly process the MID pass through an automated optical inspection station (AOI) from the Viscom company. The AOI is a standard configuration from printed-circuit board assembly modified to include 3D capability. The final step is electrical testing to check functionality and signal integrity.

Economic and Organizational Aspects

Production volume is forecast to remain high, so a dedicated assembly and test facility was set up. The high degree of automation enables cost-effective production of this specialized component. The automotive industry places high quality requirements on process capability. These high standards are met by the automated metallization unit and the assembly and testing facility already mentioned.

9.5 Pressure Sensor

Manufacturing process	Two-shot molding
Material	LCP/LCP
Process chain ▪ Injection molding ▪ Metallization ▪ Assembly	 ▪ n. a. ▪ n. a. ▪ Robert Bosch GmbH
No. of units	> 10 million p. a.
Start of series production	2009
Area of application	Automotive

Purpose of the Product

The pressure sensor is part of the Electronic Stability Control (ESC) safety system for motor vehicles. It transforms hydraulic braking pressure into an electric signal. The MID is the carrier for the printed-circuit board. It also interfaces the electrical connections from the sensor element to the printed-circuit board and from the PCB to the customer's contacts. The MID has interfaces to the sensor cell, to the printed-circuit board, to testers for the appropriate tests, and to the periphery (Fig. 9.5).

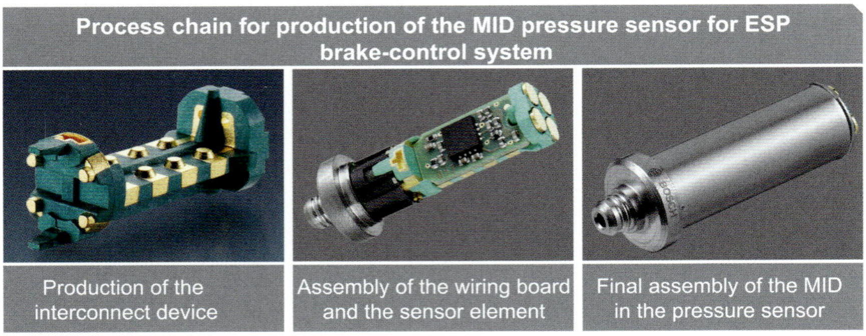

FIGURE 9.5 Process chain for production of a miniaturized pressure-sensor system using MID technology (graphics courtesy of PKT Präzisions-Kunststoff-Teile/Robert Bosch)

Technological Aspects

The pressure sensor is a two-shot molding. Production volume is on the high side of 10 million units per year. The printed-circuit board is contacted on the MID by conductive-adhesive gluing (isotropic conductive adhesive). Nonconductive adhesive is also applied for mechanical location. The contact pins of the MID are tiny bumps to which the conductive adhesive is applied in the dispensing process. [143]

Economic and Organizational Aspects

MID technology reduces the space occupied by the pressure sensor by 50%. The former solution measured 16.2 mm in diameter. The diameter of the new solution is a mere 10 mm. The single most important benefit of the new solution is that the printed-circuit board can be set upright in the pressure sensor, whereas formerly it had to lie flat. On account of this miniaturization, the pressure sensor's structural principle can be used in all new brake control systems.

9.6 MULTI LED

Manufacturing process	Laser direct structuring
Material	LCP
Process chain ▪ Injection molding ▪ Structuring ▪ Metallization ▪ Assembly	 ▪ 2E mechatronic GmbH & Co. KG ▪ 2E mechatronic GmbH & Co. KG ▪ MID-Solutions GmbH/LaserMicronics GmbH ▪ 2E mechatronic GmbH & Co. KG
No. of units	70,000 p. a.
Start of series production	2009
Area of application	Medical technology

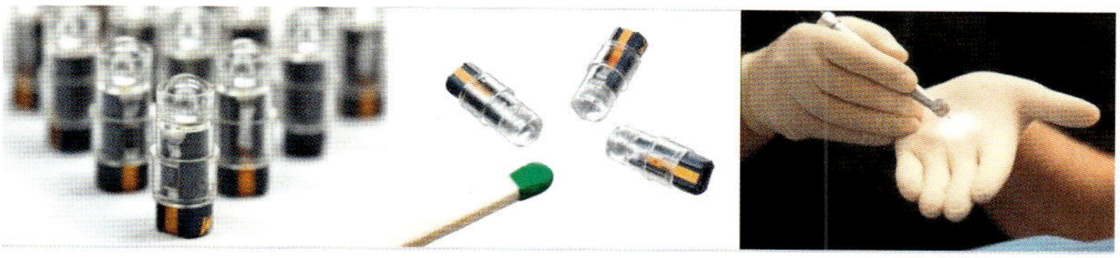

FIGURE 9.6 MULTI LED for medical technology (graphics courtesy of 2E mechatronic, KaVo Dental)

Purpose of the Product

The MULTI LED module supersedes the high-pressure lamps formerly used in treatment units for dentistry. The 1 : 1 replaceability means that existing components are easily changed over on dentists' premises. The MULTI LED, moreover, is considerably more durable and produces better-quality light.

Technological Aspects

The objective was to develop an LED solution to replace the existing high-pressure lamp. This meant that the MULTI LED had to be located and contacted by an existing lamp socket. The LDS process was eminently suitable because a 3D miniaturized solution was needed on account of the limited space available and also because optimum lighting was a must. Initial samples were milled from the LCP material chosen for the finished product. The advantage of milling is that no injection-molding tools are needed for the samples and the down-line steps are very close to the later series-production process. These parts are also suitable for close-to-series testing. Laser activation takes place on four planes normal to each other. SMDs are positioned on two of these planes after metallization (Fig. 9.6). These are an LED on the end face and a series resistor at one side. The 3D placement is done on an MID assembly line set up in the framework of the BMBF-funded AHMID research project. The line includes a dispenser, an SMD pick & place robot, and a vapor-phase soldering unit. Such 3D placement also requires the inclusion of a suitable workpiece carrier system. Final installation of sleeve and lens is a manual operation. Quality has received the solid confirmation of fault-free operation in the field.

Economic and Organizational Aspects

Not primarily economic in nature, the major factor in implementation of this project was the need for a solution that combined durability with optimum lighting. An LED solution is eminently suitable for these terms of reference, and this is the reason for the choice of LED and MID technology. The initiator was the KaVo Dental company, which also bore development responsibility for the MID lighting element. The project was implemented under considerable time pressure. Only seven months elapsed between initial contact and series introduction. Series production got off to a successful start in December 2009.

9.7 Insulin Pump

Manufacturing process	Two-shot molding
Material	PC/PC-ABS
Process chain ▪ Injection molding ▪ Metallization	 ▪ Phillips Plastic ▪ SelectConnect Technologies
No. of units	> 3 million p. a.
Start of series production	2009
Area of application	Medical technology

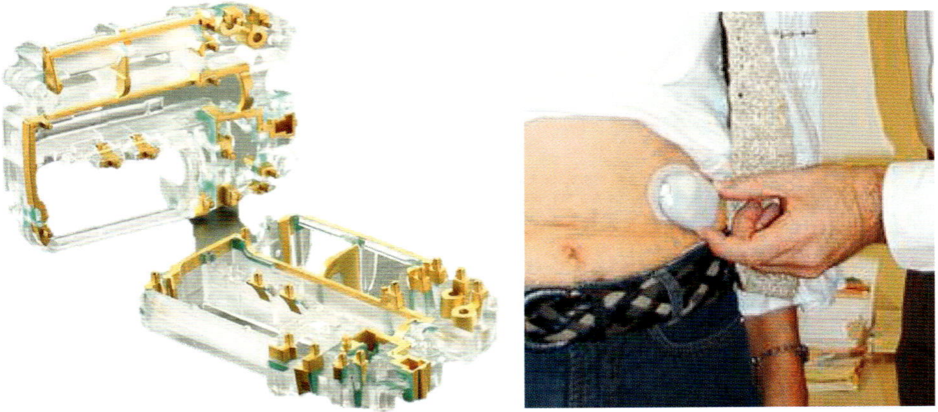

FIGURE 9.7 Two-shot molded body for an insulin pump (graphics courtesy of Insulet/SelectConnect Technologies, Wikipedia)

Purpose of the Product

This medical device is a small, programmable pump that administers insulin via the infusion set, consisting of a cannula and a needle. It is designed to be used for insulin pump therapy (continuous subcutaneous insulin infusion, CSII). The patch pump is a variant that combines the infusion set with the pump in a common housing (usually referred to as the pod). This unit is applied to the skin as an adhesive patch and replaced after two or three days. It is controlled by a separate unit (the personal diabetes manager, PDM) with a wireless link to the pod. Patch pumps were introduced in the USA around 2007 and have been available in Europe since mid-2010. The MID is the electrical chassis for the single-use insulin management system (Fig. 9.7).

Technological Aspects

The MID is manufactured by two-shot molding. The metallizable material PC/ABS (polycarbonate, acrylonitrile butadiene styrene) of the first shot is enclosed in the second shot of PC (polycarbonate). The PC/ABS forms what will become the conductors, and the PC is the package. The pins contact the PCB and have to be dimensionally stable, so precision injection molding is a must. The PC/ABS is chemically metallized with Cu-Ni-Au.

Economic and Organizational Aspects

Two-shot molding and selective metallization of the conductors integrated into the package are the key to a solution for high-volume production that connects the battery to the PCB logic.

9.8 Passive UHF RFID Transponder

Manufacturing process	Laser direct structuring
Material	LCP
Process chain ▪ Injection molding ▪ Structuring ▪ Metallization ▪ Assembly	 ▪ HARTING AG Mitronics ▪ HARTING AG Mitronics ▪ HARTING AG Mitronics ▪ HARTING AG Mitronics
No. of units	10,000 p. a.
Start of series production	2009

FIGURE 9.8 Passive UHF RFID transponder (graphics courtesy of HARTING Mitronics)

Purpose of the Product

RFID tags are used increasingly in industrial processes for tracking goods and goods flows. UHF RFID applications are encountered in medical technology and in production and transport logistics. They must evince high read ranges even when affixed in close proximity to liquids and metals. Usable temperature ranges from −65 °C to +85 °C and even 210 °C are possible when the appropriate plastics are used. The housings have IP 64, IP 67, and IP 69K protection ratings, with a number of assembly options. The passive UHF RFID transponders manufactured by HARTING AG Mitronics have read ranges of 3 to 6 m, depending on their shape.

Technological Aspects

The antenna is structured and the RFID chip is placed directly on the housing of the RFID tag (Fig. 9.8). This dispenses with the need for a separate printed-circuit board. The material is LCP, capable of meeting the high requirements for temperature exposure common to industrial applications.

Economic and Organizational Aspects

RFID chip technology is developing at a tremendous rate, and by using laser direct structuring the manufacturer has the flexibility to adapt the antenna structure rapidly to the requirements of new RFID components. This translates into a significant edge over the competition in terms of reaction speed and also economically vis-à-vis conventional technologies. It is also possible to produce curved 3D structures to lift the antenna a defined distance clear of the metal surface (for tagging items as diverse as skeleton containers, say, or hospital beds). Read range is reduced dramatically if the antenna is too close to the metal of the tagged object. So using MID technology has afforded both production-related and application-specific benefits.

9.9 LED Camera Module

Manufacturing process	Laser direct structuring
Material	LCP
Process chain ▪ Development ▪ Structuring ▪ Metallization ▪ LED placement	 ▪ HARTING AG Mitronics ▪ HARTING AG Mitronics ▪ HARTING AG Mitronics ▪ HARTING AG Mitronics
No. of units	5,000 p. a.
Start of series production	2007
Area of application	Industrial automation

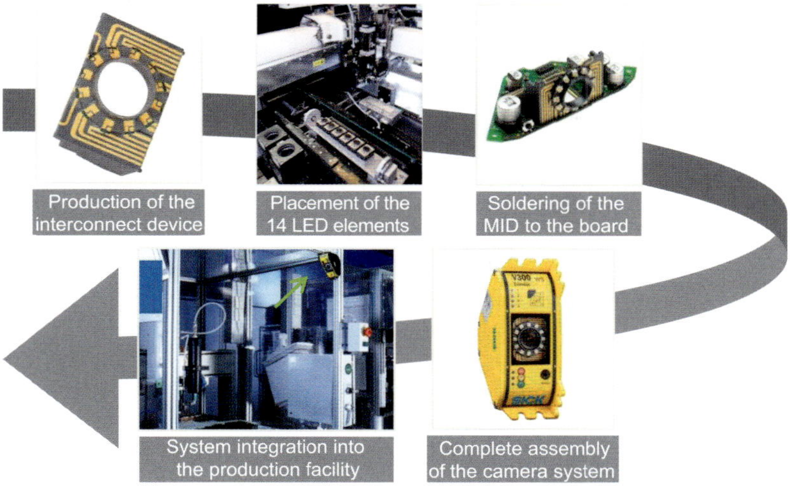

FIGURE 9.9 MID camera module on its own and integrated into the electronics module (graphics courtesy of HARTING Mitronics/SICK)

Purpose of the Product

The MID is used as a 3D carrier for the LEDs in a camera-based safety imaging system (V300 Work Station) manufactured by the Sick company. The lighting module assists state-of-the-art camera technology with a wide aperture angle. Light in the nonvisible part of the spectrum emitted by the LED unit is reflected for the purposes of detection. This optical monitoring system has no blind spots and is unaffected by ambient conditions. All the components, including the analyzer, are integrated into a single, compact housing (Fig. 9.9).

Technological Aspects

Three-dimensional orientation of the LEDs was essential for this new lighting module for a camera-based safety imaging system, necessitating the development of an MID application. The 3D layout also facilitates straightforward mechanical and electrical contacting to the FR4 printed-circuit board. Along with these technological aspects, the cost factor was also definitive for the choice of MID technology. The challenge was to implement a solution that would correspond to the user's design requirements. Laser direct structuring is the process used in manufacturing this product. The LDS-compatible material LCP is used. Metallization is Cu-Ni-Au, a plating typical for MID. The biggest challenges for 3D production were the design and the series-production maturity of the jigs for development: four sides of the part had to be laser-structured and many different angles have to be taken into account for placement of the LEDs. Consequently, pivoting arms for the lasering process and for component placement were added to the existing assembly configuration. There have been no reports of failure of an MID part from the field.

Economic and Organizational Aspects

The economic edge over alternative solution concepts was a major factor in implementation of the MID application. A second concept was initially pursued in parallel during the early stages of development. The intention for this second line of development was to integrate a printed-circuit board into the plastic molding. The Sick company was the initiator for MID development. HARTING AG Mitronics elaborated the concept with the support of Sick. The project was launched in March 2006. The first prototypes were delivered in early summer 2007. Series production started that same year in November. The process chain for production is not very complex; this is a standard MID process with a very limited number of participating partners.

9.10 3D Switching Module

Manufacturing process	Laser direct structuring
Material	PA6/6T
Process chain ▪ Injection molding ▪ Structuring ▪ Metallization ▪ Assembly	 ▪ n. a. ▪ Kromberg & Schubert GmbH & Co. KG ▪ n. a. ▪ Kromberg & Schubert GmbH & Co. KG
No. of units	Approx. 350,000 p. a.
Start of series production	2007
Area of application	Automotive

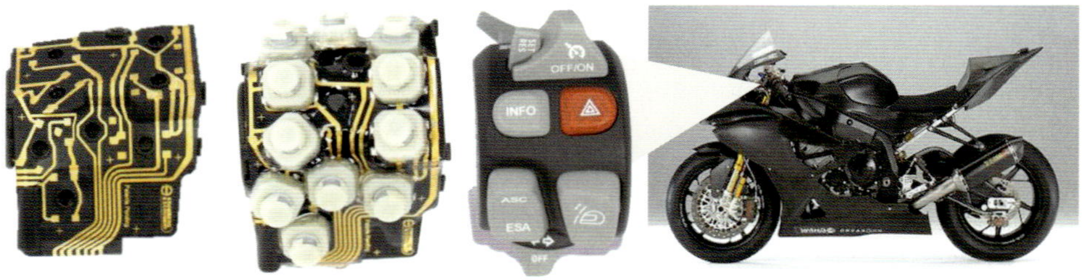

FIGURE 9.10 Multifunction switch for motorcycle handlebars (graphics courtesy of Kromberg & Schubert)

Purpose of the Product

The multifunction switch for mounting on the handlebars of a motorcycle is used in the automotive industry, for example by BMW on its R and K series models, the F800 and S1000RR. The MID integrates fourteen switches, connectors, and other electronic components. All functions including lights, horn, navigation, turn signals, and radio are operated by the multifunction switch (Fig. 9.10). Four components, namely right and left for standard handlebars and right and left for special vehicles (police, emergency medical assistance, etc.) are produced in about fifty different versions. The MID discharges the electrical function of signal carrier with currents < 1 A and ensures the requisite stability under changing mechanical loading with forces of up to 20 N when switches are actuated by the rider.

Technological Aspects

The part is laser direct structured in polyamide with a melting point of 295 °C and excellent mechanical and thermal properties. The material is qualified for use on motorcycles with vibrations, impacts, dirt, and dampness in widely varying temperature conditions. Data matrix coding is applied with the laser structuring so that the varieties of multifunction switches can be distinguished, particularly in the course of subsequent assembly processes. Following one-shot molding, structuring, and metallization, the electronic components are placed in a 3D assembly process that includes all the necessary testing. The assembly system for the multifunction switch is shown in Fig. 9.11. Lead-free soldering takes place in a vapor-phase

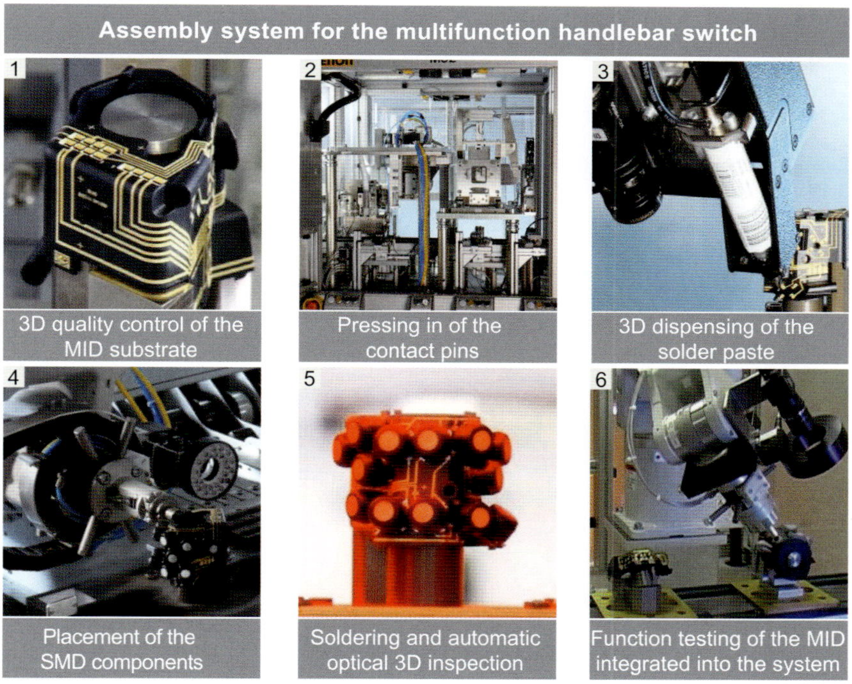

FIGURE 9.11 Process chain for assembly of the electronic components (graphics courtesy of Kromberg & Schubert/XENON Automatisierungstechnik)

soldering unit. All switch functions are tested in the final function check in the separate testing station. [116]

Paint has to be applied in a uniform thickness for added protection against moisture and other external influences. The extremely three-dimensional geometry of the multifunction switch makes applying the paint difficult, but drips and runs would be wholly unacceptable.

Economic and Organizational Aspects

Implementation of the MID application in series production necessitated the development of a fully automated assembly solution by a manufacturer of special-purpose machines. The 3D placement of the SMDs, the switch elements, and the contact pins is only one of the functions discharged by the system. Others include incoming-goods inspection, electrical testing of the conductor tracks, dispensing the solder paste with optical process monitoring, and final inspection. [116]

9.11 Security Caps

Manufacturing process	Laser direct structuring
Material	PBT
Process chain ▪ Injection molding ▪ Structuring ▪ Metallization ▪ Assembly	 ▪ HARTING AG Mitronics ▪ HARTING AG Mitronics ▪ HARTING AG Mitronics ▪ By customer (various)
No. of units	100,000 p. a.
Start of series production	2006

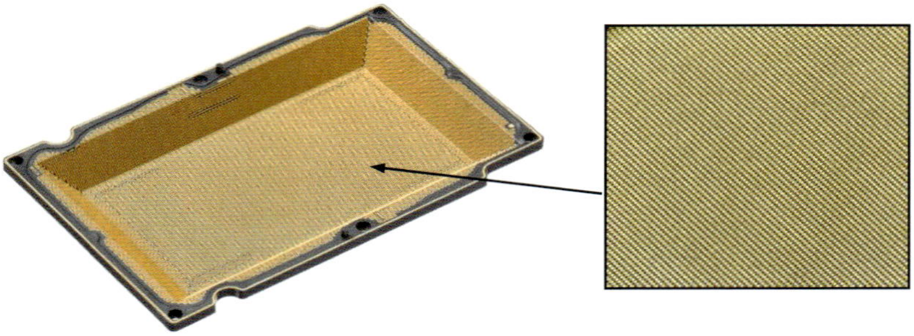

FIGURE 9.12 Security caps for electronic counting systems (graphics courtesy of HARTING Mitronics)

Purpose of the Product

Counting systems that process the data of credit card, debit card, and other forms of electronic monetary transactions have to meet very high security requirements. No critical information is present in these systems when they are idle. The critical phase for data theft and subsequent fraudulent use is only when credit card data are being read and transmitted during the transaction. It is essential to prevent a sensor from being infiltrated that could copy this data while the transaction is in progress. These security caps are like hoods over the electronics and are designed to eliminate this risk. The underside of the cap is structured with two long conductor tracks and metallized. If these conductor tracks are interrupted, for example if a hole is drilled through the cap, the electronics are disabled and the card data can be neither read nor copied. Current security standards require that any attempt to drill holes as small as 200 μm in diameter will trip this protective mechanism.

Technological Aspects

Detecting a drill with a diameter of 200 μm calls for a closely meshed conductor network with a pitch of 300 μm (150 μm conductor width and 150 μm center-to-center distance). PBT was chosen as the material because the heat-resistance requirements for these security caps are relatively low. Contacting is by conductive rubber strips; soldering is not necessary.

Economic and Organizational Aspects

The big advantages of MID security caps over conventional solutions (glued flex-print films) are very easy installation and the avoidance of quality problems in the field, often the result of failure of the adhesive. Production volume is high, so the product can be manufactured economically on frame carriers. The process parameters have to be closely monitored because the resistance range acceptable for the conductors is very closely toleranced.

9.12 Solar Sensor

Manufacturing process	Two-shot molding
Material	LCP/LCP
Process chain ▪ Injection molding ▪ Structuring / metallization ▪ 3D assembly ▪ Systems technology for assembly	 ▪ HARTING AG Mitronics ▪ HARTING AG Mitronics ▪ HELLA Fahrzeugkomponenten GmbH ▪ Häcker Automation GmbH
No. of units	1.5 million p. a.
Start of series production	2005
Area of application	Automotive

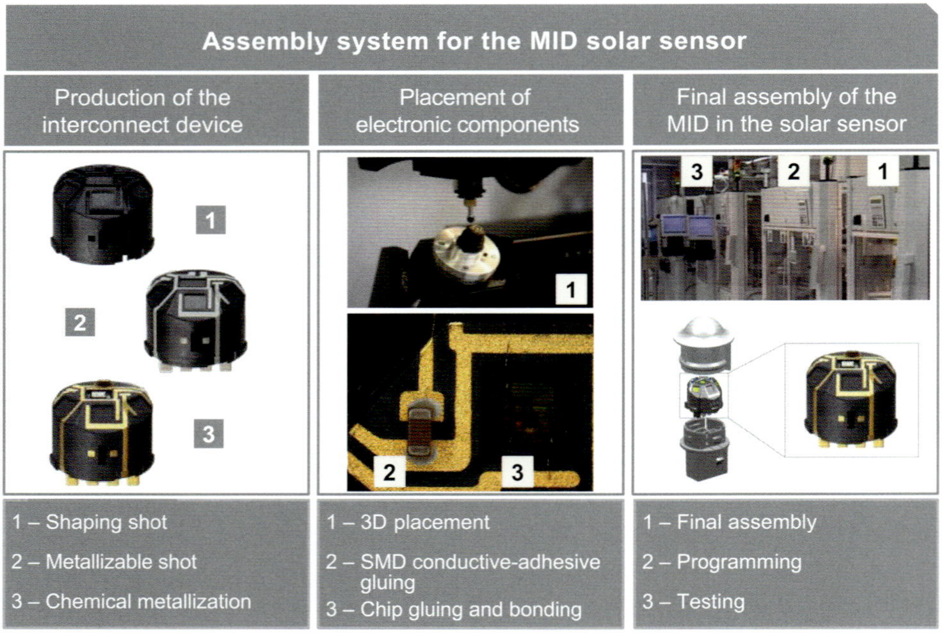

FIGURE 9.13 MID solar sensor for in-car air-conditioning control (graphics courtesy of HARTING Mitronics/HELLA Fahrzeugkomponenten)

Purpose of the Product

The MID solar sensor assists in-car air-conditioning control (Fig. 9.13). Along with temperature, the sensor detects the intensity and the direction of incidence of the sun's rays, helping to minimize energy consumption overall. Miniaturization was achieved and reliability increased by the positioning of multiple optical ASICs integrating both the opto-sensitive element and the electronic analyzer. If the sun's rays are more intense on the passenger side, the automatic increase in cooling or ventilation on that side enabled by the solar sensor counters localized overheating. The MID base body is a two-shot molding with two different types of LCP. [79]

Technological Aspects

Wire bonding is the technology for contacting the optical sensors. The silicon chips are first located mechanically on the substrate by an adhesive. Metallized plastic pins produced just like the conductor tracks directly in the injection-molding process are the contacts to the periphery.

Economic and Organizational Aspects

Two-shot molding was chosen for economy, on account of the high-volume production rate of some 1.5 million units per year. Price per unit is therefore low, despite high investment outlay for the complex molding tool. A 3D assembly line was set up for component placement on the MID blanks and final assembly of the module.

9.13 Microphone Carrier for Hearing Aid

Manufacturing process	Laser direct structuring
Material	LCP
Process chain ▪ Injection molding ▪ Structuring ▪ Metallization ▪ Assembly	 ▪ HARTING AG Mitronics ▪ HARTING AG Mitronics ▪ HARTING AG Mitronics ▪ Siemens Audiologische Technik GmbH
No. of units	10,000 p. a.
Start of series production	2005
Area of application	Medical technology

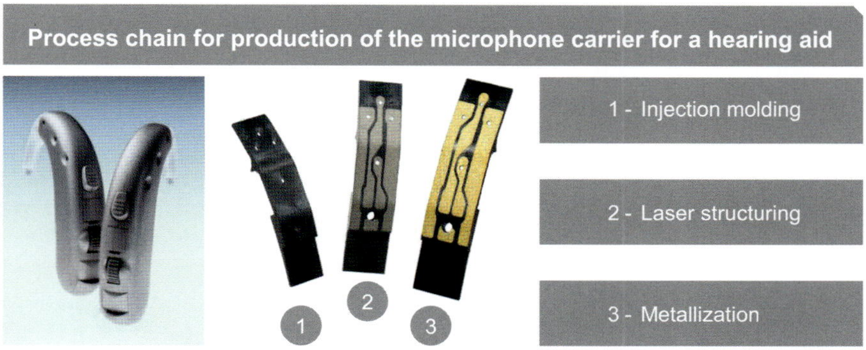

FIGURE 9.14 Microphone carrier for a behind-the-ear (BTE) hearing aid (graphics courtesy of HARTING Mitronics)

Purpose of the Product

The MID microphone carrier is part of a behind-the-ear (BTE) hearing aid. There are three microphones in all, positioned and contacted on the MID. The MID also makes the electrical connections between the microphones and the mainboard. Precision positioning of the microphones is essential for correct operation of the hearing aid, because this is what enables directional audio perception for the wearer (i.e. the ability to identify noises as coming from in front or behind).

Technological Aspects

The MID is highly miniaturized (dimensions: length 16 mm, width 4 mm). Special precautions have to be taken to avoid damage in the production process. The curved shape necessitates the construction of special holders for the laser-structuring process. Features that merit special mention are the two-sided structuring and metallization of the MID, made possible by the integration of what are known as vias, vertical interconnects extending right through from front to rear. The vias are produced directly in the injection-molding process. The substrate material is LCP, chosen because the down-line processes include soldering.

Economic and Organizational Aspects

The directional positioning of the microphones is the single most important benefit of MID technology for this particular application. It dispenses with a second carrier that would have to be connected by wires to the mainboard.

9.14 Seat-Adjustment Switch

Manufacturing process	Hot embossing
Material	PA
Process chain ▪ Structuring ▪ Metallization ▪ Assembly	 ▪ 2E mechatronic GmbH & Co. KG ▪ 2E mechatronic GmbH & Co. KG ▪ 2E mechatronic GmbH & Co. KG
No. of units	20,000 p. a.
Start of series production	2004
Area of application	Automotive

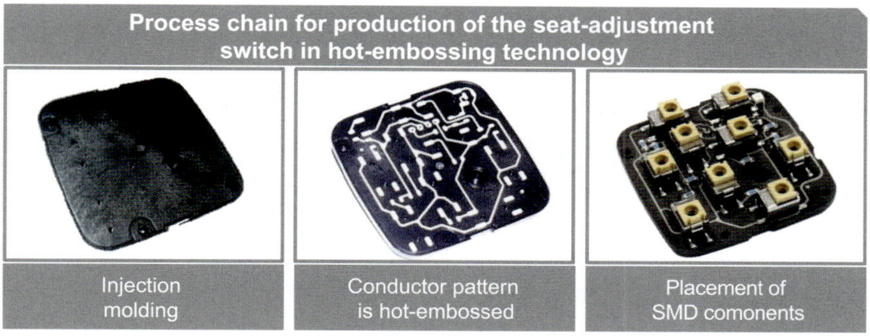

FIGURE 9.15 Seat-adjustment switch for the automotive industry, manufactured by hot embossing (graphics courtesy of 2E mechatronic)

Purpose of the Product

Mercedes-Benz uses this seat-adjustment switch in its S and M class models.

Technological Aspects

With the exception of the injection-molding step, production is on the line for hot-emboss MID set up in the framework of the AHMID research project funded by the BMBF. The line includes a dispenser, an SMD pick & place robot, and a vapor-phase soldering unit. The preassembled interconnect devices are set in groups of twenty on a purpose-built workpiece carrier and checked. The high level of quality is confirmed by what is now eight years of fault-free operation in the field.

Economic and Organizational Aspects

In 2004, the seat-adjustment switch was the first hot-emboss MID used in series production inside the passenger compartment of a luxury-class automobile. The application continues in series production today. Once the line for hot-emboss MID had been set up, this product development marked the start of series production of MID for the 2E mechatronic company in 2004.

9.15 LED Light

Manufacturing process	Hot embossing
Material	PA 66-GF
Process chain ▪ Injection molding ▪ Hot embossing ▪ Assembly	 ▪ Eaton Industries GmbH ▪ Eaton Industries GmbH ▪ Eaton Industries GmbH
No. of units	Several million p. a.
Start of series production	2000
Area of application	Industrial automation

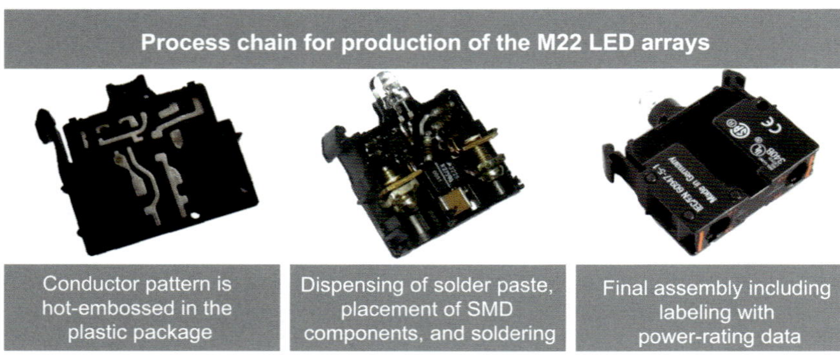

FIGURE 9.16 M22 LED arrays (graphics courtesy of Eaton Industries)

Purpose of the Product

The MID is used in the RMQ Titan series of instrumentation and control devices. Annual production volume runs to several million units.

Technological Aspects

The two parts of the housing are one-shot 66-GF polyamide injection moldings. The conductor layout is hot-embossed on to the substrate. The hot-embossing film has a conductive copper coating 30 µm thick and a tin surface finish 3 µm thick. After hot embossing, a dispenser applies solder paste to the planar process surface. A gantry handler presses the contacts into the semifinished module. There is a choice of two connection technologies. Screw connection is the variant of choice, with cage clamp as the alternative.

The next step is component placement with an SMD placement robot manufactured by the MIMOT company. Operating at a very fast cycle time, this unit places various SMDs and one LED. The components are soldered in a pass process by a vapor-phase soldering unit from the IBL Löttechnik company. The terminal screws complete with plates are then inserted, and the two parts of the housing are riveted together by ultrasonics. [45]

Economic and Organizational Aspects

The fully automated production and assembly line can handle 35 variants of the basic design. There are seven different conductor layouts. The short cycle times achievable with this process and the flexibility for manufacturing variants of the basic design are the key to economical high-volume production.

Using MID technology means that there is no need for a printed-circuit board in the conventional meaning of the term. Customers benefit from the extremely long durability and high reliability of LED as the light source. There is a choice of connection options (screw or cage clamp), colors, and operating voltage for the LED elements. Front or bottom attachment are both possible, so the customer can pick the version suitable for a given application from the existing product range, without having to fall back on custom solutions.

10 List of Abbreviations

1S	one-shot
2S	two-shot
ABS	acrylonitrile butadiene styrene
ACA	anisotropic conductive adhesives
ACC	adaptive cruise control
ACT	assembly and connection technology
AIT	advanced interconnect technology
AOI	automated optical inspection
ASIC	application-specific integrated circuit
AXI	automated X-ray inspection
BGA	ball grid array
BMBF	Bundesministerium für Bildung und Forschung (German Federal Ministry of Education and Research)
BREP	boundary representation
BTE	behind-the-ear hearing aid
C4NP	controlled collapse chip connection new process
CAD	computer-aided design
CAM	computer-aided manufacturing
CAP	computer-aided planning
CCD	charge coupled device
CFD	computational fluid dynamics
CFU	capillary flow underfill
CIGS	copper, indium, gallium, sulfur, and selenium
CNC	computerized numerical control

CONSENS	conceptual design specification technique for the engineering of complex systems
CR	chip resistor
CSII	continuous subcutaneous insulin infusion
CSP	chip scale package
CTE	coefficient of thermal expansion
CVD	chemical vapor deposition
DIN	Deutsches Institut für Normung (German Standards Institute)
DLC	diamond-like carbon
DMA	dynamic-mechanical analysis
DMLS	direct metal laser sintering
DNP	distance to neutral point
DOD	drop-on-demand
DP	duroplastics
DRC	design rule check
DSC	differential scanning calorimetry
ECAD	electronic computer-aided design
EDA	electronic design automation
EDX	energy dispersive X-ray spectroscopy
EMC	electromagnetic compatibility
EMS	electronic manufacturing services
EN	European standard
EOL	End-of-life
EP	epoxide (EP resins)
ESC	electronic stability control
EU	European Union
FEM	finite element method
FDM	fused deposition modeling
GF	glass fiber
GPS	global positioning system
GRP	glass-fiber reinforced plastic
GSM	global system for mobile communications
HDL	hardware description language

HF	high frequency
HT	high temperature
ICA	isotropic conductive adhesives
IDF	intermediate data format
IEC	International Electrotechnical Commission
IGES	initial graphics exchange specification
IMD	in-mold decoration
IP	international protection
IPC	Association Connecting Electronics Industries, formerly: Institute for Printed Circuits
IR	infrared
ISO	International Organization for Standardization
LASER	light amplification by stimulated emission of radiation
LCP	liquid crystal polymer
LDS	laser direct structuring
LED	light-emitting diode
LGA	land grid array
LS	laser sintering
LTE	long-term evolution
MCAD	mechanical computer-aided design
MEMS	micro electro-mechanical system
MID	molded interconnect device
MIDIA	Molded Interconnect Devices International Association
MIPTEC	microscopic integrated processing technology
MRC	manufacturing rule checks
NCA	non-conductive adhesive
Nd: YAG	neodymium-doped yttrium aluminum garnet
NFU	no-flow underfill
OLED	organic light-emitting diode
PA	polyamide
PBT	polybutylene terephthalate
PC	polycarbonate
PCB	printed-circuit board

PCK	printed circuit board Kollmorgen
PDM	personal diabetes manager
PEI	polyetherimide
PEEK	polyether ether ketone
PES	polyethersulfone
PET	polyethylene terephthalate
PLCC	plastic leaded chip carrier
PP	polypropylene
PPA	polyphthalamide
PPE	polyphenyl ether
ppm	parts per million
PPO	polyphenylene oxide
PPS	polyphenylene sulfide
PPSF	polyphenylsulfone
PTFE	polytetrafluorethylene
PVD	physical vapor deposition
QFN	quad flat no leads package
QFP	quad flat package
RF	radio frequency
RFID	radio frequency identification
RoHS	Directive on the restriction of the use of certain hazardous substances in electrical and electronic equipment
SAC 305	Sn/Ag 3.0/Cu 0.5
SEM	scanning electron microscope
SKW	Sankyo Kasei wiring board
SLM	selective laser melting
SLS	selective laser sintering
SM	substrate metallization
SMD	surface mount device
SMT	surface mount technology
SO	small outline
SOIC	small outline integrated circuit
SOP	small outline package

SOT	small outline transistor
sPS	syndiotactic polystyrene
STEP	standard for the exchange of product model data
STL	stereolithography/surface tesselation language
SVP	soft vapor phase
SysML	systems modeling language
TAIC	triallyl isocyanurate
THD	through-hole device
THT	through-hole technology
TI	temperature indices
TMPTA	trimethylolpropane triacrylate
TP	thermoplastics
TS	thermo-sonic
UBM	under-bump metallization
UF	underfill
UHF	ultra-high frequency
UML	unified modeling language
UMTS	universal mobile telecommunications system
US	ultrasonics
UV	ultraviolet
VDE	Verband der Elektrotechnik Elektronik Informationstechnik (German association for electrotechnology, electronics and information technology)
VDI	Verein Deutscher Ingenieure (Association of German Engineers)
VOC	volatile organic compound
WLAN	wireless local area network
WLSST	wafer level solder sphere transfer process
XML	extensible mark-up language

Chemical elements and compounds are abbreviated in accordance with the international standard, as follows:

Ag	silver
Al	aluminum
Au	gold
Bi	bismuth
C	carbon
Cu	copper
Cr	chromium
F	fluorine
H	hydrogen
N	nitrogen
Ni	nickel
O	oxygen
P	phosphorus
Pb	lead
Pd	palladium
S	sulfur
Sn	tin

11 Literature

Final reports on research projects are listed separately, as are standards, directives and guidelines.

[1] acp – advanced clean production GmbH (Ed.): CO2-Schneestrahlreinigung. http://www.acp-micron.com/5699-CO2-Schneestrahl-Reinigung.html?lang=0, retrieved 21 October 2012.

[2] Alvarez, C.: Simulationsgestützte Methoden zur effizienten Gestaltung von Lötprozessen in der Elektronikproduktion. Bamberg: Meisenbach Verlag, 2008.

[3] Amend, P. et al.: ADDIMID Technology and Aerosol Jet Printing for Functional MID Prototypes by the Use of Stereolithography. In: Proceedings: 10th International Congress Molded Interconnect Devices, Research Association Molded Interconnect Devices 3-D MID e. V., Fürth, 19–20 September 2012.

[4] Amesöder, S.: Wärmeleitende Kunststoffe für das Spritzgießen. Dissertation, Lehrstuhl für Kunststofftechnik, FAU Erlangen-Nürnberg, 2009.

[5] Bandowski, A.: Flexible Fully Automated 3D MID Assembly Systems. In: Proceedings: 10th International Congress Molded Interconnect Devices, Research Association Molded Interconnect Devices 3-D MID e. V., Fürth, 19–20 September 2012.

[6] Baraldi, L., Ernst, C., Elspass, S.: Lasergestützte MID-Technologien. In: Geiger, M. et al. (Ed.): Laser in der Elektronikproduktion & Feinwerktechnik – Tagungsband des 9. Erlanger Seminars LEF 2006, Bamberg: Meisenbach Verlag, 2006.

[7] BASF SE (Ed.): Company brochure Ultramid T4381 LOS (KT I K, F204). File: http://www.plasticsportal.net/wa/plasticsEU-de_DE/function/conversions:/publish/common/upload/engineering_plastics/Uitramid_T4381LDS.pdf, retrieved 07 July 2012.

[8] Baur, E., Brinkmann, S., Schmachtenberg, E., Osswald, T. A.: Saechtling Kunststoff Taschenbuch. Munich: Carl Hanser Verlag, 2007.

[9] Bayer AG (Ed.): Conductive nano inks für printed electronics. http://www.baytubes.com/news_and_services/pr_100217_nano-ink.html, retrieved 16 November 2010.

[10] Beer, D.: Zuverlässige Qualitätssicherung von MID-Produkten. In: Produktion von Leiterplatten und Systemen (PLUS), Vol. 13, No. 4 (2011), pp. 927–931.

[11] Beine, H.: Neue Dimensionen – Reinigen Aktivieren und Beschichten mit kaltaktivem Plasma unter Atmosphärendruck. File: http://www.all-electronics.de/track.php?p=1&ci=6576&ct=&l=http://www.all-electronics.de/media/file/7142, retrieved 20 June 2012.

[12] Bell, H.: Reflowlöten – Grundlagen, Verfahren, Temperaturprofile und Lötfehler. Bad Saulgau: Eugen G. Leuze Verlag, 2005.

[13] Bernard, D., Golubovic, D.: 3D Board Level X-Ray Inspection via Limited Angle Computer Tomography. In: Proceedings of the SMTA International Conference, Orlando, Florida, 14–28 October 2012.

[14] Birkicht, A., Meyerovich, A.: Außenstromlose Kupfer-Zinn-Abscheidung für die 3D-MID Technologie. File: www.ivam.de/download/1850_inno48%20for%20web.pdf, retrieved 24 October 2012.

[15] Borges, M.: Laser schreibt dreidimensionale Leiterbahnen. In: Elektronik Sonderausgabe Räumliche elektronische Baugruppen, 2011, pp. 12–16.

[16] Brand, A.: Prozess und Systeme zur Bestückung räumlicher elektronischer Schaltungsträger (3D MID). Bamberg: Meisenbach Verlag, 1997.

[17] Brinkmann, T.: Entwicklungsverkürzung mittels durchgängigem CAD Einsatz. In: Tagungsband: Seminar zur Kunststoffverarbeitung: Optimierte Formteilentwicklung – Schneller am Markt mit durchgängiger Rechnerunterstützung, Institut für Kunststoffverarbeitung, Aachen, 1–2 December 1998.

[18] Brocka, Z., Ehrenstein, G. W.: Strahlenvernetztes PA 6 für Schaltungsträger. In: Kunststoffe, Vol. 94, No. 11 (2004), pp. 88–94.

[19] Brose, A., Leneke, T., Hirsch, S, Schmidt, B.: Aerosol Deposition of Catalytic Ink to Fabricate Fine Pitch Metallizations for Molded Interconnect Devices (MID). In: Proceedings: Electronic System-Integration Technology Conference (ESTC) 2010, Institute of Electrical and Electronic Engineers (IEEE), Berlin, 13–16 September 2010.

[20] Cadence Design Systems, Inc. (Ed.): Products & Solutions. http://www.cadence.com/products/pages/default.aspx, retrieved 27 October 2012.

[21] CadSoft Computer GmbH (Ed.): EAGLE PCB Software. http://www.cadsoft.de/eagle-pcb-design-software/product-overview/?language=de, retrieved 27 October 2012.

[22] Campus materials database: Data sheet. http://www.campusplastics.com/campus/en/datasheet/Pocan%C2%AE+DP+ T7140+LDS/LANXESS+Deutschland+GmbH/99/80f9c97a, retrieved 07 August 2012.

[23] Chai, K., Wu, L.: The underfill processing technologies for flip chip packaging. In: Proceedings of 1st International IEEE Conference on Polymers and Adhesives in Microelectronics and Photonics (Polytronic 2001), Institute of Electrical and Electronic Engineers (IEEE), Potsdam, 21–24 October 2001.

[24] Chappell, J., Hutt, D. A., Conway, P. P.: Controlling the Feature Sizes of an Inkjet Printed Optical Waveguide Applicable Material. In: Proceedings: 2nd International Symposium on Photonic Packaging, Institute of Electrical and Electronic Engineers (IEEE), Munich, 13 November 2008.

[25] Coombs, C. F. (Ed.): Printed Circuits Handbook. 5th ed., New York: McGraw Hill, 2001.

[26] CustomPartNet (Ed.): Stereolitography. http://www.custompartnet.com/wu/stereolithography, retrieved 14 November 2011.

[26a] CustomPartNet (Ed.): Stereolitography. http://www.custompartnet.com/wu/stereolithography, retrieved 14 November 2011.

[26b] CustomPartNet (Ed.): Selective Laser Sintering. http://www.custompartnet.com/wu/selective-laser-sinterning, retrieved 14 November 2011.

[26c] CustomPartNet (Ed.): Fused Deposition Modeling. http://www.custompartnet.com/wu/fused-deposition-modeling, retrieved 14 November 2011.

[27] Dassault Systèmes (Ed.): CATIA – Das digitale Produkterlebnis. http://www.3ds.com/de/products/catia/, retrieved 27 October 2012.

[28] De Gans, B.-J., Duineveld, P. C., Schubert, U. S.: Inkjet Printing of Polymers: State of the Art and Future Developments. In: Journal of Advanced Material, Vol. 16, No. 3, 2004, pp. 203–213.

[29] Dressler, M.: Reliability Study of Stud Bump Bonding Flip Chip Assemblies on Molded Interconnect Devices. Dissertation, Technische Universität Berlin, 2010.

[30] Eberhardt, W., Kück, H., Weser, S.: Von der Deko in die Industrie. In: Elektronik Sonderausgabe Räumliche elektronische Baugruppen, 2011, pp. 24–27.

[31] Ehrenstein, G. W., Amesöder, S.: Gerichteter Wärmetransport – eine besondere Bedeutung im Thermal Management. In: Fachtagung Wärmeleitende Kunststoffe, Lehrstuhl für Kunststofftechnik, Erlangen, 2007.

[32] Ehrenstein, G. W., Geiger, M., Münstedt, H.: Bereitstellung und Qualifizierung von laseraktivierten Thermoplastwerkstoffen für die Additive Metallisierung von MIDs. Erlangen: Bayerisches Laserzentrum, 2004.

[33] Ehrenstein, G. W.: Mit Kunststoffen konstruieren. 3rd ed., Munich: Carl Hanser Verlag, 2007.

[34] Ehrenstein, G. W.: Polymer-Werkstoffe. Munich: Carl Hanser Verlag, 1999.

[35] Ehrenstein, G. W.: Polymer-Werkstoffe: Struktur – Eigenschaften – Anwendung. 3rd ed., Munich: Carl Hanser Verlag, 2011.

[36] Ehrenstein, G. W., Riedel, G., Trawiel, P.: Praxis der Thermischen Analyse von Kunststoffen. Munich: Carl Hanser Verlag, 2003.

[37] Eisenbarth, M.: Beitrag zur Optimierung der Aufbau- und Verbindungstechnik für mechatronische Baugruppen. Bamberg: Meisenbach Verlag, 2003.

[38] Eitrich, O.: Prozeßorientiertes Kostenmodell für die entwicklungsbegleitende Vorkalkulation. Karlsruhe: Universität Karlsruhe, 1996.

[39] Elektroniknet.de (Ed.): Kunststoffe umweltverträglich metallisieren und beschichten. http://www.elektroniknet.de/e-fertigung/news/article/29386/0/Kunststof/, retrieved 19 October 2012.

[40] Elsner, P. et al. (Ed.): Domininghaus – Kunststoffe: Eigenschaften und Anwendungen. 8th ed., Berlin: Springer Verlag, 2012.

[41] Endres, B.: Funktionale Metallschichten für MID. In: Tagungsband: Innovative Anwendungen der MID-Technologie, Workshop HSG-IMAT und Wirtschaftsministerium Baden-Württemberg, Stuttgart, 09 October 2009.

[42] EPP Elektronik Produktion & Prüftechnik (Ed.): Verfahren zur chemiefreien Metallisierung und Beschichtung. http://www.epp-online.de/archiv/-/article/32536724/32624659/Verfahren-zur-chemiefreien-Metallisierung-und-Beschichtung/art_co_INSTANCE_0000/maximized (as of: 02 November 2009), retrieved 19 October 2012.

[43] Essemtec AG (Ed.): Hydra – 3D MID Dosier- & Bestückungssystem. http://www.essemtec.de/product.aspx/SMD-Bestueckungssysteme/3D-MID-Dosier-Bestueckungssystem/?ArtNr=Hydra, retrieved 07 January 2013.

[44] Essemtec AG (Ed.): Essemtec präsentiert 3D-Dispens- und Bestückungssystem. http://www.essemtec.de/news.asp?id=10502 (As at: 07 December 2011), retrieved 10 January 2013.

[45] Feldmann, K., Kunze, A.: Schaltungsträger als 3D-Teile gefertigt. In: Maschinenmarkt Plastics – Spritzgießen und Formenbau, Issue 2/3 (2003), pp. 18–21.

[46] Fischer, C., Franke, J., Feldmann, K.: Two Approaches for the Design of Molded Interconnect Devices (3D-MID). In: G. Huang et al. (Ed.): Proceedings of the 6th CIRP-Sponsored International Conference on Digital Enterprise Technology (DET 2009). Berlin: Springer Verlag, 2010, pp. 67–78.

[47] FIT Fruth Innovative Technologien GmbH (Ed.): DMLS Spritzgusswerkzeug. http://www.pro-fit.de/info.php?leistung=0166, retrieved 04 November 2012.

[48] Forschungsvereinigung Räumliche Elektronische Baugruppen 3-D MID e. V. (Ed.): 3D-MID Technologie: Räumliche elektronische Baugruppen: Herstellungsverfahren, Gebrauchsanforderungen, Materialkennwerte. Munich: Carl Hanser Verlag, 2004.

[49] Franke, J. et al.: Integration von Makro-MID-Technologie im PKW. In: Gausemeier, J. et al. (Ed.): 8. Paderborner Workshop – Entwurf mechatronischer Systeme, HNI-Verlagsschriftenreihe, Paderborn, Vol. 294, 2011, pp. 39–55.

[50] Franke, J., Gausemeier, J., Goth, C., Dumitrescu, R.: MID-Studie 2011 – Markt- und Technologieanalyse. Eine Studie im Auftrag von Forschungsvereinigung Räumliche Elektronische Baugruppen 3-D MID e. V. (Ed.), Erlangen, 2011.

[51] Franke, J.: Integrierte Entwicklung neuer Produkt- und Produktionstechnologien für räumliche spritzgegossene Schaltungsträger (3-D MID). Munich: Carl Hanser Verlag, 1995.

[52] Fraunhofer IZM (Ed.): Flip Chip Adhesive Bond Technologies. http://www.izm.fraunhofer.de/de/abteilungen/system_integrationinterconnectiontechnologies/leistungsangebot/prozess-_und_produktentwicklung/flip_chip_adhesivebondtechnologies.html, retrieved 10 October 2012.

[53] Freiermuth, M.: 3D Pick & Place with Integrated Robotics. In: Proceedings: 9th International Congress Molded Interconnect Devices, Research Association Molded Interconnect Devices 3-D MID e. V., Fürth, 29–30 September 2010.

[54] Friedrich, J.: Untersuchungen zur Strombelastbarkeit von Leiterbahnen auf metallisierten Thermoplasten. Fürth, 2006.

[55] Fuchs, M., Ranft, F., Drummer, D.: Thermal Conductive-Modified Polyamide Compounds for Manufacturing MID and Film Substrates. In: Proceedings: 10th International Congress Molded Interconnect Devices, Research Association Molded Interconnect Devices 3-D MID e. V., Fürth, 19–20 September 2012.

[56] Gaida, B., Aßmann, K.: Technologie der Galvanotechnik. 1st ed., Saulgau: Eugen G. Leuze Verlag, 1996.

[57] Gajski, D. D.: Principles of Digital Design. Upper Saddle River: Prentice Hall, 1997.

[58] Gausemeier, J., Binger, V., Peitz, T.: Chancen und Grenzen für den Einsatz der Technologie MID: Markt- und Technologieanalyse. Eine Studie im Auftrag von Forschungsvereinigung Räumliche Elektronische Baugruppen 3-D MID e. V. (Ed.), Paderborn, 2003.

[59] Gausemeier, J., Feldmann, K.: Integrative Entwicklung räumlicher elektronischer Baugruppen. Munich: Carl Hanser Verlag, 2006.

[60] Gausemeier, J., Peitz, T.: MID-Studie 2006: MID-Markt: Deutschland-Analyse von MID-Projekten. Eine Studie im Auftrag von Forschungsvereinigung Räumliche Elektronische Baugruppen 3-D MID e. V. (Ed.), Paderborn, 2006.

[61] Gebhardt, A.: Generative Fertigungsverfahren: Rapid Prototyping – Rapid Tooling – Rapid Manufacturing. 3rd ed., Munich: Carl Hanser Verlag, 2007.

[62] Gevatter, H.-J., Grünhaupt, U. (Ed.): Handbuch der Mess- und Automatisierungstechnik in der Produktion. Berlin: Springer Verlag, 2006.

[63] Goth, C.: Analyse und Optimierung der Entwicklung und Zuverlässigkeit räumlicher Schaltungsträger (3D-MID). Bamberg: Meisenbach Verlag, 2013.

[64] Goth, C., Franke, J., Putzo, S.: Aerosol Jet Printing on Rapid Prototyping Materials for Fine Pitch Electronic Applications. In: Proceedings: 61st Electronic Components and Technology Conference (ECTC 2011), Institute of Electrical and Electronic Engineers (IEEE), Buena Vista, Florida, 31 May – 3 June 2011, pp. 1211–1216.

[65] Goth, C., Putzo, S., Hörber, J., Franke, J.: Aerosol Jet Printing and Rapid Prototyping Methods for Flexible Structuring of 3D Interconnect Devices. In: Proceedings: International Conference on Electronics Packaging (ICEP 2011), Nara (Japan), 13–15 April 2011, pp. 324–329.

[66] Grellmann, W., Seidler, P. (Ed.): Kunststoffprüfung. 2nd ed., Munich: Carl Hanser Verlag, 2011.

[67] Gu, Y., Nakamura, T.: Interfacial delamination and fatigue life estimation of 3D solder bumps in flip-chip packages. In: Microelectronics Reliability, Vol. 44, Issue 3 (2004), pp. 471–483.

[68] Häcker Automation GmbH (Ed.): Assembly of 3-Dimensional Components Using a Special 3D Substrate Carrier. In: Proceedings: 9th International Congress Molded Interconnect Devices, Research Association Molded Interconnect Devices 3-D MID e. V., Fürth, 29–30 September 2010.

[69] Häcker Automation GmbH (Ed.): Datenblatt VICO 520 M Multi-Chip + 3D-Mikro-Montageanlage. File: http://www.haecker-automation.de/fileadmin/user_upload/Datenblaetter/520M_d.pdf, retrieved 07 January 2013.

[70] Harting AG: Gestaltungsrichtlinie 3D-MID LDS-Prozess, HARTING AG Mitronics, Biel, 2012.

[71] Heger, A: Technologie der Strahlenchemie von Polymeren. Munich: Carl Hanser Verlag, 1990.

[72] Heicks Parylene Coating GmbH (Ed.): Grundlagen der Parylenebeschichtung. http://www.heicks.de/grundlagen-parylene.html, retrieved 27 September 2012.

[73] Heininger, N., John, W., Boßler, H.-J.: Fertigung von MID-Bauteilen vom Rapid Prototyping bis zur Serie mit innovativer LDS-Technologie. Firmenschrift LPKF Laser & Electronics AG, Garbsen, 2004.

[74] Heininger, N., John, W., Boßler, H.-J.: Manufacturing of Molded Interconnect Devices from Prototyping to Mass Production with Laser Direct Structuring. In: Proceedings: 6th International Congress Molded Interconnect Devices, Research Association Molded Interconnect Devices 3-D MID e. V., Fürth, 22–23 September 2004.

[75] Heininger, N.: Laserstrukturierte Schaltungsträger. Firmenschrift LPKF Laser & Electronics AG, Garbsen, 2008.

[76] Heininger, N.: LPKF-LDS® wird zum Standard. LPKF Laser & Electronics AG (Ed.): http://www.lpkf.de/_mediafiles/1540-newsletter-mid-10-2009.htm, retrieved 17 October 2012.

[77] Heinle, C.: Simulationsgestützte Entwicklung von Bauteilen aus wärmeleitenden Kunststoffen. Dissertation, Lehrstuhl für Kunststofftechnik, FAU Erlangen-Nürnberg, 2012.

[78] Hellerich, W., Harsch, G., Haenle S.: Werkstoff-Führer Kunststoffe: Eigenschaften, Prüfungen, Kennwerte. Munich: Carl Hanser Verlag, 2004.

[79] Henke, T. et al.: Three-dimensional combination sensor in MID technology. In: Proceedings: 7th International Congress Molded Interconnect Devices, Research Association Molded Interconnect Devices 3-D MID e. V., Fürth, 27–28 September 2006.

[80] Hering, E., Martin, R.: Photonik: Grundlagen, Technologie und Anwendung. Berlin: Springer Verlag, 2006.

[81] Hunziker, W.: Chipmontage auf MID (Molded Interconnect Device) – Ein Weg zu Chipmodulen höherer Funktionalität. In: HARTING KGaA (Ed.): HARTING tec. News Issue 13 (2005), pp. 20–25. File: http://www.harting.de/aktuelles/kundenzeitschrift-tec-news/, retrieved 15 August 2011.

[82] Jaroschek, C.: Spritzgießen für Praktiker. 2nd ed., Munich: Carl Hanser Verlag, 2008.

[83] Jillek, W., Schmitt, E.: Inkjet gedruckte Schaltungen. In: Produktion von Leiterplatten und Systemen (PLUS), Vol. 13, No. 1 (2011), pp. 151–162.

[84] Johannaber, F., Michaeli, W.: Handbuch Spritzgießen. 2nd ed., Munich: Carl Hanser Verlag, 2004.

[85] John, W.: New Possibilities for MID Engineers based on LPKF-LDS ProtoPainting. In: Proceedings: 9th International Congress Molded Interconnect Devices, Research Association Molded Interconnect Devices 3-D MID e. V., Fürth, 29–30 September 2010.

[86] John, W.: Rapid Prototyping für 3D-Schaltungsträger – Neues Verfahren ermöglicht schnellen und kostengünstigen 3D-Prototypenbau. In: Elektronik, No. 8 (2011).

[87] John, W.: Rapid-Prototyping mechatronischer Baugruppen mittels LDS-Technologie. Hanser-Seminar: "Rapid Prototyping für mechatronische Produkte" directed by Dr.-Ing. Rudolf Meyer, Augsburg, 10 May 2006.

[88] Kaiser, I.: Systematik zur Entwicklung mechatronischer Systeme in der Technologie MID (Molded Interconnect Devices), Heinz Nixdorf Institut, Universität Paderborn, HNI-Verlagsschriftenreihe, Vol. 248, Paderborn, 2009.

[89] Kannengießer, U., Stober, S.: Lötbeständiges PBT. In: Kunststoffe, Vol. 91, No. 8 (2001), pp. 108–111.

[90] Kester, Inc. (Ed.): Lead-free Solutions. http://www.kester.com/sidemenu/leadfreesolutions/tabid/145/default.aspx, retrieved 04 October 2012.

[91] King, B., Renn, M.: Aerosol Jet® Direct Write Printing for Mil-Aero Electronics Applications. File: http://www.optomec.com/Additive-Manufacturing-Downloads/3D-Printed-Electronics, retrieved 12 October 2012.

[92] Klein-Wassink, R. J.: Weichlöten in der Elektronik. 2nd ed., Bad Saulgau: Eugen G. Leuze Verlag, 1991.

[93] Kohlhepp, K: Der richtige Werkstoff für Zukunftsmärkte. In: Kunststoffe, Vol. 94, No. 8 (2004), pp. 26–31.

[94] Krebs, T.: Integration elektromechanischer CA-Anwendungssysteme über einem STEP-Produktmodell. Bamberg: Meisenbach Verlag, 1996.

[95] Krebs, T., Franke, J.: Konstruktionswerkzeuge für elektronisch/mechanisch integrierte Produkte – ECAD- und MCAD-Funktionen in einem dreidimensionalen Entwicklungssystem integriert. In: Elektronik, No. 18 (2005), pp. 60–66.

[96] Lee, D. W.: Formteile durch Vernetzen mit ionisierenden Strahlen veredeln. In: Kunststoffe, Vol. 83, No. 2 (1993), pp. 127–130.

[97] Lee, D. W.: Permanent vernetzt. In: Kunststoffe, Vol. 91, No. 6 (2001), pp. 78–80.

[98] Leicht, H. W.: Vapor Phase Soldering for MID High Volume Production. Chances and Risks. In: Proceedings: 10th International Congress Molded Interconnect Devices, Research Association Molded Interconnect Devices 3-D MID e. V., Fürth, 19–20 September 2012.

[99] Lienig, J.: Layoutsynthese elektronischer Schaltungen – Grundlegende Algorithmen für die Entwurfsautomatisierung. Berlin: Springer Verlag, 2006.

[100] LPKF Laser & Electronics AG (Ed.): Große Bauteile im LDS-Verfahren – LPKF Fusion3D 1500. File: http://www.lpkf.de/_mediafiles/2444-fusion3d-1500.pdf, retrieved 19 November 2012.

[101] LPKF Laser & Electronics AG (Ed.): High Volume Production with Multiple Laser Heads – LPKF Fusion3D 6000. File: http://www.lpkf.de/_mediafiles/1319-mid-high-volume-production.pdf, retrieved 19 November 2012.

[102] LPKF Laser & Electronics AG (Ed.): Laser-Direkt-Strukturierung (LPKF-LDS) – Herstellung von 3D-Schaltungsträgern (MIDs). http://www.lpkf.de/produkte/mid/index.htm, retrieved 06 December 2012.

[103] LPKF Laser & Electronics AG (Ed.): Laser-Direkt-Strukturierung von mechatronischen Schaltungsträgern – LPKF MicroLine 3D. File: http://www.lpkf.de/_mediafiles/1205-lpkf-microline-3d.pdf, retrieved 19 November 2012.

[104] LPKF Laser & Electronics AG (Ed.): LDS-MID Designregeln – Technische Information. Version 2.0, as at 10 November 2010, Garbsen, 2010.

[105] LPKF Laser & Electronics AG (Ed.): LPKF ProtoPaint LDS – Herstellung LDS-fähiger Kunststoffteile. http://www.lpkf.de/produkte/mid/protopaint.htm, retrieved 04 November 2012.

[106] LPKF Laser & Electronics AG (Ed.): LPKF stellt einfache Metallisierung für 3D-Schaltungsträger vor. http://www.lpkf.de/presse/pressemitteilungen/909/334.htm, retrieved 22 October 2012.

[107] LPKF Laser & Electronics AG (Ed.): Neuer Markt: LDS-Antennen in Laptops. http://www.lpkf.de/newsletter/mid/2011/maerz/271.htm, retrieved 25 Setptember 2012.

[108] LPKF Laser & Electronics AG (Ed.): Wirtschaftlicher Einstieg in die Laser Direkt Strukturierung (LDS) – LPKF Fusion3D 1100. File: http://www.lpkf.de/_mediafiles/2250-fusion3d-1100-broschuere.pdf, retrieved 19 November 2012.

[109] Luchs, R.: Einsatzmöglichkeiten leitender Klebstoffe zur zuverlässigen Kontaktierung elektronischer Baulemente in der SMT. Bamberg: Meisenbach Verlag, 1998.

[110] Mäntysalo M. et al.: Capability of Inkjet Technology in Electronics Manufacturing. In: Proceedings: 2009 Electronic Components and Technology Conference, San Diego, 26–29 May 2009.

[111] Mathee, U.: ECAD/MCAD-Collaboration – Effizienter Tauschhandel. In: CAD CAM, Issue 9–10 (2009), pp. 10–12.

[112] MECADTRON GmbH (Ed.): System NEXTRA®. http://www.mecadtron.com/produkte/nextra.de.php, retrieved 06 December 2012.

[113] Meier, R.: Strategien für einen produktorientierten Einsatz räumlicher spritzgegossener Schaltungsträger (3-D MID). Bamberg: Meisenbach Verlag, 2002.

[114] Menges, G., Michaeli, W., Mohren, P.: Spritzgießwerkzeuge. Munich: Hanser Verlag, 2007.

[115] Mentor Graphics Corporation (Ed.): The EDA Technology Leader. http://www.mentor.com/products/, retrieved 27 October 2012.

[116] N. N.: Mechatronik im Fokus. In: Produktion von Leiterplatten und Systemen (PLUS), Vol. 13, No. 11 (2011), pp. 2655–2658.

[117] Novacentrix (Ed.): Advanced Curing for Printed Electronics. File: http://www.novacentrix.com/sites/all/themes/novacentrix/downloads/PF_Brochure_4pg.pdf, retrieved 14 October 2012.

[117a] Nowottnick, M.: Hochtemperatur-Lotwerkstoffe. In: Reichl, H. (Ed.) Systemintegration in der Mikroelektronik Fertigung elektronischer Baugruppen für die Automobilelektronik. Berlin: VDE Verlage, 2008, pp. 107–124.

[118] ODU-Steckverbindungssysteme GmbH & Co. KG (Ed.): Einzelkontakte im ODU Springtac®-Kontaktsystem und ODU-Lamellen-Kontaktsystem. File: http://www.odu.de/fileadmin/template/pdf/einzel/Einzelkontakte.deu.pdf, retrieved 09 October 2012.

[119] Oettel, H., Schumann, H. (Ed.): Metallografie – Mit einer Einführung in die Keramografie. 15th ed., Weinheim: Wiley-VCH Verlag, 2011.

[120] Orrock, J.: Rapid LDS Prototyping Made Possible by Fused Deposition Modeling. In: Proceedings: 10th International Congress Molded Interconnect Devices, Research Association Molded Interconnect Devices 3-D MID e. V., Fürth, 19–20 September 2012.

[121] Orthmann, K.: Kleben in der Elektronik – elektrisch leitende Klebstoffe – Anwendung, Aufbau, Applikation, Qualität, Zuverlässigkeit. Kontakt & Studium, Bd. 472, Renningen-Malmsheim: expert Verlag, 1995.

[122] Pahl, G., Beitz, W.: Konstruktionslehre: Grundlagen erfolgreicher Produktentwicklung – Methoden und Anwendung. 7th ed., Berlin: Springer Verlag, 2007.

[123] Panasonic Corporation (Ed.): MIPTEC 3D Packaging Technology. http://www3.panasonic.biz/ac/e/tech/miptec/index.jsp, retrieved 07 October 2012.

[124] Panasonic Corporation (Ed.): MIPTEC. File: http://www3.panasonic.biz/ac/e_download/tech/miptec/catalog/con_eng_miptec.pdf?f_cd=401242, retrieved 07 October 2012.

[125] Parametric Technology Corporation (Ed.): PTC Creo – Konstruktionssoftware, die Potenzial nutzbar macht. http://de.ptc.com/product/creo/, retrieved 27 October 2012.

[126] Peitz, T.: Methodik zur Produktoptimierung mechanisch elektronischer Baugruppen durch die Technologie MID (Molded Interconnect Devices), Heinz Nixdorf Institut, Universität Paderborn, Paderborn: HNI-Verlagsschriftenreihe, Vol. 221, 2008.

[127] Perelaer, J. et al.: Plasma and Microwave Flash Sintering of a Tailored Silver Nanoparticle Ink, Yielding 60% Bulk Conductivity on Cost-Effective Polymer Foils. In: Journal of Advanced Material, Vol. 24, Issue 29 (2012), pp. 3993–3998.

[128] Pfeffer, M., Goth, C., Craiovan, D., Franke, J.: 3D-Assembly of Molded Interconnect Devices with Standard SMD Pick & Place Machines Using an Active Multi Axis Workpiece Carrier. In: Proceedings: 2011 IEEE International Symposium on Assembly and Manufacturing, ISAM 2011, Tampere (Finland), 25–27 May 2011.

[129] Pöhlau, F.: Entscheidungsgrundlagen zur Einführung räumlicher spritzgegossener Schaltungsträger (3-D MID). Bamberg: Meisenbach Verlag, 1999.

[130] Pongratz, S., Ehrenstein, G. W.: Lötbeständigkeit von Kunststoffen. In: Kunststoffe, Vol. 88, No. 7 (1998), pp. 1025-1028.

[131] Pongratz, S., Ehrenstein, G. W.: Long-Time Behaviour of Soldered Plastics. In: Proceedings, ANTEC 1999 Plastics: Bridging the Millennia, Society of Plastics Engineers, New York City, 2-6 May 1999.

[132] Prinz, U., Schnoor, A.: Recent Developments for Plating of LDS Serial Parts and Prototypes Based on One Step Copper Technology In: Proceedings: 10th International Congress Molded Interconnect Devices, Research Association Molded Interconnect Devices 3-D MID e. V., Fürth 19-20 September 2012.

[133] ProSTEP iViP Association (Ed.): ECAD/MCAD Collaboration - Recommendation. File: http://www.prostep.org/fileadmin/freie_downloads/Empfehlungen-Standards/ProSTEP_iViP/ProSTEP-iViP_Recommendation_ECAD-MCAD_5_2-0_01.pdf, retrieved 06 December 2012.

[134] Pscherer, C.: Functional Prototypes for 3D Molded Interconnect Device Applications by the Use of Direct Rapid Prototyping Technologies. In: Proceedings: 9th International Congress Molded Interconnect Devices, Research Association Molded Interconnect Devices 3-D MID e. V., Fürth, 29-30 September 2010.

[135] Ranft, F., Drummer, D., Heinle, C.: Wärmemanagement mit hochgefüllten Kunststoffen. Erlanger-Bayreuther-Kunststofftage, Erlangen 2012.

[136] Raschig GmbH (Ed.): Data Sheets. http://www.raschig.de/Kunststoffe, retrieved 18 January 2013.

[137] Reichl, H. (Ed.): Direktmontage: Handbuch für die Verarbeitung ungehäuster ICs. Berlin: Springer Verlag, 1998.

[138] Reinhardt, A., Franke, J., Goth, C.: Reliability of Micromechatronic Systems with Chip on Molded Interconnected Devices and Flexible Substrates. In: Proceedings: 12th Electronics Packaging Technology Conference (EPTC 2010), Institute of Electrical and Electronic Engineers (IEEE), Singapore, 8-10 December 2010, pp. 79-83.

[139] Reinhausen Plasma GmbH (Ed.): Ein kostengünstiges und temperaturschonendes Verfahren zur lösungsmittelfreien Metallisierung. Training, Regensburg, 2012.

[140] Reinhausen Plasma GmbH (Ed.): plasmadust® - Die Revolution der Beschichtungstechnik. File: www.reinhausen.com/de/XparoDownload.ashx?raid=21258, retrieved 19 October 2012.

[141] Remer, U.: From Prototyping to Mass Production of a MID Based Flow Sensor. In: Proceedings: 10th International Congress Molded Interconnect Devices, Research Association Molded Interconnect Devices 3-D MID e. V., Fürth, 19-20 September 2012.

[142] Richter, H., Lapper, S., Eberhardt, W., Kück, H.: Heiße Technik. In: Elektronik Sonderausgabe Räumliche elektronische Baugruppen 2011, pp. 17-20.

[143] Rohde, H.: MID Drucksensor für die Fahrzeugtechnik. In Tagungsband: Innovative Anwendungen der MID-Technik, Hahn-Schickard-Gesellschaft Institut für Mikroaufbautechnik (HSG-IMAT), Stuttgart, 5 October 2011.

[144] Rösch, M.: Potenziale und Strategien zur Optimierung des Schablonendruckprozesses in der Elektronikproduktion. Bamberg: Meisenbach Verlag, 2011.

[145] Rose, M.: Leiterplattenentwurf. Heidelberg: Hüthig Verlag, 1992.

[146] Rouif, S.: Radiation cross-linked polymers: Recent developments and new applications. Nuclear Instruments and Methods in Physics Research, B Volume 236, Issue 1–4 (2005), pp. 68–72.

[147] RTP Company (Ed.): Data Sheets. http://www.rtpcompany.com/info/data/index.html, retrieved 07 August 2012.

[148] Scheel, W., Wittke, K., Nowottnick, M.: Die Lotverbindung – Buch 1. Templin: Verlag Dr. Markus A. Detert, 2007.

[149] Schierbock, P.: Formeln und Tabellen für Metalltechnische Berufe. Cologne: Bildungsverlag Eins, 2007.

[150] Schimpf, C.: Optimierung von Zuverlässigkeitsuntersuchungen, Prüfabläufen und Nacharbeitsprozessen in der Elektronikproduktion. Bamberg: Meisenbach Verlag, 2009.

[151] Schmid, B.: Leiterplatte und Leiterbahnen im Spritzguss. In: Elektronik Sonderausgabe Räumliche elektronische Baugruppen 2011, pp. 9–11.

[152] Schober, M., Grözinger, T., Kück, H.: Numerical Reliability Analysis of MID. In: Proceedings: 10th International Congress Molded Interconnect Devices, Research Association Molded Interconnect Devices 3-D MID e. V., Fürth, 19–20 September 2012.

[153] Schüßler, F.: Verbindungs- und Systemtechnik für thermisch hochbeanspruchte und miniaturisierte elektronische Baugruppen. Bamberg: Meisenbach Verlag, 2010.

[154] Schulz, U.: Advances in Assembly and Packaging Technology for 3D MID. In: Proceedings: 10th International Congress Molded Interconnect Devices, Research Association Molded Interconnect Devices 3-D MID e. V., Fürth, 19–20 September 2012.

[155] Sendler, U.: Das PLM-Kompendium – Referenzbuch des Produkt-Lebenszyklus-Managements. Berlin: Springer Verlag, 2009.

[156] Siemens Industry Software GmbH & Co. KG (Ed.): Explore NX and discover your solution. http://www.plm.automation.siemens.com/de_de/products/nx/, retrieved 27 October 2012.

[157] Specialty Coating Systems, Inc. (Ed.): Parylene Knowledge. http://scscoatings.com/what_is_parylene/index.aspx, retrieved 19 December 2012.

[158] Speedmold GmbH (Ed.): Rapid Prototyping. http://www.speedmold.de/index.php?p=rp, retrieved 04 November 2012.

[159] Stojek, M., Stommel, M, Korte, W.: FEM für die mechanische Auslegung von Kunststoff- und Elastomerbauteilen. 1st ed., Düsseldorf: Springer-VDI-Verlag, 1998.

[160] Suppa, M.: Schutzlacke für elektronische Baugruppen: Anwendungsgebiete, Anforderungsprofile, Merkmale, Verarbeitung. Lackwerke Peters, 2010.

[161] Süß-Wolf, R., De Paolis, M.: LEONI FLAMECON® – Strukturierte Metallisierung von der Idee zur technischen Umsetzung. In: Produktion von Leiterplatten und Systemen (PLUS), Vol. 9, No. 2 (2007), pp. 321–323.

[162] Tröger, L.: 3D MID Function Specimen in the Rapid Prototyping Process. In: Proceedings: 10th International Congress Molded Interconnect Devices, Research Association Molded Interconnect Devices 3-D MID e. V., Fürth, 19–20 September 2012.

[163] Türke, A.: Synthese und Charakterisierung von druckbaren Metalltinten. Institut für Photonische Mikrosysteme Fraunhofer Gesellschaft, 2009.

[164] Uhl, K.: PTC Pro/Engineer Wildfire 4. In: CAD CAM, Issue 1–2 (2008), pp. 60–62.

[165] VDI Technologiezentrum GmbH (Ed.): Evaluierung Plasmatechnik. File: https://www.techportal.deuploadspublications320EvaluierungPlasmatechnik.pdf, retrieved 15 May 2012.

[166] Vollhardt, K. P. C., Schore, N. E.: Organische Chemie. Weinheim: Wiley-VCH Verlag, 2012.

[167] Vyncolit NV & Sumitomo Bakelite Co.,Ltd. (Ed.): Material Data Sheets. http://www.vyncolit.net/index.php?option=com_rsfiles&task=files.display&path=/Home/DataSheets&Itemid=69, retrieved 07 August 2012.

[168] Weckenmann, A., Krämer, P.: Multisensorielle Prüfung eingebetteter Baugruppen – Teilprojekt C2. In: Weckenmann, A. (Ed.): DFG Sonderforschungsbereich 694 – Integration elektronischer Komponenten in mobile Systeme – Abschlussbericht (01 January 2006 – 31 December 2010). Aachen: Shaker, 2010, pp. 299–318.

[169] Christoph, R. Neumann, H. J.: Multisensor-Koordinatenmesstechnik. Landsberg: Verlag Moderne Industrie, 2003.

[170] Wikipedia: Flamecon – Wikipedia. Wikimedia Foundation Inc. (Ed.): http://de.wikipedia.org/wiki/Flamecon (as of: 31 October 2012), retrieved 17 October 2012.

[171] WirtschaftsWoche Online (Ed.): Die Finalisten des Innovationspreises. http://www.wiwo.de/technologie/wettbewerb-die-finalisten-des-innovationspreises-seite-6/5225504-6.html, retrieved 18 August 2012.

[172] Wißbrock, H.: Laser-Direkt-Strukturieren von Kunststoffen – Ein neues Verfahren im Spiegel eingeführter MID-Technologien. In: Kunststoffe, Vol. 92, No. 11 (2002), pp. 101–105.

[173] XENON Automatisierungstechnik GmbH (Ed.): Datenblatt 3D Präzisionsmontagemodul Space 400. http://www.xenon-dresden.de/index.php?ILNK=automotive_montageanlagen_3d_mid&iL=1&PHPSESSID=g7ss4s0bvlqpkjpjoob3d0223mg93527, retrieved 07 January 2013.

[174] XENON Automatisierungstechnik GmbH (Ed.): Datenblatt 3D-MID Bestückungs- und Montageanlagen. http://www.xenon-dresden.de/index.php?ILNK=automotive_montageanlagen_3d_mid&iL=1&PHPSESSID=g7ss4s0bvlqpkjpjoob3d0223mg93527, retrieved 07 January 2013.

[175] Zeh, K.: Selection and Qualification of a Test Method for Determining Adhesion of MID Metallization in the LDS Technology. In: Proceedings: 10th International Congress Molded Interconnect Devices, Research Association Molded Interconnect Devices 3-D MID e. V., Fürth, 19–20 September 2012.

[176] Zhang, Z., Wong, C. P.: Recent advances in flip-chip underfill: materials, process, and reliability. In: IEEE Transactions on Advanced Packaging Vol. 27, No. 3 (2004), pp. 515–524.

[177] Zhuo, Y.: Entwurf eines rechnergestützten integrierten Systems für Konstruktion und Fertigungsplanung räumlicher spritzgegossener Schaltungsträger (3D-MID). Bamberg: Meisenbach Verlag, 2007.

[178] Zimmermann, M.: Aufbau- und Verbindungstechnik auf LDS-MID. In: Produktion von Leiterplatten und Systemen (PLUS), Vol. 11, No. 3 (2009), pp. 622–632.

[179] ZSW Zentrum für Sonnenenergie- und Wasserstoff-Forschung Baden-Württemberg (Ed.): PV-Materialien. http://www.zsw-bw.de/themen/pv-materialforschung/pv-materialien.html, retrieved 19 October 2012.

[180] Zuken GmbH (Ed.): Software-Lösungen für die Leiterplattenentwicklung. http://www.zuken.com/de/products/pcb-design, retrieved 27 October 2012.

Final reports on research projects

[181] Final report on AiF project 11454 N "Zweikomponentenspritzgießen von MIDs aus Kombinationen hochtemperaturbeständiger Thermoplaste," 2000.

[182] Final report on AiF project 12417 N "Untersuchungen zur Einsetzbarkeit der Ultraschall-Drahtbondtechnik zur Kontaktierung in miniaturisierten MID-Gehäusen für die Mikrosystemtechnik – Drahtbonden auf MID," 2002.

[183] Final report on AiF project 13258 "Untersuchungen zur Einsetzbarkeit der Flipchip-Technik für die Kontaktierung von Nacktchips in miniaturisierten MID Gehäusen in der Mikrosystemtechnik – Flip-Chip auf MID," 2004.

[184] Final report on AiF project 15257 N "Untersuchungen zur Verbesserung der Stempelstandzeit für die Heißpräge-MID-Technik – Heißprägestempel," 2009.

[185] Final report on AiF project 15664 N "Untersuchungen zur Verbesserung der Kontaktierung von LDS-MID durch Planarisierung – Kontaktierung LDS-MID," 2009.

[186] Final report on AiF project 15684 N "Herstellung funktionaler Schaltungsträger mittels Rapid Prototyping für MID-Anwendungen – PROMID," 2010.

[187] Final report on AiF project 15839 N "Simulationsunterstützte Auslegung von Kontaktelementen zur SMT-kompatiblen Montage von 3D-MID Packages auf Leiterplatte – MID auf Leiterplatte," 2011.

[188] Final report on AiF project 16591 N "Einhausung von MID-Baugruppen im Montagespritzguss für den Einsatz unter extremen Umweltbedingungen – MID-Einhausung," 2012.

[189] Final report on AiF project 16683 N "Untersuchungen zum umweltfreundlichen und wirtschaftlichen Reinigungsverfahren für laserstrukturierte 3D-MID-Baugruppen – Reinigung Laser-MID," 2012.

[190] Final report on AiF project 259 ZN "Untersuchungen zur Optimierung der Temperaturfestigkeit von Heißpräge-MID-Baugruppen – HT MID," 2009.

[191] Final report on AiF project 303 ZBG "Zuverlässigkeit mikromechatronischer Systeme mit Chip auf MID und flexiblen Substraten – ZUMIKROSYS," 2011.

[192] Final report on BMBF collective project "Automatisierte Fertigungslinie für Heißpräge-MID-Baugruppen – AHMID," 2002.

[193] Final report on BMBF collective project "Mess- und Prüftechnik für laserbasierte Fertigung von multifunktionalen 3D-Packages," 2007.

[194] Final report on BMWi project IN 6025 "Modellunterstützte Optimierung der Stiftkontaktierung bei MID – MOSKOM," 2009.

[195] Final report on research project AZ 928-10 "Funktionalisierung von Kunststoffen im Inkjet- und Aerosoljetdruck – FKIA" funded by the Bayerische Forschungsstiftung (Bavarian Research Foundation), 2012.

Standards, directives, and guidelines

ANSI/IPC J-STD-005A
Association Connecting Electronics Industries: Requirements for Soldering Pastes. Arlington, VA, Northbrook, IL, US, 2012.

ASTM D 648
ASTM International: Determination of heat deflection temperatures of plastics. Berlin: Beuth-Verlag, 2007.

DIN 16742
Deutsches Institut für Normung e. V. (DIN): Plastics moldings – Tolerances and acceptance conditions. Berlin: Beuth-Verlag, 2012.

DIN 32513-1
Deutsches Institut für Normung e. V. (DIN): Soldering pastes – Part 1: Composition, technical specifications. Berlin: Beuth-Verlag, 2005.

DIN 60617
Deutsches Institut für Normung e. V. (DIN): Graphical symbols for diagrams. Berlin: Beuth-Verlag, 1997.

DIN 60825
Deutsches Institut für Normung e. V. (DIN): Safety of laser products. Berlin: Beuth-Verlag, 2008.

DIN EN 60068
Deutsches Institut für Normung e. V. (DIN): Environmental testing. Berlin: Beuth-Verlag, 1995.

DIN EN 60352
Deutsches Institut für Normung e. V. (DIN): Solderless connections. Berlin: Beuth-Verlag, 2012.

DIN EN 61082
Deutsches Institut für Normung e. V. (DIN): Preparation of documents used in electrotechnology. Berlin: Beuth-Verlag, 2007.

DIN EN 61340
Deutsches Institut für Normung e. V. (DIN): Electrostatics. Berlin: Beuth-Verlag, 2003.

DIN EN ISO 75
Deutsches Institut für Normung e. V. (DIN): Plastics – Determination of temperature of deflection under load. Berlin: Beuth-Verlag, 2004.

DIN EN ISO 178
Deutsches Institut für Normung e. V. (DIN): Plastics – Determination of flexural properties. Berlin: Beuth-Verlag, 2011.

DIN EN ISO 179
Deutsches Institut für Normung e. V. (DIN): Determination of Charpy impact properties. Berlin: Beuth-Verlag, 2010.

DIN EN ISO 294-3
Deutsches Institut für Normung e. V. (DIN): Plastics – Injection molding of test specimens of thermoplastic materials – Part 3: Small plates. Berlin: Beuth-Verlag, 2003.

DIN EN ISO 527
Deutsches Institut für Normung e. V. (DIN): Plastics – Determination of tensile properties. Berlin: Beuth-Verlag, 2012.

DIN EN ISO 868
Deutsches Institut für Normung e. V. (DIN): Plastics and ebonite – Determination of indentation hardness by means of a durometer (Shore hardness). Berlin: Beuth-Verlag, 2003.

DIN EN ISO 2039-1
Deutsches Institut für Normung e. V. (DIN): Plastics – Determination of hardness – Part 1: Ball indentation test. Berlin: Beuth-Verlag, 2003.

DIN EN ISO 2039-2
Deutsches Institut für Normung e. V. (DIN): Plastics – Determination of hardness – Part 2: Rockwell hardness. Berlin: Beuth-Verlag, 2000.

DIN EN ISO 2409
Deutsches Institut für Normung e. V. (DIN): Paints and varnishes – Cross-cut test. Berlin: Beuth-Verlag, 2007.

DIN EN ISO 2578
Deutsches Institut für Normung e. V. (DIN): Plastics – Determination of the time-temperature limits after prolonged exposure to heat. Berlin: Beuth-Verlag, 1998.

DIN EN ISO 3167
Deutsches Institut für Normung e. V. (DIN): Plastics – Multipurpose test specimens. Berlin: Beuth-Verlag, 2003.

DIN EN ISO 3497
Deutsches Institut für Normung e. V. (DIN): Metallic coatings – Measurement of coating thickness – X-ray spectrometric methods. Berlin: Beuth-Verlag, 2000.

DIN EN ISO 4624
Deutsches Institut für Normung e. V. (DIN): Paints and varnishes – Pull-off test for adhesion. Berlin: Beuth-Verlag, 2003.

DIN EN ISO 6507
Deutsches Institut für Normung e. V. (DIN): Metallic materials – Vickers hardness test. Berlin: Beuth-Verlag, 2006.

DIN EN ISO 8256
Deutsches Institut für Normung e. V. (DIN): Plastics – Determination of tensile impact strength. Berlin: Beuth-Verlag, 2005.

DIN EN ISO 10350
Deutsches Institut für Normung e. V. (DIN): Plastics – Acquisition and presentation of comparable single-point data. Berlin: Beuth-Verlag, 2010.

DIN EN ISO 10724-2
Deutsches Institut für Normung e. V. (DIN): Plastics – Injection molding of test specimens of thermosetting powder molding compounds (PMC) – Part 2: Small plates. Berlin: Beuth-Verlag, 2000.

DIN EN ISO 11359
Deutsches Institut für Normung e. V. (DIN): Plastics – Thermo-mechanical analysis (TMA). Berlin: Beuth-Verlag, 1999.

DIN ISO 857-2
Deutsches Institut für Normung e. V. (DIN): Welding and allied processes – Vocabulary – Part 2: Soldering and brazing processes and related terms (ISO 857-2:2005). Berlin: Beuth-Verlag, 2007.

DIN ISO 12165
Deutsches Institut für Normung e. V. (DIN): Tools for molding – Components of compression and injection molds and die-casting dies – Terms and symbols (ISO 1265:2000). Berlin: Beuth-Verlag, 2002.

ISO 16750
International Organization for Standardization: Road vehicles — Environmental conditions and testing for electrical and electronic equipment. Geneva: 2006.

2002/95/EC
Directive 2002/95/EC of the European Parliament and of the Council of 27 January 2003 on the restriction of the use of certain hazardous substances in electrical and electronic equipment. Brussels: 2003.

IPC-2221A
IPC – Association Connecting Electronics Industries: Generic Standard on Printed Board Design. Northbrook, Illinois: 2003.

VDI 2206
Verein Deutscher Ingenieure (VDI): Design Methodology for Mechatronic Systems. Berlin: Beuth-Verlag, 2004.

VDI 2221
Verein Deutscher Ingenieure (VDI): Systematic Approach to the Design of Technical Systems and Products. Berlin: Beuth-Verlag, 1993.

VDI 2234
Verein Deutscher Ingenieure (VDI): Basic economical information for design engineers. Berlin: Beuth-Verlag, 1990.

VDI 4005
Verein Deutscher Ingenieure (VDI): Einflüsse von Umweltbedingungen auf die Zuverlässigkeit technischer Ergebnisse. (Effects of environmental conditions on the reliability of technical products) Berlin: Beuth-Verlag, 1983.

VDI/VDE 2251
Verein Deutscher Ingenieure (VDI), Verband der Elektrotechnik Elektronik Informationstechnik (VDE): Precision-engineered components; Joints; Overview. Berlin: Beuth-Verlag, 1991.

VDI/VDE 2421
Verein Deutscher Ingenieure (VDI), Verband der Elektrotechnik Elektronik Informationstechnik (VDE): Plastic surface treatment in precision engineering – Survey. Berlin: Beuth-Verlag, 1980.

VDI/VDE 2422
Verein Deutscher Ingenieure (VDI), Verband der Elektrotechnik Elektronik Informationstechnik (VDE): Systematical development of devices controlled by microelectronics. Berlin: Beuth-Verlag, 1994.

VDI/VDE 2424
Verein Deutscher Ingenieure (VDI), Verband der Elektrotechnik Elektronik Informationstechnik (VDE): Industrial Design. Berlin: Beuth-Verlag, 1986.

VDI/VDE 2428
Verein Deutscher Ingenieure (VDI), Verband der Elektrotechnik Elektronik Informationstechnik (VDE): Equipment engineering. Berlin: Beuth-Verlag, 1989.

VDI/VDE 2628
Verein Deutscher Ingenieure (VDI), Verband der Elektrotechnik Elektronik Informationstechnik (VDE): Automated visual inspection; description of the testing task. Berlin: Beuth-Verlag, 1989.

VDI/VDE 3712
Verein Deutscher Ingenieure (VDI), Verband der Elektrotechnik Elektronik Informationstechnik (VDE): Placement of printed circuit boards; determination of placement accuracy and performance of SMD pick & place machines. Berlin: Beuth-Verlag, 1991.

VDI/VDE 3715
Verein Deutscher Ingenieure (VDI), Verband der Elektrotechnik Elektronik Informationstechnik (VDE): SMD – Surface Mount Devices – Process measurement and testing for printed circuit boards (PCB). Berlin: Beuth-Verlag, 1995.

12 Authors and Contributors

12.1 Editor

Jörg Franke
is a professor at the Friedrich-Alexander-University Erlangen-Nuremberg, where since March 2009 he has been Chair of the Institute for Factory Automation and Production Systems (FAPS). He also chairs the board of the Forschungsvereinigung Räumliche Elektronische Baugruppen (3-D MID) e. V., or Research Association Molded Interconnect Devices (3-D MID), and he is director of the E|Drive Center, the Bavarian Technology Center for Electrical Drives. Professor Franke worked in the industry for many years with companies that include McKinsey, Robert Bosch GmbH, ZF Lenksysteme (ZFLS) GmbH, and INA Schaeffler KG. In his most recent industrial capacity he was chairman of the management board of ABM Greiffenberger Antriebstechnik GmbH.

12.2 Authors

Peter Buckmüller
studied microsystem and precision-engineering technology at the University of Stuttgart; since 2007 he has worked as a research assistant at the Hahn-Schickard Institute for Micro Assembly Technology in Stuttgart. Since 2011 he has been in charge of the institute's Design, Toolmaking, and Injection Molding group.

Payam Daneschwar
studied business administration at the Württembergische Verwaltungs- und Wirtschafts-Akademie school of administrative and economic sciences, and mechanical engineering at the Friedrich-Alexander-University Erlangen-Nuremberg. Since 2012 he has been a research assistant at the Institute for Factory Automation and Production Systems of the University of Erlangen-Nuremberg.

Roman Dumitrescu
studied mechatronics at the Friedrich-Alexander-University Erlangen-Nuremberg. He then took up a post as research assistant at the Faculty for Product Engineering of the Heinz Nixdorf Institute, University of Paderborn. He worked there under Professor Dr.-Ing. Jürgen Gausemeier and was awarded his doctorate in 2010 in the field of systems engineering for intelligent mechatronic systems. Since then he has been director of the Product Engineering department in the Mechatronics System Design project group of the Fraunhofer Institute for Production Technology IPT. He is also managing director of the "Intelligent Technical Systems OstWestwalenLippe" (it's OWL) leading-edge cluster.

Wolfgang Eberhardt
studied chemistry at the University of Tübingen, where he received his doctorate from the Institute for Organic Chemistry. Since 1998 he has been a research assistant at the Hahn-Schickard Institute for Micro Assembly Technology in Stuttgart, specializing in the metallization of MID. He is currently head of the Technology department at the Hahn-Schickard Institute.

Christian Fischer
studied computer sciences at the Friedrich-Alexander-University Erlangen-Nuremberg. Since 2008 he has been a research assistant in the Systems Engineering group at the Institute for Factory Automation and Production Systems of the University of Erlangen-Nuremberg, working mainly on software solutions relating to the design of mechatronic products in general and 3D-MID in particular.

Michael Fuchs
studied production engineering and management at the Friedrich-Alexander-University Erlangen-Nuremberg. In the period 2006 to 2013 he has been a research assistant in the Plastics in Mechatronics department of the Institute of Polymer Technology of the University of Erlangen-Nuremberg.

Christian Goth
studied production engineering and management at the Friedrich-Alexander-University Erlangen-Nuremberg. Since 2007 he has been a research assistant at the Institute for Factory Automation and Production Systems of the University of Erlangen-Nuremberg. In the period 2007 to 2011 he was managing director of the Research Association Molded Interconnect Devices (3-D MID). In 2012, working under Professor Franke, he was awarded his doctorate for his work on "Analysis and optimization of the development and reliability of three-dimensional molded interconnect devices (3D-MID)."

Johannes Hörber
studied mechatronics at the Friedrich-Alexander-University Erlangen-Nuremberg. Since 2008 he has been a research assistant in the Electronics Production specialist group at the Institute for Factory Automation and Production Systems of the University of Erlangen-Nuremberg. He is currently researching assembly and connection technologies for molded interconnect devices, with a primary focus on Aerosol-Jet® printing.

Christoph Jürgenhake
studied aerospace technology, specializing in lightweight fiber composites and adaptronics, at the Technische Universität Braunschweig. On completing his studies he accepted a post in industry, specializing in the field of requirements management with a European aircraft manufacturer. Since 2011 he has been a research assistant at the Fraunhofer Institute for Production Technology. Within the project group Mechatronics System Design he is responsible for the field of intelligent electromechanical structures.

Heinz Kück
obtained his doctorate from the Physics Institute of the University of Bonn in 1982 and worked for a little more than two years at the Deutsches Elektronen-Synchrotron accelerator center in Hamburg. He then moved to the Fraunhofer Institute for Microelectronic Circuits and Systems in Duisburg, where he stayed for seven years. In 1991 he joined the FhG-IMS Dresden (now the FhG-IPMS), where as head of the Components and Technology department and vice-director he was instrumental in establishing and expanding the institute. Since 1998 he has been director of the Institute for Chronometry, Precision Engineering, and Micro-integration at the University of Stuttgart and the Hahn-Schickard Institute for Micro Assembly Technology.

Thomas Kuhn
studied production engineering and management at the Universities of Applied Sciences of Ulm and Neu-Ulm. Since 2012 he has been a research assistant at the Institute for Factory Automation and Production Systems of the University of Erlangen-Nuremberg and managing director of the Research Association Molded Interconnect Devices (3-D MID).

Michael Pfeffer
studied Electrical Engineering, Electronics, and Information Technology at the Friedrich-Alexander-University Erlangen-Nuremberg. Since 2006 he has been a research assistant in electronics production at the Institute for Factory Automation and Production Systems of the University of Erlangen-Nuremberg. His fields are component placement technologies and assembly solutions for three-dimensional interconnect devices.

Thomas Schierbaum
studied production engineering and management, specializing in mechanical engineering, at the University of Paderborn. Since 2010 he has been a research assistant at the Institute for Product Development, working under Professor Dr.-Ing. Jürgen Gausemeier at the Heinz Nixdorf Institute.

René Schramm
studied mechatronics at the Friedrich-Alexander-University Erlangen-Nuremberg. Since 2011 he has been a research assistant at the Institute for Factory Automation and Production Systems of the University of Erlangen-Nuremberg and has worked as director of the MID Applications Center (MIDAZ).

12.3 Consulting Experts

While working on the individual chapters, the authors received invaluable assistance in the form of specialist input from the following members of the research council of the Research Association Molded Interconnect Devices (3-D MID):

Albert Birkicht, HARTING AG Mitronics

Prof. Dr. Dietmar Drummer, Friedrich-Alexander-University Erlangen-Nuremberg, Institute of Polymer Technology

Prof. Dr. Jörg Franke, Friedrich-Alexander-University Erlangen-Nuremberg, Institute for Factory Automation and Production Systems

Dr. Wolfgang John, LPKF Laser & Electronics AG

Dr. Ingo Kriebitzsch, BMW AG

Dr. Andreas Pojtinger, 2E mechatronic GmbH & Co. KG

Hartmut Rohde, Robert Bosch GmbH

Robert Süß-Wolf, Kromberg & Schubert GmbH & Co. KG

13 Addresses

13.1 Research Association Molded Interconnect Devices (3-D MID)

Forschungsvereinigung Räumliche Elektronische Baugruppen 3-D MID e. V.
Fürther Straße 246b
D-90429 Nürnberg, Germany
☎ +49 911 5302-9101
🖷 +49 911 5302-9102
info@3dmid.de
www.3dmid.de

13.2 Members of the Research Association Molded Interconnect Devices (3-D MID)

2E mechatronic GmbH & Co. KG
Dr. Andreas Pojtinger
Maria-Merian-Straße 29
D-73230 Kirchheim unter Teck, Germany
☎ +49 7021 9301-63
🖷 +49 7021 9301-70
a-pojtinger@2e-mechatronic.de
www.2e-mechatronic.de

Atotech Deutschland GmbH
Felix Bozsa
Erasmusstraße 20
D-10553 Berlin, Germany
☎ +49 30 34985-239
📠 +49 30 34985-469
felix.bozsa@atotech.com
www.atotech.com

BLZ Bayerisches Laserzentrum GmbH
Philipp Amend
Konrad-Zuse-Straße 2–6
D-91052 Erlangen, Germany
☎ +49 9131 97790-0
📠 +49 9131 97790-11
p.amend@blz.org
www.blz.org

BMW AG
Dr. Ingo Kriebitzsch
Knorrstraße 147
D-80788 München, Germany
☎ +49 89 382-48315
📠 +49 89 382-76085
ingo.kriebitzsch@bmw.de
www.bmw.de

Buss-Werkstofftechnik GmbH & Co. KG
Volker Zippmann
Butzbacher Straße 6
D-35516 Münzenberg-Gambach, Germany
☎ +49 6033 9879-17
📠 +49 6033 9879-45
volker.zippmann@buss-werkstofftechnik.de
www.buss-werkstofftechnik.de

Castsolut Sales & Service GmbH
Helmut Lutz
Ludwig-Lutz-Straße 10
D-73479 Ellwangen, Germany
☎ +49 7961 9149-0
📠 +49 7961 9149-49
info@castsolut.de
www.castsolut.de

Centro Ricerche Plast-Optica S.p.A.
Franco Marcori
Via J. Linussio 1
IT-33020 Amaro (UD), Italy
☎ +39 0433 4875-12
🖷 +39 0433 4875-55
franco.marcori@magnetimarelli.com
www.crpo.it

Cicor Technologies Ltd.
Nouhad Bachnak
Route de l'Europe 8
CH-2017 Boudry, Switzerland
☎ +41 32 84305-04
🖷 +49 32 84305-24
nbachnak@cicorel.ch
www.cicor.com

Collini AG
Dr. Christian Ernst
Ringstrasse 9
CH-8600 Dübendorf, Switzerland
☎ +41 44 8241-157
🖷 +41 44 8241-112
cernst@collini.eu
www.collini.eu

Continental AG
Wolfgang Fey
Peter-Dornier-Straße 10
D-88131 Lindau, Germany
☎ +49 8382 9699-374
🖷 +49 8382 969922-374
wolfgang.fey@continental-corporation.com
www.temic.com

Dage Deutschland GmbH
Armin Struwe
Kelterstraße 69
D-73265 Dettingen/Teck, Germany
☎ +49 7021 95069-0
🖷 +49 7021 82149
sales.de@nordsondage.com
www.dage.de

Doduco GmbH
Mike Liphardt
Breite Seite 1
D-74889 Sinsheim, Germany
☎ +49 7261 409-225
Fax +49 7231 602-12139
mliphardt@doduco.net
www.doduco.net

DOW Electronic Materials
Martin Höynck
Am Kronberger Hang 4
D-65824 Schwalbach, Germany
☎ +49 711 553650-0
Fax +49 711 553650-77
mhoynck@dow.com
www.electronicmaterials.rohmhaas.com

DSM Engineering Plastics B. V.
Patrick Duis
Urmonderbaan 22
NL-6167 RD Geleen, Netherlands
☎ +31 46 4764 892
Fax +31 6 1303 8767
patrick.duis@dsm.com
www.dsmep.com

Eaton Industries GmbH
Harald Maus
Hein-Moeller-Straße 1
D-56357 Holzhausen a. d. Haide, Germany
☎ +49 6772 911-140
Fax +49 6772 911-120
haraldmaus@eaton.com
www.eaton.com

EMS-Chemie AG
Sabine Bertram
Via Innovativa 1
CH-7013 Domat/Ems, Switzerland
☎ +41 81 632-6567 or -6276
Fax +41 81 632-7410
sabine.bertram@emsgrivory.com
www.emsgrivory.com

Ensinger

Ensinger Compounds, HP Polymer GmbH
Jürgen Fraer
Werkstraße 3
A-4860 Lenzing, Austria
☎ +49 151 108 65643
🖷 +43 7672 96865
fraer@ensinger-compounds.com
www.ensinger-online.com/de/compounds

enthone

Enthone GmbH
Dr. Joachim Heyer
Elisabeth-Selbert-Straße 4
D-40764 Langenfeld, Germany
☎ +49 2173 8490-800
🖷 +49 2173 8490-208
jheyer@cookson.com
www.enthone.com

essemtec swiss made

Essemtec AG
Florian Schildein
Mosenstrasse 20
CH-6287 Aesch, Switzerland
☎ +41 41 919-9044
🖷 +41 41 919-6050
fls@essemtec.com
www.essemtec.com

EVONIK INDUSTRIES

Evonik Industries AG
Volker Strohm
Paul-Baumann-Straße 1
D-45772 Marl, Germany
☎ +49 2365 49-9022
🖷 +49 2365 49-809022
volker.strohm@evonik.com
www.evonik.com

FED
Ihr Fachverband
für Design, Leiterplatten-
und Elektronikfertigung

FED e. V.
Michael Ihnenfeld
Alte Jakobstraße 85/86
D-10179 Berlin, Germany
☎ +49 30 844 714 45
🖷 +49 30 834 18 31
m.ihnenfeld@fed.de
www.fed.de

Festo AG & Co. KG
Dr. Peter Post
Ruiter Straße 82
D-73734 Esslingen, Germany
☏ +49 711 347-3144
🖷 +49 711 347-3613
pos@de.festo.com
www.festo.com

FMB Technik GmbH
Falko A. Eidner
Ferdinand-von-Steinbeis-Ring 21
D-75447 Sternenfels, Germany
☏ +49 7045 200 64-0
🖷 +49 7045 200 64-29
f.a.eidner@fmb-technik.de
www.fmb-technik.de

Fraunhofer IFAM
Uwe Maurieschat
Wiener Straße 12
D-28359 Bremen, Germany
☏ +49 421 2246-491
🖷 +49 421 2246-430
uwe.maurieschat@ifam.fraunhofer.de
www.ifam.fraunhofer.de

Fraunhofer IPT – Entwurfstechnik Mechatronik
Dr. Roman Dumitrescu
Zukunftsmeile 1
D-33102 Paderborn, Germany
☏ +49 5251 5465-101
🖷 +49 5251 5465-102
roman.dumitrescu@ipt.fraunhofer.de
www.ipt.fraunhofer.de

Fraunhofer IST
Dr. Ralf Bandorf
Bienroder Weg 54E
D-38108 Braunschweig, Germany
☏ +49 531 2155-602
🖷 +49 531 2155-900
ralf.bandorf@ist.fraunhofer.de
www.ist.fraunhofer.de

Fraunhofer IZM

Fraunhofer IZM
Dr. Frank Ansorge
Argelsrieder Feld 6
D-82234 Oberpfaffenhofen, Germany
☎ +49 8153 9097-500
📠 +49 8153 9097-511
frank.ansorge@oph.izm.fraunhofer.de
www.izm.fraunhofer.de

FRITSCH
COMPLETE AND FLEXIBLE SMT SOLUTIONS

Fritsch GmbH
Markus Fritsch
Kastler Straße 11
D-92280 Kastl, Germany
☎ +49 9625 9210-0
📠 +49 9625 9210-49
mf@fritsch-smt.com
www.fritsch-smt.com

HÄCKER Automation

Häcker Automation GmbH
Gerrit Häcker
Inselsbergstraße 17
D-99891 Schwarzhausen, Germany
☎ +49 36259 300-0
📠 +49 36259 300-29
gerrit.haecker@haecker-automation.com
www.haecker-automation.com

HARTING®
Pushing Performance

HARTING AG Mitronics
Ellen McMillan
Leugenestrasse 10
CH-2500 Biel 6, Switzerland
☎ +41 32 344-2137
📠 +41 32 344-2102
ellen.mcmillan@harting.com
www.HARTING.com

HASEC
ELEKTRONIK GMBH

HASEC-Elektronik GmbH
Marco Zimmermann
Röberstraße 5
D-99848 Wutha-Farnroda, Germany
☎ +49 36921 300-0
📠 +49 36921 300-199
m.zimmermann@hasec.de
www.hasec.de

Heicks Industrieelektronik GmbH
Rudolf Heicks
Am Schwarzen Weg 25–31
D-59590 Geseke, Germany
☎ +49 2942 979 26-0
🖷 +49 2942 979 26-150
r.heicks@heicks.de
www.heicks.de

Henkel AG & Co. KGaA
Andreas Mende
Gutenbergstraße 3
D-85748 Garching, Germany
☎ +49 89 9268-0
🖷 +49 89 9268-1583
andreas.mende@henkel.com
www.henkel.com/electronics

Hochschule Aalen
Prof. Dr. Manfred Bartel
Beethovenstraße 1
D-73430 Aalen, Germany
☎ +49 7361 576-4107
manfred.bartel@htw-aalen.de
www.htw-aalen.de

Hochschule Ostfalia – IfR
Prof. Dr. Achim Schmiemann
Heinenkamp 16
D-38444 Wolfsburg, Germany
☎ +49 5308 939-910
🖷 +49 5308 939-9120
a.schmiemann@ostfalia.de
www.ostfalia.de

Hoffmann + Krippner GmbH
Martin Gehrig
Siemensstr. 1
D-74722 Buchen, Germany
☎ +49 6281 5200-71
🖷 +49 6281 5200-90
m.gehrig@tastatur.de
www.tastatur.de

HSG-IMAT
Prof. Dr. Heinz Kück
Allmandring 9B
D-70569 Stuttgart, Germany
☎ +49 711 685-83712
🖷 +49 711 685-83705
kueck@hsg-imat.de
www.hsg-imat.de

IBL-Löttechnik GmbH
Andreas Thumm
Messerschmittring 61–63
D-86343 Königsbrunn, Germany
☎ +49 8231 95889-0
🖷 +49 8231 95889-30
a.thumm@ibl-loettechnik.de
www.ibl-tech.com

IMO Oberflächentechnik GmbH
Wolfgang Westphal
Remchinger Straße 5
D-75203 Königsbach-Stein, Germany
☎ +49 7232 3006-503
🖷 +49 7232 3006-8503
w.westphal@imo-gmbh.com
www.imo-gmbh.com

Infotech AG
Ernest Fischer
Vogelherdstrasse 4
CH-4500 Solothurn, Switzerland
☎ +41 32 626 36-60
🖷 +41 32 626 36-69
ernest.fischer@infotech-automation.com
www.infotech-automation.com

INOTECH Kunststofftechnik GmbH
Josef Gleixner
Boschstraße 3
D-92507 Nabburg, Germany
☎ +49 9433 2400-11
🖷 +49 9433 2400-40
gleixner@inotech.de
www.inotech.de

Jaudas GmbH & Co. KG
Gustav Jaudas
Erlenbrunnenstraße 4
D-72411 Bodelshausen, Germany
☎ +49 7471 9599-12
🖷 +49 7471 9599-65
gustav.jaudas@jaudas.de
www.jaudas.de

Karlsruher Institut für Technologie (KIT) – wbk
Prof. Dr. Volker Schulze
Kaiserstraße 12
D-76131 Karlsruhe, Germany
☎ +49 721 608-42440
🖷 +49 721 608-45004
volker.schulze@kit.edu
www.wbk.kit.edu

LaserMicronics GmbH
Lars Ederleh
Osteriede 9a
D-30827 Garbsen, Germany
☎ +49 5131 90811-0
🖷 +49 5131 90811-29
l.ederleh@lasermicronics.de
www.lasermicronics.de

Laser Zentrum Hannover e. V.
Dr. Dietmar Kracht
Hollerithallee 8
D-30419 Hannover, Germany
☎ +49 511 2788-0
🖷 +49 511 2788-100
info@lzh.de
www.lzh.de

LEONI AG
Joachim Neubert
Flugplatzstraße 74
D-97318 Kitzingen, Germany
☎ +49 9321 304-758
🖷 +49 9321 304-435
joachim.neubert@leoni.com
www.leoni.com

LPKF Laser & Electronics AG
Stephan Krause
Osteriede 7
D-30827 Garbsen, Germany
+49 5131 7095-1759
+49 5131 7095-90
stephan.krause@lpkf.com
www.lpkf.de

Lüberg Elektronik GmbH & Rothfischer KG
Dr. Ulrich Rothfischer
Marienstraße 20
D-90402 Nürnberg, Germany
+49 911 99534-0
+49 911 505137
info@lueberg.com
www.lueberg.de

MacDermid Inc.
Lisa Guerrera
245 Freight Street
Waterbury, CT 06702, USA
+1 203 575-5743
+1 203 575-7990
lguerrera@macdermid.com
www.macdermid.com/electronics

MECADTRON GmbH
Dr. Thomas Krebs
Allersberger Straße 185
D-90461 Nürnberg, Germany
+49 911 462369-10
+49 911 462369-11
thomas.krebs@mecadtron.de
www.mecadtron.de

Mektec Europe GmbH
Dr. Wolfgang Bochtler
Im Technologiepark 1
D-69469 Weinheim, Germany
+49 6201 80-4191
+49 6201 80-7863
wolfgang.bochtler@mektec.de
www.mektec.de

MID Solutions GmbH
Manfred Hellmich
Bahnhofstraße 3b
D-37534 Gittelde, Germany
☎ +49 5327 859-335
📠 +49 5327 859-372
mhellmich@mid-solutions.de
www.mid-solutions.de

MID-TRONIC Wiesauplast GmbH
Karl Görmiller
Max-Planck-Weg 7
D-95676 Wiesau, Germany
☎ +49 9634 88-810
📠 +49 9634 88-7810
k.goermiller@mid-tronic.de
www.mid-tronic.de

MEP Europe B.V. - Technical Center
Group of Mitsubishi Engineering-Plastics Corporation

Mitsubishi Engineering-Plastics Europe B. V.
Francis van Vehmendahl
Urmonderbaan 22
NL-6167 RD Geleen, Netherlands
☎ +31 464 761 221
📠 +31 464 106 999
francis.vehmendahl-van@mepeu.nl
www.xantar.com

Neotech Services MTP
Dr. Martin Hedges
Petzoltstraße 3
D-90443 Nürnberg, Germany
☎ +49 911 274-5501
📠 +49 911 274-5502
info@neotechservices.com
www.neotechservices.com

Otto Dunkel GmbH
Gerhard Nicklbauer
Pregelstraße 11
D-84453 Mühldorf am Inn, Germany
☎ +49 8631 6156-1314
📠 +49 8631 6156-4314
gerhard.nicklbauer@odu.de
www.odu.de

Panasonic Electric Works Europe AG
Michael Immle
Rudolf-Diesel-Ring 2
D-83607 Holzkirchen, Germany
+49 8024 648-254
+49 8024 648-700
michael.immle@eu.panasonic.com
www.panasonic-electric-works.com

PEP – Pôle Européen de Plasturgie
Maël Moguedet
2, rue Pierre et Marie Curie
FR-01100 Bellignat, France
+33 474 8181-10
+33 474 8192-61
mael.moguedet@poleplasturgie.com
www.poleplasturgie.com

PKT Präzisions-Kunststoff-Teile GmbH
Thomas Kaupp
Daimlerstraße 5–7
D-75233 Tiefenbronn, Germany
+49 7234 9550-18
+49 7234 1010
t.kaupp@pkt-gmbh.de
www.pkt-gmbh.de

Plasmatreat GmbH
Peter Langhof
Dornierstraße 4
D-75127 Birkenfeld, Germany
+49 7231 280-9941
+49 7231 280-9944
peter.langhof@plasmatreat.de
www.plasmatreat.de

PLEXPERT GmbH
Thomas Mann
Pfromäckerstraße 21
D-73432 Aalen, Germany
+49 7361 97535-20
+49 7361 97535-80
t.mann@plexpert.de
www.plexpert.de

pretema GmbH
Dietmar Kurzeja
Enztalstraße 6
D-75223 Niefern, Germany
☎ +49 7233 69-211
🖶 +49 7233 69-258
dietmar.kurzeja@pretema.com
www.pretema.com

Raschig GmbH
Markus Walcher
Mundenheimerstraße 100
D-67061 Ludwigshafen, Germany
☎ +49 621 5618-317
🖶 +49 621 5618-661
mwalcherr@raschig.de
www.raschig.de

Reinhausen Plasma GmbH
Daniel Fisch
Weidener Straße 16
D-93057 Regensburg, Germany
☎ +49 941 4090-9100
🖶 +49 941 4090-9101
d.fisch@reinhausen-plasma.com
www.reinhausen-plasma.com

RF Plast GmbH
Dr. Simon Amesöder
Weinstraße 8
D-91710 Gunzenhausen, Germany
☎ +49 9831 6196-0
🖶 +49 9831 6196-297
simon.amesoeder@rf-plast.de
www.rf-plast.de

RKT Rodinger Kunststoff-Technik GmbH
Werner Schneider
Ernst-Abbe-Straße 3
D-93426 Roding, Germany
☎ +49 9461 954-160
🖶 +49 9461 954-180
werner.schneider@rkt.de
www.rkt.de

Robert Bosch GmbH
Dr. Andre Zimmermann
Postfach 1131
D-71301 Waiblingen, Germany
☎ +49 7151 503-2413
🖷 +49 7151 503-2664
andre.zimmermann@de.bosch.com
www.bosch.com

Salcon International
Markus Thamm
Im Höllengrund 6
D-69118 Heidelberg, Germany
☎ +49 172 7449 526
markus.thamm@salconinternational.de
www.salconinternational.de

Seuffer GmbH & Co. KG
Gerhard Tritt
Bärental 26
D-75365 Calw, Germany
☎ +49 7051 6001 17
🖷 +49 7051 6002 17
gerhard.tritt@seuffer.de
www.seuffer.de

Siemens AG
Joseph Sauer
Gebbertstraße 125
D-91058 Erlangen, Germany
☎ +49 9131 308-3326
🖷 +49 9131 308-3359
joseph.sauer@siemens.com
www.hearing.siemens.com

TDK-EPC AG & Co. KG
Oliver Bard
Nobelstraße 2–6
D-12057 Berlin, Germany
☎ +49 30 68291-429
🖷 +49 30 68291-505
oliver.bard@epcos.com
www.epcos.de

TE Connectivity
Christian Köhler
Lengefelder Straße 3
D-09125 Chemnitz, Germany
☎ +49 371 514522
📠 +49 371 514597
christian_koehler@te.com
www.te.com

Technische Hochschule Nürnberg – KAM
Dr. Jens Helbig
Hohfederstraße 40
D-90489 Nürnberg, Germany
☎ +49 911 5880-1751
📠 +49 911 5880-5750
jens.helbig@ohm-hochschule.de
www.ohm-hochschule.de/kompetenzzentren

Technische Universität Chemnitz – IFK
Agnieszka Kalinowska
Reichenhainer Straße 70
D-09126 Chemnitz, Germany
☎ +49 371 531-39544
📠 +49 371 531-839544
agnieszka.kalinowska@mb.tu-chemnitz.de
www.tu-chemnitz.de/mb/KunstStTechn

Ticona GmbH
Dr. Alexandra Jacobs-Hattwig
Am Unisys-Park 1
D-65843 Sulzbach, Germany
☎ +49 69 45009-1826
📠 +49 69 45009-81826
jacobs@ticona.de
www.ticona.de

TNO
Barend Vermeulen
De Rondom 1
NL-5612 AP Eindhoven, Netherlands
☎ +31 40 265-0534
📠 +31 401 265-0850
barend.vermeulen@tno.nl
www.tno.nl

University of Akron – Mechanical Engineering
Dr. Jae-Won Choi
244 Summer Street
44325 Akron, OH, USA
☎ +1 330 972-5276
🖷 +1 330 972-6027
palagent@gmail.com
www.uakron.edu

Universität Erlangen-Nürnberg – FAPS
Prof. Dr. Jörg Franke
Egerlandstraße 7–9
D-91058 Erlangen, Germany
☎ +49 9131 85-27569
🖷 +49 9131 302528
franke@faps.uni-erlangen.de
www.faps.uni-erlangen.de

Universität Erlangen-Nürnberg – LKT
Prof. Dr. Dietmar Drummer
Am Weichselgarten 9
D-91058 Erlangen, Germany
☎ +49 9131 85-29700
🖷 +49 9131 85-29709
drummer@lkt.uni-erlangen.de
www.lkt.uni-erlangen.de

Photonische Technologien

Universität Erlangen-Nürnberg – LPT
Prof. Dr. Michael Schmidt
Paul-Gordan-Straße 3
D-91052 Erlangen, Germany
☎ +49 9131 85-23456
🖷 +49 9131 85-23234
michael.schmidt@lpt.uni-erlangen.de
www.lpt.uni-erlangen.de

Universität Erlangen-Nürnberg – LSP
Prof. Dr. Dirk W. Schubert
Martensstraße 7
D-91058 Erlangen, Germany
☎ +49 9131 85-27752
🖷 +49 9131 85-28321
polymer@ww.uni-erlangen.de
www.lsp.uni-erlangen.de

Universität Erlangen-Nürnberg – WW3
Prof. Dr. Andreas Roosen
Martensstraße 5
D-91058 Erlangen, Germany
☎ +49 9131 85-27547
🖷 +49 9131 85-28311
ww3@ww.uni-erlangen.de
www.glass-ceramics.uni-erlangen.de

Universität Hannover – HFT
Dr. Bernd Geck
Appelstraße 9a
D-30167 Hannover, Germany
☎ +49 511 762-5163
🖷 +49 511 762-4010
geck@hft.uni-hannover.de
www.hft.uni-hannover.de

Universität Magdeburg – IMOS
Prof. Dr. Bertram Schmidt
Postfach 4120
D-39016 Magdeburg, Germany
☎ +49 391 67-18399
🖷 +49 391 67-12609
bertram.schmidt@ovgu.de
www.uni-magdeburg.de/imos

Universität Paderborn – HNI
Prof. Dr. Jürgen Gausemeier
Fürstenallee 11
D-33102 Paderborn, Germany
☎ +49 5251 60-6266
🖷 +49 5251 60-6268
juergen.gausemeier@hni.upb.de
www.hni.upb.de

Universität Weimar – Interface Design Group
Prof. Dr. Jens Geelhaar
Marienstraße 5, Raum 105
D-99421 Weimar, Germany
☎ +49 3643 583-763
🖷 +49 3643 583-806
jens.geelhaar@uni-weimar.de
www.uni-weimar.de

Vipem Hackert GmbH
Jürgen Hackert
Hammerbrücker Straße 4
D-08223 Grünbach, Germany
☎ +49 3745 7805-0
🖷 +49 3745 7805-44
juergen.hackert@vipem.de
www.vipem.de

Viscom AG
Detlef Beer
Carl-Buderus-Straße 9–15
D-30455 Hannover, Germany
☎ +49 511 94996-0
🖷 +49 511 94996-900
detlef.beer@viscom.de
www.viscom.de

Walter Söhner GmbH & Co. KG
Oskar Schütz
Daimlerstraße 13
D-74193 Schwaigern, Germany
☎ +49 7138 22-224
🖷 +49 7138 22-242
oskar.schuetz@walter.soehner.com
www.soehnergruppe.de

WWS Technik in Form und Kunststoff GmbH
Michael Hahnau
Im Seelenwinkel 4
D-71570 Oppenweiler, Germany
☎ +49 7191 36788-0
🖷 +49 7191 4225
michael.hahnau@wws-gmbh.com
www.wws-gmbh.com

XENON Automatisierungstechnik GmbH
Dr. Hartmut Freitag
Heidelberger Straße 1
D-01189 Dresden, Germany
☎ +49 351 40209-100
🖷 +49 351 40209-109
hartmut.freitag@xenon-automation.com
www.xenon-automation.com

Zollner Elektronik AG
Ulrich Niklas
Industriestraße 2–14
D-93499 Zandt, Germany
☏ +49 9944 201-167
🖷 +49 9944 201-9388
ulrich_niklas@zollner.de
www.zollner.de

Index

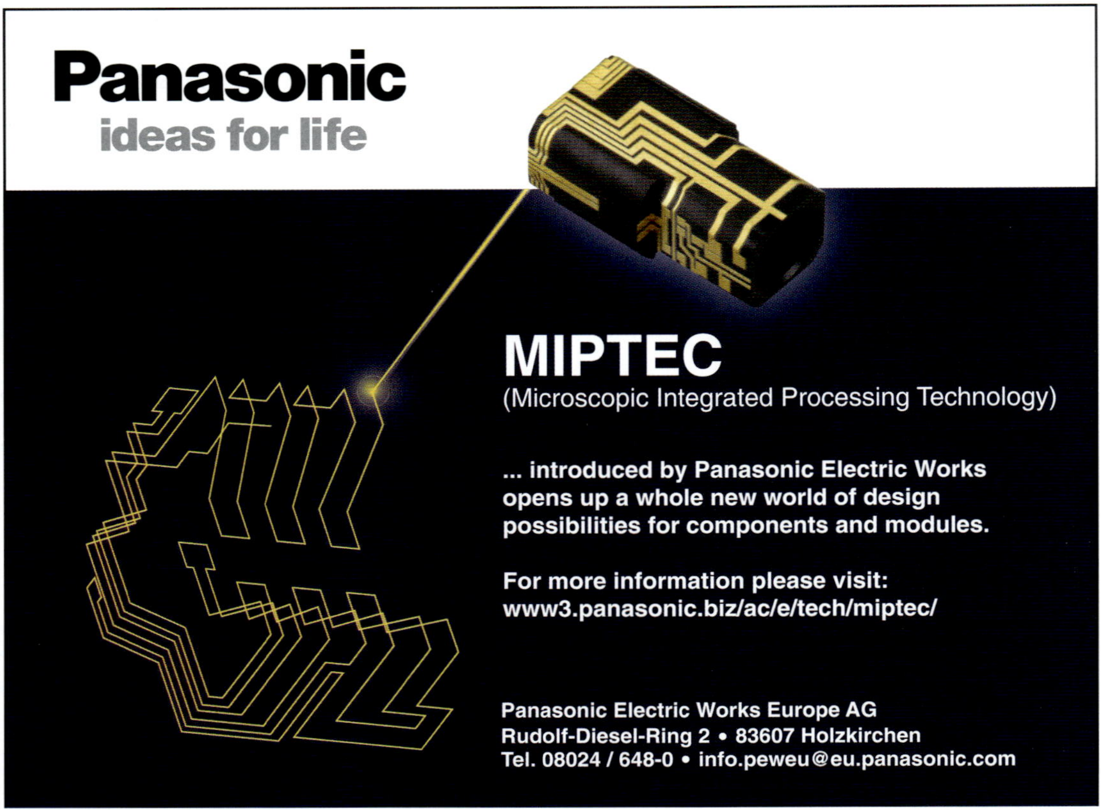
Panasonic
ideas for life
MIPTEC
(Microscopic Integrated Processing Technology)
... introduced by Panasonic Electric Works opens up a whole new world of design possibilities for components and modules.
For more information please visit:
www3.panasonic.biz/ac/e/tech/miptec/
Panasonic Electric Works Europe AG
Rudolf-Diesel-Ring 2 • 83607 Holzkirchen
Tel. 08024 / 648-0 • info.peweu@eu.panasonic.com

Contact us
USA – Frank Basile
Phone: +1-408-234-7737
email: fjb@te.com
EMEA – Christian Köhler
Phone: +49-371-51-45-22
email: christian_koehler@te.com
Asia Pacific – Matt Tsai
Phone: +886-2-8768-2788 (Ext. 239)
email: matt.tsai@te.com
MID Technology
Customized Solutions. Continued Innovation
antenna.te.com
TE connectivity
EVERY CONNECTION COUNTS

2E mechatronic –
The source for
MID solutions
Idea – Development –
Production
2E mechatronic GmbH & Co. KG
D-73230 Kirchheim unter Teck
+49 7021 93010
info@2e-mechatronic.de
www.2e-mechatronic.de
mechatronic

Laser Direct Structuring
Taking the Heat!
with VESTAMID® HTplus
The latest generation of VESTAMID® HTplus (PPA) is precisely tailored to the requirements of the LDS process. Suitable grades are a PA10T with a particularly wide processing window, which is also 50% biobased, and a PA6T/X with higher thermal resistance.
Technical advantages
• Very small components are possible
• High dimensional stability
• Good processability
• Very good chemical resistance
• Very good temperature resistance
www.vestamid.com
Evonik. Power to create.
EVONIK
INDUSTRIES